CONTROLLING CONDUCTED EMISSIONS BY DESIGN

CONTROLLING CONDUCTED EMISSIONS BY DESIGN

John C. Fluke, Sr.

Industrial Control Services Co.
Kirkland, Washington

VNR VAN NOSTRAND REINHOLD
New York

Library of Congress Catalog Card Number 90-38647
ISBN 0-442-23904-1

Printed in the United States of America

Van Nostrand Reinhold
115 Fifth Avenue
New York, New York 10003

Chapman and Hall
2-6 Boundary Row
London, SE1 8HN, England

Thomas Nelson Australia
102 Dodds Street
South Melbourne 3205
Victoria, Australia

Nelson Canada
1120 Birchmount Road
Scarborough, Ontario MIK 5G4, Canada

16 15 14 13 12 11 10 9 8 7 6 5 4 3 2 1

Library of Congress Cataloging in Publication Data
Fluke, John C.
Controlling conducted emissions by design / by John C. Fluke, Sr.
p. cm.
Includes index.
ISBN 0-442-23904-1
1. Electronic circuit design--Data processing. 2. Computer-aided design.
3. Electromagnetic compatibility. Title.
TK7867.F59 1991
621.381'5--dc20 90-38647
CIP

To Daniel
I hope this is an inspiration

CONTENTS

Preface xi

PART I. FUNDAMENTALS OF CONDUCTED EMISSION DESIGN xv

1. Designing for EMC 1
- 1.1 Noise (EMI) 2
- 1.2 EMI Source, Path, and Victim 5
- 1.3 Conductive Paths 8
- 1.4 Conduction or Radiation? 18
- 1.5 Design to Control Conducted Emissions 20

2. EMI Spectrum 23
- 2.1 Time and Frequency Domains 23
- 2.2 Description of FFT Software 24
- 2.3 Data Interpretation 25
- 2.4 Bare Bones FFT 26
- 2.5 Methods of Inputting Data to FFT 29
- 2.6 An Enhanced Version of FFT 32
- 2.7 Examples of FFT Conversions from Time to Frequency Domains 37
- 2.8 Some Possible Pitfalls 44
- 2.9 Subharmonics 45

3. Capacitor Modeling 49
- 3.1 The Capacitor Model 53
- 3.2 Parasitic Elements of Capacitors 54
- 3.3 Capacitor Types 59
- 3.4 Capacitor Voltage Ratings 61

4. Inductor Modeling 65
- 4.1 Inductor Losses 67
- 4.2 Inductor Capacitance 68
- 4.3 Air Core with Conductor Near Experiment 70
- 4.4 Inductor Cores Form Capacitive Paths 70
- 4.5 Inductor Impedance Curve 73
- 4.6 Parasitic Elements of Inductors 75
- 4.7 Simulation 75

5. Balun Modeling **83**
5.1 Differential Mode Flux 84
5.2 Common Mode Flux 85
5.3 The Truth about Windings on Inductor Cores 87
5.4 Coupling *K* Factor 87
5.5 Differential Balun Inductance 91
5.6 Common Mode Balun Inductance 93
5.7 Effects of Load and Source Resistances on Attenuation 94
5.8 Balun Driving Impedance 96
5.9 Balanced Circuits 97
5.10 Design Criteria 101
5.11 Model 102

6. Filters **105**
6.1 Parasitic Inductances and Capacitances 105
6.2 Academic *LC* Filter 106
6.3 Simple Real World *LC* Filter 106
6.4 Control Parasitics by Design 107
6.5 Parasitics Caused by Circuit Layout 110
6.6 Filter Circuit Design 113
6.7 Characteristic Impedance of *LC* Filters 115
6.8 Parallel Capacitors to Lower the ESR 117
6.9 *LC* Filter 121
6.10 Line Impedance Stabilization Networks 123
6.11 Filter Layout and Packaging Design 124

7. Grounding Electronic Circuits **133**
7.1 Grounding 133
7.2 Safety Grounds 138
7.3 Ground Geometries 138
7.4 Ground Design for Packaging Electronic Circuitry 145
7.5 Shielding 153

8. EMI Analysis **163**
8.1 EMI Modeling 164
8.2 EMI Analysis Using SPICE 169

PART II. ADVANCED CONDUCTED EMISSION DESIGN **179**

9. EMC Regulations **181**
9.1 FCC 182
9.2 VDE 184
9.3 MIL-STD-461 185
9.4 Voltage/LISN Measurement Method 189
9.5 Current/Capacitor Measurement Method 191
9.6 A Comparison of Some of the RF Conducted Emissions Standards 192

10. Switch Mode Power Supplies **197**
10.1 Typical Power Supply Block Diagram 197
10.2 Typical Switch Mode Power Supply EMI Problem Areas 202
10.3 EMI Simulation and Laboratory EMI Test Setup 205
10.4 SMPS EMI Design Example 208
10.5 Model the Problem 210
10.6 Simulation Problems 213
10.7 Back to Fundamental Model 217
10.8 Identify the Players 224
10.9 Other Types of EMI Modeling for SMPS 230
10.10 Conclusion 233

11. Transistor and Diode Packaging Problem for EMI **235**
11.1 New Semiconductor Device Packages 235
11.2 Common Mode Shorting Screens 236
11.3 Typical System with Power Conversion 236
11.4 Common Mode Current Paths 237
11.5 Conducted Emissions Reduction by Choice of Package 244

12. Circuit Examples **245**
12.1 Example 1 245
12.2 Example 2 254
12.3 Example 3 (FFT) 265

13. Computers and Digital Logic Circuitry **269**
13.1 Conducted Emissions Coupling Paths 269
13.2 Sequential Logic and Clocks 272
13.3 Example of Internal Conducted Emissions 275
13.4 What Is the Best Bypass Capacitor? 280
13.5 Power Entry Capacitor 288

14. What This Analysis Method Is Not **291**
14.1 Diagnostics 291
14.2 Fields 292
14.3 Radiation 292
14.4 Characteristic Impedances of Common Pairs of Conductors 294
14.5 Shortcomings of EMI Test Simulation as Described Herein 294

15. Magnetic Saturation Modeling **301**
15.1 The Polarization of Magnetic Domains 301
15.2 Device, Core, and Material Properties 302
15.3 Core Geometry Effects 303
15.4 Effects of Cores Made of Two Different Materials 305
15.5 Some Crucial Parameters to Model Saturation 306
15.6 Methods of Integrating Voltage 306
15.7 Dr. Lauritzen's Saturation Model 309
15.8 The Core Geometry and Material Porosity Region of the *B-H* Loop 313
15.9 Curve Fitting versus Parametric Models 315
15.10 Conclusion 316

Appendix. BASIC FFT **319**

Index **329**

PREFACE

This book presents a useful way to "design in" electromagnetic compatibility (EMC). EMC design considerations are often an addendum to the design. These Band-Aid fixes are not the best approach most of the time but are all that is possible at a late stage in the design and development process. This book is not the classic "EMI fix cookbook"; it is intended for all electronics design engineers. The analytical tools presented enable the designer to address EMC considerations early in the design process. Power conversion engineers will find the enclosed information especially important because of the inherent conducted emissions problems in power conversion equipment. Switching power supplies are commonly the most significant noise generators in electronic systems.

In most design work, if the conducted emission problem is addressed, good layout and packaging will ensure that the conducted and radiated electromagnetic interference (EMI) requirements are met. The EMI process involves three components: source, path, and victim. These elements are easily modeled on the computer. The methods of modeling and analysis on the computer are the essence of this book. The EMI source is analyzed using the FFT and the results are applied to a computer model of the path and victim (test setup). The resulting currents are measured and compared to a standard.

Switching power supplies and computers are discussed in specific detail because of their proliferation and special considerations. The fundamentals apply in all electronic circuits and are easily extended to any area of concern. In addition to the use of the computer as a simulation tool for achieving EMC, important aspects of layout and packaging are discussed. Understanding the effects of layout and packaging on electronics is crucial to being able to create accurate models and analyses with the computer. The reader will be able to use the computer as a tool to expand his or her knowledge of EMC beyond any existing limits in any electronics field he or she chooses to pursue.

The book is divided into two parts. Part 1 provides a brief overview of the EMC design problem and covers the fundamentals of computer simulation and modeling.

Chapter 1 is an overview of the conducted emissions problem. It provides a background of understanding as a starting point to begin the process of putting together the analysis tools and methods necessary to quantify and refine the design parameters that affect EMC performance. Chapter 2 presents the fast Fourier transform, which can be used to convert time domain waveforms to the frequency domain. The frequency domain view of circuit waveforms is generally most useful in EMI analysis (also, the measurements are usually done in the frequency domain).

Chapters 3 through 5 present detailed descriptions of how to model passive components along with their parasitic elements. Chapter 6 presents the important design aspects and parameters in filter applications using the passive models developed in Chapters 3 through 5. Chapter 7 presents a basic understanding of grounding electronic circuits. EMI problems are sometimes caused by poor grounding, which usually results from the circuit designer's lack of understanding of the differences between grounds and returns. Chapter 8 develops a way to do EMI analysis using SPICE. This is a very unique use of SPICE that was originally developed by Ralph Carpenter.

Part 2 begins with a detailed description of the conducted emission limits for some of the common EMC standards. These limits, when required, are additional design constraints to the functional design requirements. Many applications using advanced simulation techniques are presented.

Chapter 9 is an overview of the most common or important international standards for controlling EMI in electronic circuits. The conducted emissions measurement method is a major difference in the standards and is the focus of this chapter. Chapter 10 gives a full-scale conducted emissions analysis of a typical switching power supply circuit problem. A model of a circuit and a test setup according to Mil-Std-462 are used in a simulation of an EMI measurement to predict the EMC performance. First the simulation is performed in an as-designed configuration. The results are compared to the EMI measured in the laboratory to verify the accuracy of the model. Then the computer model of the circuit is modified to improve EMI performance. The circuit improvements can then be implemented in the breadboard and a retest in the laboratory can be performed to verify the circuit modifications.

Chapter 11 focuses on a major problem in electronic circuits that causes a great deal of common mode noise problems. The classic diode and transistor packages create capacitive common mode paths and are the major problem in switching power supply design in regard to EMI. Retrofit solutions and the use of the new TO-247 and TO-254 packages are presented. The new packages enable great reductions in the capacitance of the common mode paths to the chassis. The reduction of common mode capacitance to chassis reduces conducted emissions.

Chapter 12 consists of three examples of conducted emission problems in power conversion and their solutions. Chapter 13 focuses on the use of presented analysis methods to solve functional problems in computer circuits. This application of EMI simulation is not done to meet EMI standards but is used to solve a functional problem that is caused by internally conducted emissions. These internal emissions are not controlled by EMI standards and testing, but they affect the circuit performance.

Chapter 14 is possibly the most important chapter in the book. It presents a discussion of the limitations of the analysis methods. If these limitations are not understood, there is a good chance that the analysis results will mislead the designer. The successful use of any tool depends upon the proper application. Chapter 15 presents information on modeling magnetic saturation. This information will help a designer to deal with modeling problems such as ferro-resonant regulators and magnetic amplifiers, both of which functionally depend upon saturation. Also, unexpected magnetic saturation in a circuit or filter may cause an EMI problem. Magnetic saturation modeling is an important tool when needed.

The appendix contains a listing of the entire FFT program described in Chapter 2.

I wish to thank Ralph Carpenter for taking me under his wing many years ago and teaching me useful methods of design in power conversion. His help will have an everlasting impact on my career, and I hope this book will have an enduring impact upon the career of every reader. I want to thank John Folsum for his help and advise.

PART I

FUNDAMENTALS OF CONDUCTED EMISSION DESIGN

1

DESIGNING FOR EMC

This chapter provides an overview of an approach to electronic design that has proven to be very effective on many projects. The methods and techniques presented will help designers to include electromagnetic compatibility (EMC) as an important design criteria or parameter. It gives a fundamental understanding of electromagnetic interference (EMI) and outlines methods and techniques for developing design tools that use computer simulation and modeling. This is accomplished with high-level electronic circuit analysis programs such as SPICE (Simulation Program, Integrated Circuit Emphasis), which is commonly used by most electronic equipment manufacturers.

SPICE has many versions, supported by various companies, that provide enhanced capabilities beyond the public domain version. Most of the enhancements in the different versions of SPICE are the input and output capabilities such as schematic capture and graphic output plots. Some analysis capabilities are also improved in its commercial versions. SPICE is very powerful when hosted on number crunching mainframe computers. Personal computer versions of SPICE are available for under $100.

ECAP (Electronic Circuit Analysis Program) and MCAP (Microcomputer Circuit Analysis Program) electronic circuit analysis programs are designed specifically for the personal computer. They are generally not as powerful as SPICE but can be very useful. SABRE is a relatively new circuit analysis program that offers features not found in SPICE. SABRE is becoming much more commonly used and may someday replace SPICE as the "standard" circuit analysis program. The simulation fundamentals developed in this book are applicable regardless of the electronic circuit analysis software being used. Also, the fundamental simulation and modeling techniques presented here can be used for solving many circuit problems in addition to EMI analysis.

1.1 NOISE (EMI)

"Noise" is a term used for many purposes and has many different meanings. The signal is the desired portion of a waveform, and noise is the undesired portion of a waveform. EMI is quite often just called "noise"; of course, this is "electronic" noise. One person's noise can be another person's bread and butter. Radio frequency interference (RFI) is one example. Some older stereo amplifiers, made with vacuum tubes, would peak detect and amplify a neighbor's citizen band (CB) conversation. On a quiet evening while listening to a record album of a soft passage of Mozart's music, one may have experienced a loud voice over the stereo saying "ten four, Roger and out." This not only interfered with a music listening session but could also ruin the speakers or cause an amplifier to fail from the unexpected overdriving. Today, audio interference is rarely a problem because of technical advances and EMI regulations imposed by the FCC.

Navy ships have many radar, communication, and ordnance (weapons) systems that work in different frequency ranges (close together) that must not interfere with one another. Frequency allocation and system planning on board military ships are a major endeavor.

Noise and Ripple

"Noise and ripple" are parameters that are time domain measures of waveforms usually associated with dc voltages from power supplies. Ripple (see Fig. 1.1) is usually defined as the resulting waveform, after filtering, that comes from a power source such as sine waves from ac sources or rectangular and triangular waveforms from switching power supplies. Ripple is the residual ac voltage or current that is not averaged to smooth dc voltage.

The term "noise," in this context, applies to the high frequency perturbation caused by fast waveform transitions. In low-frequency ac power circuits these fast transitions are caused by rectifier diodes changing from OFF to ON and from ON to OFF states. A switching power supply creates noise from the switch and output diodes changing states. These changes in current induce large voltages in circuit inductances that ring with circuit capacitances. The resulting waveforms are damped sinusoids that decay because of the circuit losses (resistances). Noise and ripple are measured

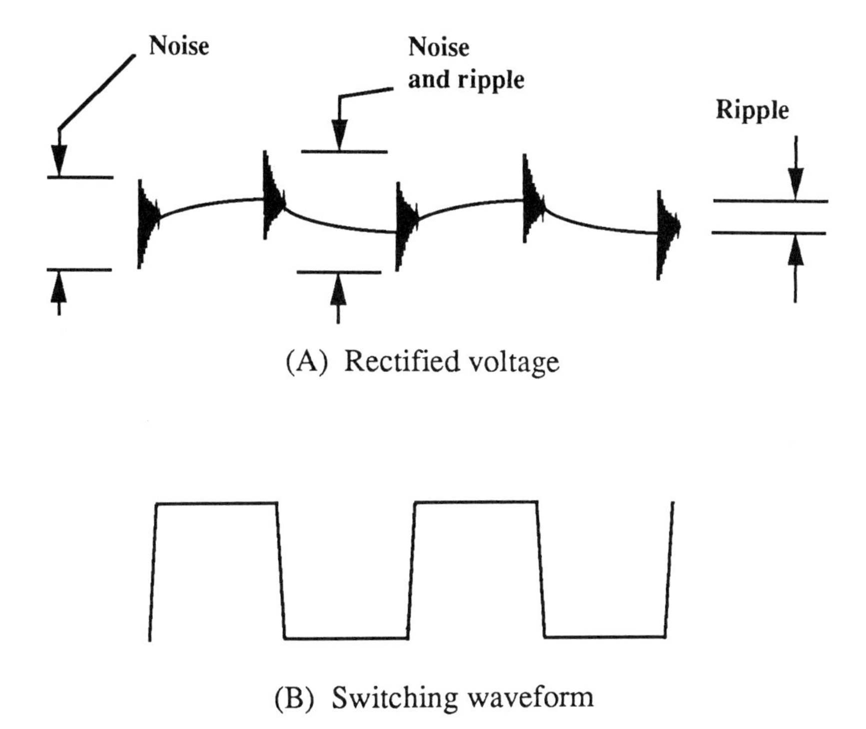

(A) Rectified voltage

(B) Switching waveform

Figure 1.1 Noise and ripple.

differentially. The purpose of these time domain parameters is to specify an acceptable amount of interference so as to not upset the equipment being powered from a given source.

Sources of Noise

There are many causes of differential noise, some of which are listed here:

1. Switching noise, differential mode noise as signal "by-products"
2. Shot noise and all sorts of "microscopic" noise mechanisms
3. Rectifier noise
4. Crosstalk
5. Magnetic loop coupling
6. Shared common impedances in the returns

Differential noise is also a by-product of many signals such as clock waveforms in computers or rectangular switching waveforms in power supplies. These are major sources of EMI. Clock waveform filters can spoil the fast rise and fall times, which will cause circuit upsets in the logic circuits. Good layout practices are the best way to deal with the EMI in clock waveforms. Power supply switching waveforms must be filtered for both functional and EMI reasons. The filter must perform at least two functions, low-frequency averaging to get direct current and high-frequency EMI filtering.

Shot and electron quantum noise are two types of atomic or molecular noise mechanisms with amplitudes in the microvolt or microamp range. The frequencies are generally random (white noise). The frequency content is broadband and in the high-frequency range. The atomic noise amplitudes are small and are not generally considered EMI except in low-level instrumentation or applications of high amplification.

Rectifier noise arises from the fast change of states (ON to OFF or OFF to ON) that cause large and/or fast changes in current. These current changes induce large voltages in the diode leads and associated wiring or circuit traces. These inductances resonate with the capacitances of the diode, the circuit, and/or the printed wiring board (PWB) traces. Because of the circuit resistances, inductances, and capacitances, the damped sinusoids "ring down" at all of the natural frequencies of the circuit.

Crosstalk, another source of EMI, can be coupled either magnetically or capacitively. Signals from adjacent signal lines can couple and cause circuit upsets or erroneous data communications.

Magnetic loop coupling problems are generally created by poor circuit layout. Signal and return lines should be laid out to create as small a loop as possible. The magnetic field strength is proportional to the area of the loop. Crosstalk, ground loop noise, and other unrelated noise can be coupled by this mechanism.

Noise can also be caused connectively by shared returns or by common ground impedances as they are sometimes called. Large currents from one circuit can cause voltages on the common return that are large enough to cause interference on a circuit using the same return.

Interference Is Not the Only Need for EMC

Electromagnetic compatibility means that electromagnetic emissions of the individual components (or any combination of components) of a system do not cause electronic upsets in any of the other components in the system. In other words, the components do not interfere with one another. EMI and EMC are generally system qualities that are concerned with both common mode and differential mode emissions and susceptibilities. Noise and interference are not the only aspects of EMC. Common mode currents in the hulls of ships can be hundreds of amps. These currents not only cause ground noise problems but cause electroplating in the seams of the ship hull that exacerbates corrosion and causes structural failure.

1.2 EMI SOURCE, PATH, AND VICTIM

Source

Sources of EMI are either external or internal to the equipment being designed. Generally for power conversion, the internal noise source is the greatest threat. Usually filters designed for internal power supply emissions have enough attenuation to reduce any external susceptibility threat to an acceptable level. The internal threat in computers (clock signals) is not as severe a threat as are power switches in power supplies. Computers are more susceptible to upsets than are power supplies. Incoming and outgoing EMI are more important considerations for computers than for power supplies. For low-power electronic instruments and communications, the major threat is generally susceptibility to external EMI sources. External threats are a susceptibility problem and are not at the focus of this book. However, susceptibility can be handled with the same fundamentals as emissions but in the opposite direction. The design focus for EMI is directed at internal EMI sources.

Electronic sources of EMI are any voltage or current that changes amplitude. The amplitude of the EMI created depends upon the amplitude of the changing voltages or currents. The faster the change in voltage or current, the higher the frequency of harmonic components created.

Path

There are many different paths for EMI at each of the system levels. A simple list of levels of systems, or environments, is as follows:

1. Earth
2. Buildings and vehicles
3. Equipment rack
4. Chassis
5. Printed wiring assemblies
6. Electronic components

This is obviously not the only possible list; one could go into much finer detail or expand into outer space or go into the submolecular level. Although the paths for EMI all have different physical features, the fundamentals of physics applied are the same for all. Some examples of conducted paths are:

1. Power lines with noise sources
2. Cables in systems
3. Circuits with relays and solenoids
4. Common ground impedance

Power lines often have many loads connected that emit EMI onto the power lines. The power lines serve as a conductive path for the EMI to penetrate equipment housings. Most all power lines must be filtered at the power entry point. Usually both differential and common mode filters are needed.

Cables, in many applications, unintentionally serve as a conduit for EMI. Common mode filters are often needed and differential filters can help, but they must not destroy the integrity of the incoming signal. The differential filter cutoff frequency must be above $1/\pi T_r$ or $1/\pi T_f$ (T_r and T_f are the rise and fall times, respectively).

Circuits that drive solenoids, relays, or other inductive devices inherently have paths of EMI from the loads they drive. Active devices can be upset or even suffer catastrophic failure from the voltages induced by rapid current changes in inductances. Voltage clamping and current rise and fall time control are two of the methods commonly used to avoid noise problems from inductive loads.

If a circuit (circuit A) is attached at more than one point to a ground reference that has current from another circuit flowing through the reference, circuit A will have a common mode voltage impressed upon it equal to the current times the impedance of the reference between the two or more attached points. The path is through the shared ground being used as a return.

Victim

Typical victims of EMI are:

1. Amplifiers
2. Receivers
3. Computers
4. Industrial control systems
5. Ordnance (munitions)
6. Human and animal (biological hazards)

Victim responses to EMI are:

1. No effect
2. Logic upset
3. Erroneous analog signals
4. Physical degradation
5. Catastrophic failure

A well-designed system or piece of equipment will not respond to levels of EMI that are within the specification requirements. Noncatastrophic EMI problems are circuit upsets. Some upsets will self-reset after removal of the EMI threat and some systems "latch up." In cases of latch up most circuits need to have the input power removed and reapplied to reset the circuit.

Living biological tissue can be destroyed if subjected to high levels of electromagnetic radiation. Microwave ovens use this process for cooking. Unintentional electromagnetic radiation or EMI can maim or kill human beings and animals.

1.3 CONDUCTIVE PATHS

Characteristic Impedance of Conductive Paths

The characteristic impedance of a pair of conductors is defined as:

$$Z_0 = (L/C)^{1/2} \tag{1.1}$$

The characteristic impedance is the square root of ratio of the inductance to the capacitance. A pair of foil conductors are shown in Fig. 1.2 with the following parameters:

A_C	area of the plates
d	distance between the plates
A_L	area of the circuit loop
l_m	magnetic path length

The capacitance C is proportional to A_C/d. The inductance L is proportional to A_L/l_m. Therefore the characteristic impedance is proportional to:

$$\left(\frac{A_L / l_m}{A_C / d} \right)^{1/2} \tag{1.2}$$

Reorganizing we get:

$$Z_o \text{ proportional to } \left(\frac{A_L \, d}{A_C \, l_m} \right)^{1/2} \tag{1.3}$$

If A_L or d is increased without changing A_C or l_m, the impedance is increased. If A_C or l_m is increased without changing A_L or d, the impedance is decreased. If the length (L) of the pair of conductors is increased, the area (A_C) of the conductors and the area (A_L) of the circuit loop area increase proportionately so that the ratio of inductance to capacitance does not change. The characteristic impedance is the same for any length of a pair of conductors with a given geometry in a homogeneous medium (continuous and constant permittivity and permeability). The impedance is the characteristic of the geometry of the conductors.

We generally think of water flowing in a pipe as an analogy or "picture" of electron current in a wire. This analogy is really very rough and in most cases is misleading. For instance, electrons

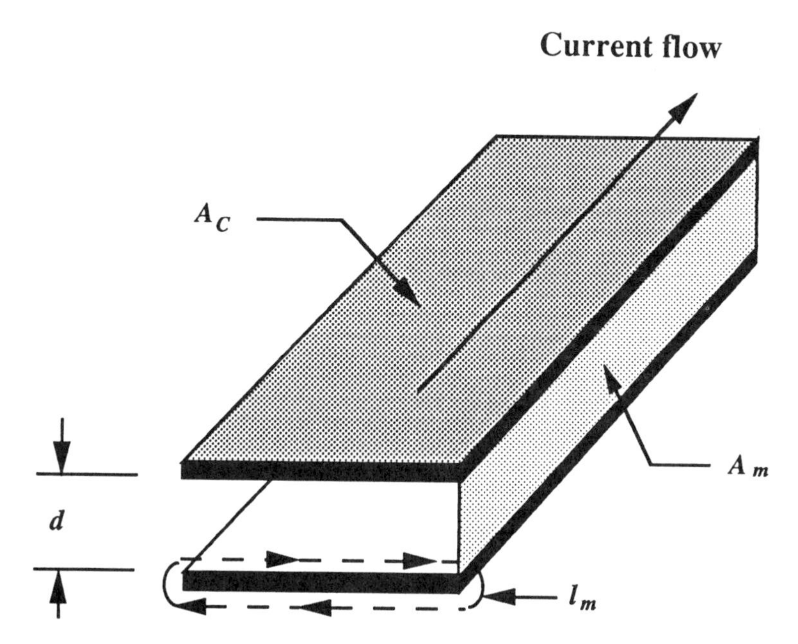

Figure 1.2 Characteristic impedance of conductors.

(negative charges) repel each other because they are of the same polarity. In direct current (zero frequency), the free conduction electrons spread out evenly in the conductor. All of the electrons flow in the path of the wire or conductor. Current will always flow in the path of lowest impedance.

At higher frequencies the path of lowest impedance is not so obvious as in direct current. End and top views of two parallel wires are shown in Fig. 1.3A. One wire is the signal line; the other is the return. On the two conductors most of the current flows on the adjacent shaded areas because that area of the conductors is the path of least impedance (L smaller and C bigger). As the capacitance gets larger, the impedance gets smaller. The capacitance between the shaded areas is larger than between the other areas.

The inductance of a loop of wire (signal and return) is proportional to the area of the loop. Fig. 1.3B is a top view of the same two parallel wires of Fig. 1.3A. The loop area of the adjacent (shaded) area is much less than the loop formed by the outer areas of the wires and less area than any other chosen path. The smallest loop area has the least inductance. The conductive path of the shaded area has the least inductance and the most

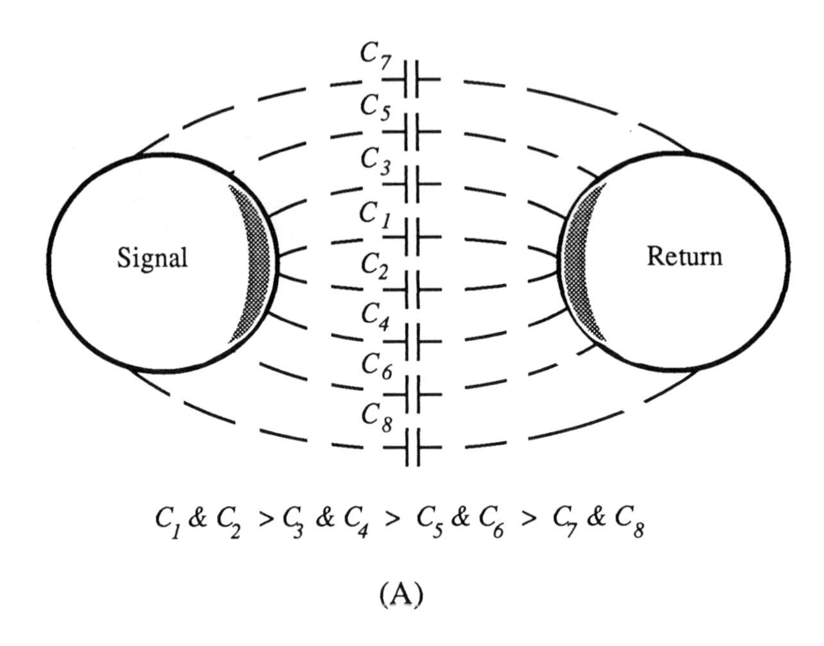

(A)

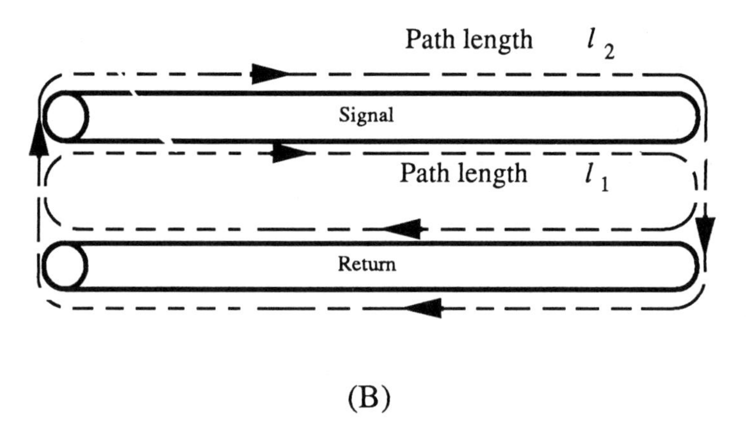

(B)

Figure 1.3 The characteristic impedance of conductors; Z_c is the square root of L/C.

capacitance, making it the path of lowest impedance. Most of the current will flow in the shaded areas.

Figure 1.4 shows a top and side view of a printed wiring board. The return plane on the bottom side of the PWB is a conductive plane that covers the whole PWB. A high-frequency sinusoidal voltage source V_s is attached between point A on the trace and point A on the return plane. A load (R_L) is attached between point B on the trace and point B on the return plane. The return current will flow in the return plane in the same snake-like path as the trace from point A to point B. The path immediately below the trace has the lowest impedance. For any other path of return current, the inductance is higher and the capacitance is lower. Inductance usually determines the path of lowest impedance. It is often hard to make the loop area as small as we would like.

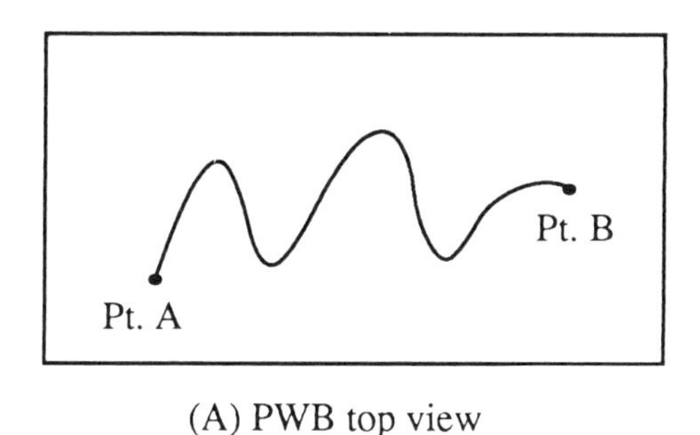

(A) PWB top view

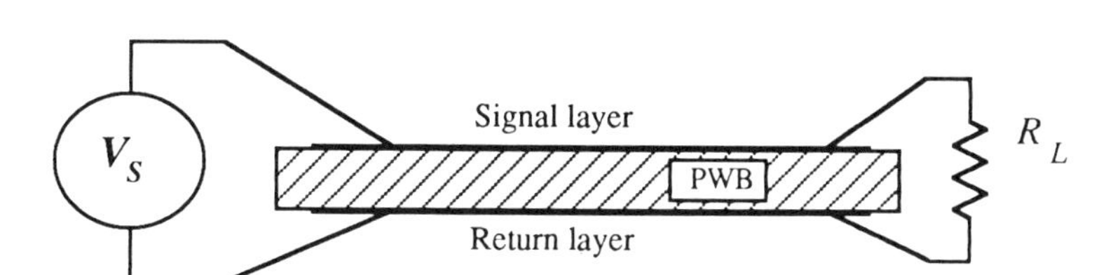

(B) PWB section view

Figure 1.4 Current paths in PWB return planes.

Differential Mode Paths

Differential noise, sometimes the by-product of signals such as clocks or power supply switching waveforms, is directly coupled into the circuit and/or is inherent to the circuit. The path is inherent to the circuit.

Crosstalk can be coupled either magnetically or capacitively. Signals from adjacent signal lines can couple magnetically from loop to loop and induce differential mode currents.

Magnetically coupled noise must have two loops to propagate magnetic fields. One current loop is required to transmit the magnetic field and one conductive loop is required to receive it. Magnetic shielding can usually be avoided in many electronic circuits. Reducing the transmitting and receiving loop areas is the most efficient method of controlling magnetic field radiation.

Low-frequency magnetic fields pass through aluminum plate. In magnetic shields, incident magnetic fields induce eddy currents in the shield that oppose the incident field.

Reduce Loop Area

Magnetic loop coupling problems are generally created by poor circuit layout. Signal and return lines should be laid out to create

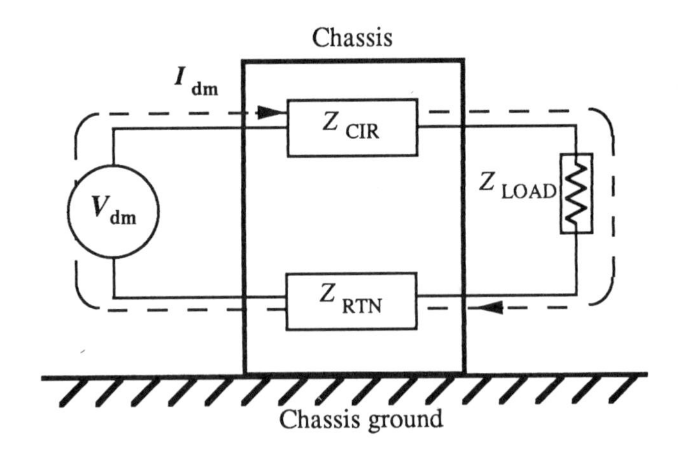

Figure 1.5 Typical differential circuit.

as small a loop as possible. The magnetic field strength is proportional to the area of the loop. Crosstalk, ground loop noise, and other types of noise can be coupled by this mechanism. Three ways to improve differential EMC are:

1. Filtering
2. Rise and fall time control (when possible)
3. Snubbing (form of filtering).

Common Mode Path

Figure 1.5 shows a basic block diagram for a differential model of a typical signal source (V_{dm}), electronic circuit (Z_{CIR} and Z_{RTN}), load (Z_{LOAD}), and chassis ground. The differential current I_{dm} flows from the source, through the electronic circuit, through the load, and then back to the source. I_{dm} is the desired signal current. When the load is grounded to the chassis, a voltage source V_{cm} placed between node A and chassis ground (as shown in Fig. 1.6) will cause common mode currents I_{cm1} and I_{cm2} to flow. The

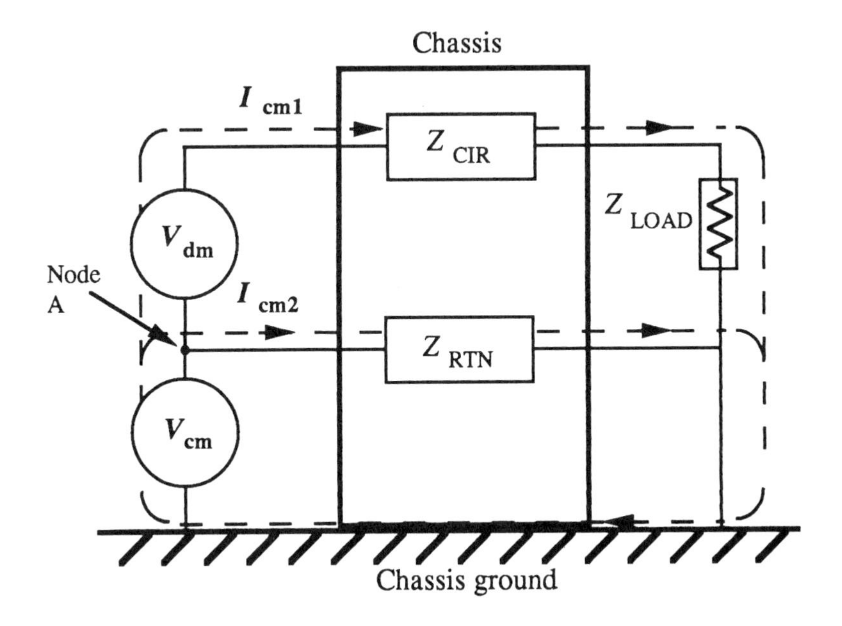

Figure 1.6 Typical common mode paths.

mechanisms creating V_{cm} will be discussed later. The total current in Z_{CIR} is $I_{cm1} + I_d$. The total current in the return is $I_{cm2} - I_d$. I_{cm2} will be greater than I_{cm1} because the impedance of loop 1 = Z_{CIR} + Z_{LOAD} and the impedance of loop 2 = Z_{RTN} ($Z_{CIR} >> Z_{RTN}$). Any impedance unbalance in the circuit will cause differential voltages to appear within the circuit.

The signal is desired to be V_{dm}, but because of the common mode currents causing $V_{noise\ cir}$ and $V_{noise\ ret}$, the signal including common mode noise is $V_{dm} + V_{noise\ cir} + V_{noise\ load} - V_{noise\ ret}$ (Fig. 1.7). The unbalances in the system cause common mode currents to be converted to differential noise voltages. Figure 1.8 is an example of a balanced circuit. These common mode currents in a balanced system do not cause differential noise. The common mode voltages created cancel each other differentially so as to create no net differential noise.

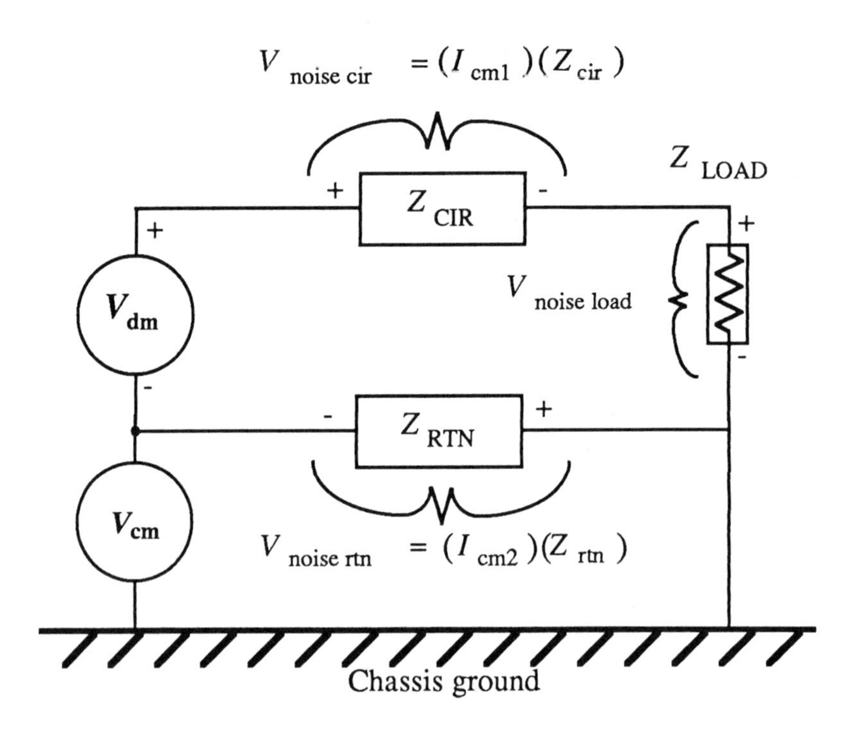

Figure 1.7 Typical common mode voltages.

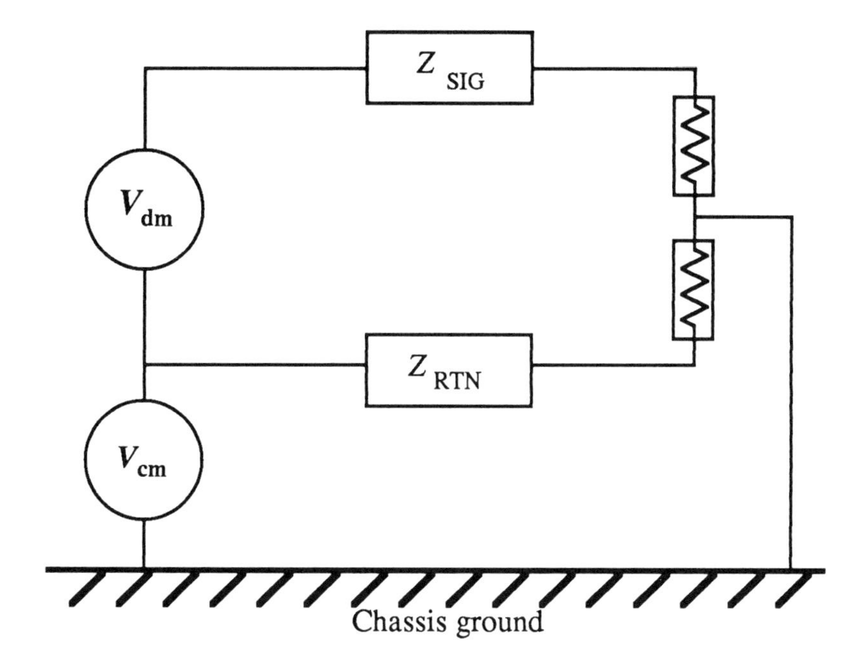

Figure 1.8 Balanced circuit.

Figure 1.9 is a typical example of circuit to chassis capacitance that we will define as common mode capacitance. Notice that there is no need to include a common mode voltage source to get common mode currents. Common mode current is injected capacitively into the chassis because of circuit voltage changes. The common mode current flows through the system or earth ground and back to the source. Again the unbalances in the circuit cause common mode currents to be converted into differential noise voltages.

Figure 1.10 is similar to the previous circuits except that a wholly separate circuit is added that passes a large current I_{ci} through a section of the system ground, but the current does not circulate in the original circuit. Because the system ground has a finite, although small, impedance, a voltage is created by I_{ci} and the common ground impedance. Common mode current will flow in the original circuit because of the common impedance voltage V_{ci}. This can be a system problem (as in this case) or can happen in a single piece of equipment with more than one connection point to the chassis for grounding.

Common mode noise can be controlled by:

1. Reducing common mode amplitudes
2. Circuit balancing
3. Avoiding shared ground impedances
4. Using isolation techniques

Common mode amplitudes can be reduced by the increasing common mode impedance. Circuit layout techniques (detailed in chapter 7) can reduce the common mode capacitance (C_{cm}). A balun (common mode inductor) can be added to the circuit to increase the impedance of the common mode loops (larger L_{cm}). The details are explained in chapter 5.

Balance the system to reduce conversion of common mode noise to differential noise. The source, the cable, and the load must all be balanced for a balanced system. Unfortunately, balanced systems are often not practical.

Layout methods to avoid shared ground impedances are included in chapter 7. A circuit with a shared ground impedance can cause noise in another circuit sharing the common ground reference.

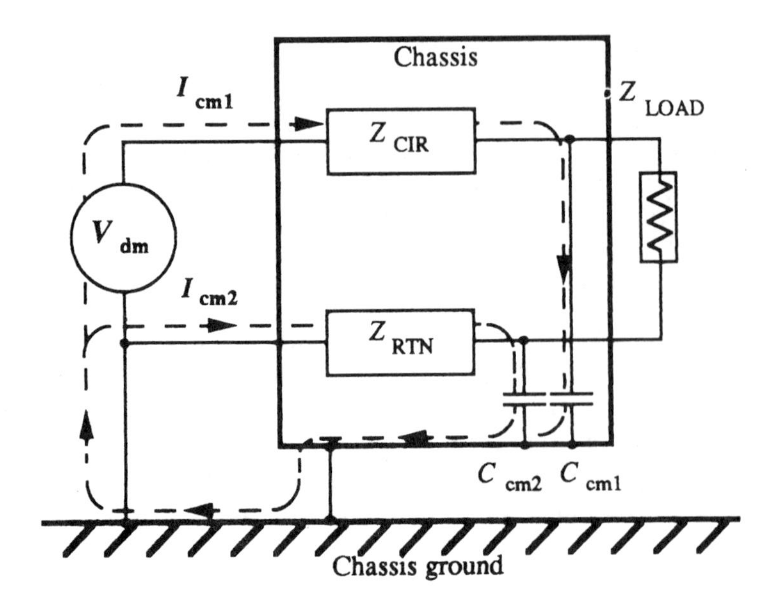

Figure 1.9 Current mode paths.

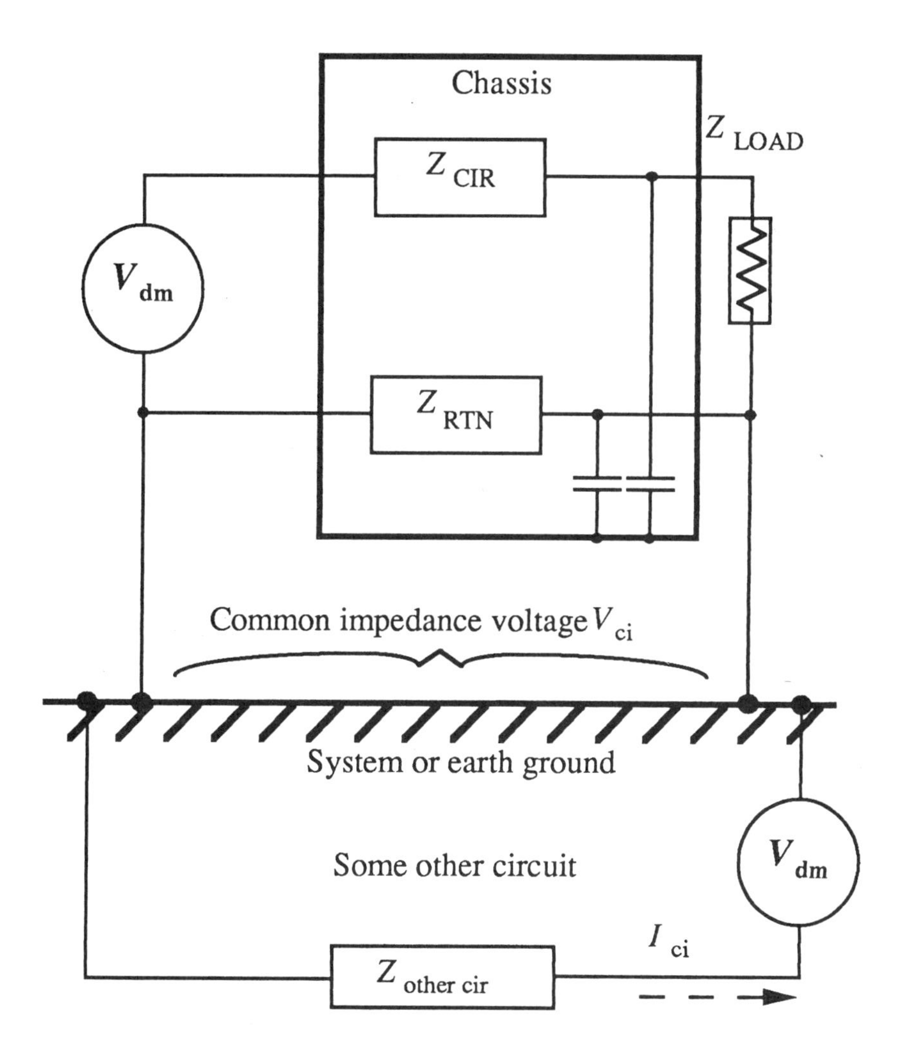

Figure 1.10 Shared common ground impedance.

Some components that are useful for isolating circuits to reduce common mode noise are:

1. Difference amplifiers
2. Isolation transformers
3. Optical couplers

1.4 CONDUCTION OR RADIATION?

There are four noise coupling mechanisms (paths):

1. Conductive	(direct contact)	Metallic path	
2. Capacitive	(electric field)	Near field	$\delta v/\delta t$
3. Inductive	(magnetic field)	Near field	$\delta i/\delta t$
4. Radiative	(electromagnetic)	Far field	$d>\lambda/2\pi$

These are also the four mechanisms for transferring electromagnetic energy. Many would say, and rightfully so, that capacitive, inductive, and radiative are all radiated energy. But for this book (and in general) we will consider conductive (metal path), capacitive, and inductive to be conducted energy. This seems to be a contradiction, and it is. This contradiction cannot be resolved, but it does not need to be. It only needs to be understood. The contradiction arises because of the definitions used to organize thoughts about physical phenomena. We usually accept that the current through a capacitor is conducted yet there is no metallic connection between the two capacitor leads. Alternating current passes through the field between the capacitor plates. This is near-field radiation because the distance between the plates is much less than the wavelength λ divided by 2π. It is primarily an electrostatic field; the magnetic field strength is very small. Noise can be induced capacitively or inductively (near-field radiation) or can be induced by far field electromagnetic radiation.

A high-voltage source (implying high impedance) can be made to radiate an electric field by using a monopole antenna (see Fig. 1.11A). In the near-field range, close to the source, the field is primarily electric. As the electric field propagates away from the source, the energy naturally converts to transverse electromagnetic waves. Some of the electrical energy is converted into a magnetic field that is perpendicular to the electric field.

A current source forcing a current through a loop works as an antenna, as shown in Fig. 1.11B. In the region close to the magnetic field source, the field is primarily magnetic. As the magnetic field propagates away from the source, the energy converts to a transverse electromagnetic wave with an electric field perpendicular to the magnetic field.

The impedance of an electromagnetic wave propagating from an antenna is a function of distance and wavelength. Figure 1.11C shows the relationship of the distance from the source and the

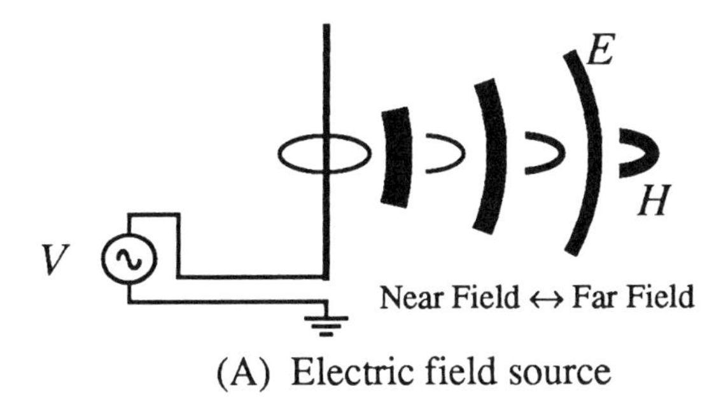

(A) Electric field source

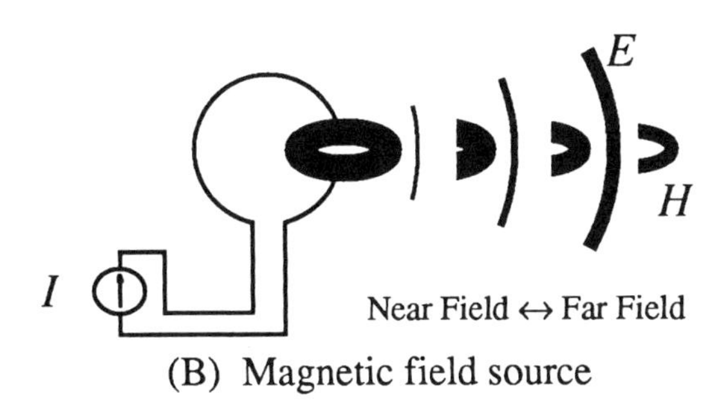

(B) Magnetic field source

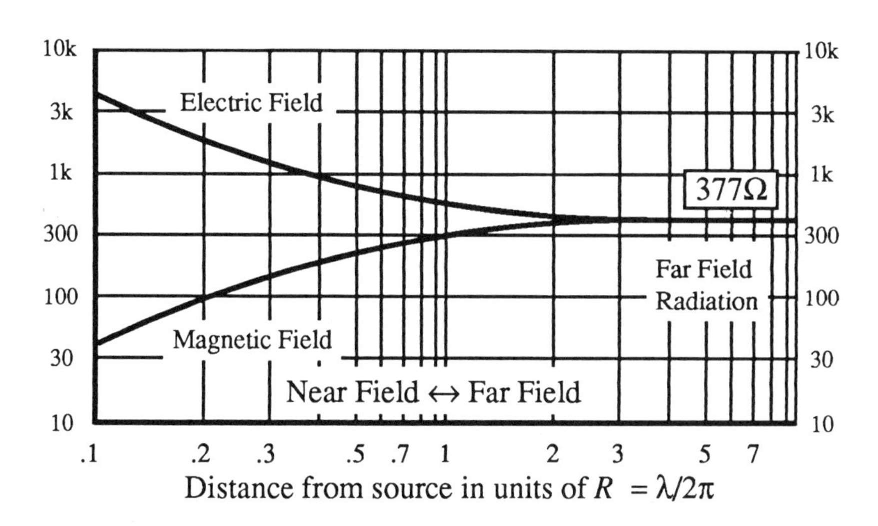

(C) Field impedance as a function of source distance.

Figure 1.11 Near-field and far-field.

wave impedance. At a distance greater than $\lambda/2\pi$, far-field, the wave impedance converges to 377 ohms, the impedance of free space. The impedance of an electromagnetic wave is the ratio of the square root of the electrostatic energy to the magnetic energy (or the ratio of the fields E/H). The electric and magnetic fields are perpendicular to each other and the direction of propagation. Far-field electromagnetic energy is the same whether it is created from a magnetic source or an electric source.

1.5 DESIGN TO CONTROL CONDUCTED EMISSIONS

Because all radiated electromagnetic energy must be conducted before any energy can be radiated, it is important in any design to reduce the conducted energy. To radiate, a magnetic antenna (loop of wire) must conduct current. An electric dipole antenna must have a modulated voltage potential across the dipole. If the conducted EMI sources could be eliminated, there would be no radiation. This is not possible, of course. We seek to develop design methods and techniques to reduce the conducted emissions to an acceptable level.

With the conducted emissions reduced to a minimum, the amount of resulting radiation depends upon the coupling from the circuit to the electromagnetic medium, which is a function of the physical geometry of the components, interconnecting circuitry, and packaging techniques. This coupling can also be minimized but is out of the scope of this book.

Most electronic equipment must be designed to meet commercial or military standards. Electronic circuit designs must be qualified to these standards by verification tests. Understanding of the purposes of the tests and the meaning of the tests is necessary for the overall design effort. The tests and requirements guide the system and equipment level design to be functional in all environments likely to be encountered. For functionality at the equipment level, EMC design practices need to be applied at the circuit and printed wiring board level in addition to the system level. Designing for EMC is needed at all levels: system, equipment, circuit, and printed wiring board.

QUESTIONS

1.1 Why is it important to distinguish types of EMI such as noise and ripple?

1.2 For what reasons do we identify common mode and differential mode currents?

1.3 Provide an additional entry for any list in this chapter.

1.4 Why does the characteristic impedance of conductors have no geometric dimensions?

1.5 Where does the majority of current flow in any circumstances?

1.6 Why should the impedance of circuit traces be matched to drivers and receivers even though the lengths of runs are much shorter than the length at which transmission line reflections are a consideration?

1.7 Why is it impossible to distinguish if a transmitter viewed electronically from deep far field is electric or magnetic?

1.8 Is the current in a capacitor conducted or radiated from the input lead to the output lead?

1.9 Contrast (when designing for conducted emissions) the process of designing to meet EMI requirements versus designing for functionality.

1.10 Discuss the differences between conducted emissions problems and conducted susceptibility (hint: upset thresholds).

2

EMI SPECTRUM

2.1 TIME AND FREQUENCY DOMAINS

The sources of conducted emissions are most easily analyzed in the frequency domain. EMI conducted emissions measurement test data are normally in the frequency domain form. Designers can mathematically convert time domain waveforms to the frequency domain. Any time domain waveform can be resolved into the sum of many sine waves. The Fourier transform or Fourier series is used to pass from the time domain to the frequency domain in order to analyze a particular waveform.

Analysis is the process of dividing an object into its parts for individual study. The attenuation of a given filter can be found at each frequency. The Fourier representation of a waveform can be applied to a filter and the resulting calculated current flow can be compared to specified limits to predict compliance or noncompliance.

The frequency content of a waveform can be used to identify noise sources and possible paths of conduction. The frequency content of the noise is also a guide to the appropriate remedies to be applied to the EMI design problem. This information can be used early in the design phase by solving predicted problems or can be used as diagnostic techniques when problems arise later in the project. We, of course, would like to solve all of the problems in the early design phases for economic and schedule reasons.

In the early stages of an electronic design the fundamental waveforms are known to the designer. An experienced designer can even predict some of the finer details. Switching power supplies often have rectangular waveforms with rise and fall times that are dependent on the type of switch used and the impedance and amplitude of the drive circuit. Knowing the amplitude, rise and fall times, pulse width, and frequency of the rectangular switching waveform, the designer can easily calculate the envelope of the frequency domain representation of the waveform. In digital circuitry the clock waveform is a primary generator of noise. The digital designer must know the shape of the waveform very well early in the design phase. The frequency spectrum can be calculated in a fashion that is similar to calculating for the switching waveform. This calculation is very simple and is described in detail in chapter 8.

The discrete Fourier transform (DFT) is a simple mathematical process that can be used to transform a time domain waveform to the frequency domain. The fast Fourier transform (FFT) algorithm can process the DFT on a personal computer. Methods to interpret the results follow and include warnings and guidance to avoid possible pitfalls.

The FFT can be processed as a forward or inverse transform time to frequency or frequency to time domain, respectively. The forward transform is most useful in EMI analysis. Early in the design phase the known EMI threat waveforms can be mathematically generated on the computer and a time domain data file created.

2.2 DESCRIPTION OF FFT SOFTWARE

It is assumed that the reader has an elementary understanding of BASIC programming. If not, it would be worth while to acquire a familiarity with BASIC to use the FFT and as a tool for solving other problems.

In the FFT presented in this book, the time domain data file is stored in the array called INPUTRE (input real). The computer first displays the time domain waveform to verify that the data are appropriate. If the time domain data are judged by the engineer to be appropriate, the FFT is processed. After the program has been run, the amplitudes of the frequency domain data are stored in the array INPUTRE. The phase data are stored in the array called INPUTIM (input imaginary). The frequency domain data are displayed graphically and may also be printed out as a data file.

In later stages of the design process, waveform analyzer data or digitized oscilloscope data from breadboards or brassboards can be loaded into the FFT data array INPUTRE and processed to verify and/or to refine the analysis.

2.3 DATA INTERPRETATION

The book *Fundamentals of FFTs* by Robert Ramirez is highly recommended to those who care to know more about the use of FFTs. Some of the fundamentals will be presented here to get started. The output of the forward FFT is frequency domain data in rectangular form (real and imaginary parts). The magnitudes are calculated by the relation:

$$\text{Mag} = (\text{Re}^2 + \text{Im}^2)^{1/2} \quad (2.1)$$

The phase angles can be calculated by the relation:

$$\phi = \arctan\,(\text{Im}/\text{Re}) \quad (2.2)$$

They are not included in the FFT program because in EMI analysis, the magnitudes are the measure to which the standards are compared. The phase information is useful for academic purposes and may be added to the FFT program if desired.

Fourier Integral and Series Interpretation

The Fourier integral is useful when analyzing a single pulse. It is assumed that for all history before the pulse and in the future after the pulse, the amplitude will remain at zero.

The Fourier series is used to analyze a continuous wave of rectangular pulses. The duty cycle for a continuous waveform is defined as the pulse width divided by the period. As the duty cycle goes to zero (the period becomes infinite), the frequency domain data becomes continuous. At the limit, as the duty cycle goes to zero, the continuous wave becomes a single pulse and the Fourier integral interpretation is more appropriate. Singular events or very small duty cycles result in broadband frequency domain data.

The frequency domain data for the series is discrete sinusoid spectra located at integer multiples of the repetition rate of the repeated rectangular pulse waveform. The discrete Fourier transform assumes a window of time domain data that is repeated to infinity or for a very long time. The first and last data points of the window must be such that if the window is repeated, the windowed waveform is the same as the continuous wave. Any discontinuities caused by windowing will cause aliasing. Aliasing is any of the errors caused by the process of the transformation from time domain to frequency domain. There are many other types of aliasing that are described thoroughly in *Fundamentals of FFTs*.

When testing to MIL-STD-461 standards for conducted emissions, both narrowband and broadband emissions are measured. The design effort for conducted emissions must focus on the noise generators internal to the equipment being designed. The two most typical noise generators, internal to our designs, are power switching waveforms and digital clocks, both of which are continuous waves. The continuous waves of sources in the frequency range of CE03 are narrowband, so it is appropriate to design to meet narrowband requirements. If the design is done well, the broadband tests will be passed as well. The broadband problems for CE03 requirements generally arise from low-frequency (60 or 400 Hz) rectification. At frequencies in the megahertz region, the harmonics of 400 hertz noise are so close together that they appear as broadband noise.

2.4 BARE BONES FFT

The bare bones version of the FFT (Fig. 2.1) is included for academic understanding of the fundamental operation of the FFT. The input and output handling sections of the whole program are much larger than the actual DFT processing portions. This version of the FFT first plots the frequency domain data in linear frequency versus linear amplitude and then plots the data in log frequency versus log amplitude. This is a modified version of an FFT developed by Robert W. Ramirez. The Ramirez version only printed a data table for output. Graphic plots have been added to the bare bones version. The log plot amplitude is scaled in decibels above a microamp or microvolt, as selected. This is generally the most useful scale and it is also used in MIL-STD-461, the military standard, making it convenient for comparison to standard limits. The inverse portions of the FFT have been removed for clarity.

Program lines 140 through 160 plot the time domain data so that integrity of the data can be verified before processing. The program stops and prompts for any key to be pressed. The time domain data can be printed by using the print screen key on the computer keyboard. The time domain plot may be included in a hardcopy file for documentation purposes. Program lines 570

```
******************************************
******************************************
10 REM
30 REM  BARE BONES FFT  TWO SINUSOID INPUT COMPONENTS
40 REM
50 PRINT "RECORD LENGTH MUST BE A POWER OF 2."
60 INPUT "INPUT RECORD LENGTH.";N
70 M=LOG(N)/LOG(2)
80 DIM INPUTRE(N-1),INPUTIM(N-1),BUFFERRE(N-1),BUFFERIM(N-1)
90 DIM MAG(N-1) , MAGLOG(N-1)
100 DIM TFRE#(N/2-1),TFIM#(N/2-1)
110 REM   ****INPUT DATA FOR XFORM HERE--REAL DATA INTO ARRAY
120 FOR J = 0 TO N-1
130 INPUTRE(J)=10*SIN(J*2*3.14159/N) +2.5*SIN(J*5*2*3.14159/N)
140 NEXT J :CLS :KEY OFF :SCREEN 2 :WINDOW (0,-20)-(270,20)
150 FOR J = 0 TO N-2 :LINE (J,INPUTRE(J)) - (J+1,INPUTRE(J+1))
160 NEXT J :INPUT TYPE$ :SIGN=(-1)
165 REM ****GENERATE TWIDDLE FACTORS****
170 PI#=3.141592653589795#:PI2#=2*PI#
180 FOR P=0 TO N/2-1 :TFRE#(P)=COS(PI2#*(-P)/N)
185 TFIM#(P)=(SIGN)*SIN(PI2#*(-P)/N)
190 NEXT P
200 REM ****COMPUTE FAST FOURIER TRANSFORM****
210 FOR I= 1 TO M :L=0: H=0 :G=(N/2^I)
220 FOR K=0 TO (N-1) STEP G :TFI=0 :TFIFLAG=(-1)^(L+1)
230 FOR J=0 TO (G-1) :TFI=J*2^(I-1) :R=K+J:S=J+H:T=J+G+H
240 IF TFIFLAG>0 THEN 270
250 BUFFERRE(R)=INPUTRE(S)+INPUTRE(T)
:BUFFERIM(R)=INPUTIM(S)+INPUTIM(T)
260 GOTO 300
270 TEMPRE=INPUTRE(S)-INPUTRE(T) :TEMPIM=INPUTIM(S)-INPUTIM(T)
280 BUFFERRE(R)=TEMPRE*TFRE#(TFI)-TEMPIM*TFIM#(TFI)
290 BUFFERIM(R)=TEMPRE*TFIM#(TFI)+TEMPIM*TFRE#(TFI)
300 NEXT J :L=L+1:H=INT(L/2)*G*2
310 NEXT K
320 FOR II=O TO N-1 :INPUTRE(II)=BUFFERRE(II)
:INPUTIM(II)=BUFFERIM(II)
330 NEXT II
340 NEXT I
```

Figure 2.1 Bare bones FFT.

```
350 FOR I=O TO N-1
360 INPUTRE(I)=INPUTRE(I)/N :INPUTIM(I)=INPUTIM(I)/N
370 NEXT I
380 REM ****BIT REVERSAL ROUTINE TO UNSCRAMBLE FFT RESULTS****
390 FOR I=O TO N-1
400 INDEX%=I :IOUT%=O
410 FOR J=1 TO M
420 TEMP%=1 AND INDEX% :IOUT%=IOUT%*2 :IOUT%=IOUT%+TEMP%
:INDEX%=INDEX%\2
430 NEXT J
440 BUFFERRE(I)=INPUTRE(IOUT%) :BUFFERIM(I)=INPUTIM(IOUT%)
450 NEXT I
460 REM ****ORDER FFT OUTPUT FOR NEG. FREQ. AT O TO
470 REM N/2-2, DC AT N/2-1, POS. FREQ. AT N/2 TO N-2,
480 REM AND NYQUIST AT N-1.****
490 FOR I=O TO N/2
500 INPUTRE(I+(N/2-1))=BUFFERRE(I) :INPUTIM(I+(N/2-1))=BUFFERIM(I)
510 NEXT I
520 FOR I=O TO N/2-2
530 INPUTRE(I)=BUFFERRE(I+(N/2+1)) :INPUTIM(I)=BUFFERIM(I+(N/2+1))
540 NEXT I
545 REM
550 REM     ****FFT OUTPUT IS HERE IN ARRAYS INPUTRE FOR REAL
560 REM     PART AND INPUTIM FOR IMAGINARY PART****
570 KEY OFF :CLS :SCREEN 2 :WINDOW (0,0)-(N+20,6)
580 FOR L=0 TO N-1
590 MAG(L) = (INPUTRE(L)^2+INPUTIM(L)^2)^(1/2) :LINE (L,0)-
(L,MAG(L))
600 NEXT L
610 INPUT TYPE$ :CLS :WINDOW (0,0)-(100,25)
620 FOR L=N/2-1 TO N-2
630 IF MAG(L) = 0 THEN MAG(L) = .000001 :MAGLOG(L) =8.68589 *
LOG(MAG(L))
640 HOG = 20 * LOG(L-N/2+2) :LINE (HOG,0)-(HOG,MAGLOG(L))
650 NEXT L
660 FOR L = N/2-1 TO N-2
670 LPRINT MAG(L) "  VOLTS      " MAGLOG(L) " DB"
680 NEXT L
690 STOP
********************************************
********************************************
```

Figure 2.1 Bare bones FFT (*Continued*).

through 650 are routines that plot the frequency domain data, linear versus linear and log versus log.

Lines 120 through 140 load sinusoidal waveform data into the array called INPUTRE. Any mathematical function can be implemented in BASIC statements and loaded into INPUTRE for processing. The linear versus linear and log versus log frequency domain data plots can be screen printed just as the time domain plots were.

2.5 METHODS OF INPUTTING DATA TO FFT

Mathematical Inputs

The FOR-NEXT loop commands can be used to process mathematical equations and store the results in array INPUTRE. Fig. 2.2 is an example of a sinusoid input. Any mathematical function can be processed in a similar fashion.

Rectangular Waveforms

A logic generated waveform with residential data is shown in Fig. 2.3. A rectangular or trapezoidal waveform is generated with the logic and mathematical commands shown. The use of resident data can be used in debugging runs so that the data for test runs need not be entered for each run. The input parameters are inserted on the appropriate program lines. Another advantage of this method of input is that a single parameter can be changed without having to put in all of the parameters for each run. This method is convenient for fast program reruns to observe the effects of changes made to a waveform. The program is modified before it is compiled, so the data variables are entered as a response to computer prompts as in Fig. 2.4. After compilation, the resident variable values cannot be changed.

```
120 FOR J = 0 TO N-1
130 INPUTRE(J)=10*SIN(J*2*3.14159/N)+2.5*SIN(J*5*2*3.14159/N)
140 NEXT J
```

Figure 2.2 FFT with mathematical data input generation.

```
95 REM * LOGIC GENERATED WAVEFORM WITH PROGRAM RESIDENT DATA
100 WW=.01:A=10:PW=.005:T=.01:TR=.001:TF=.001:TS=WW/N:XXX=TR/TS
102 SLOPETR=A/XXX:YYY=PW/TS:ZZZ=((PW+TF)/TS)-1:SLOPETF=A/(TF/TS)
105 REM WW IS THE WIDTH OF THE WINDOW IN TIME
115 REM **A IS AMPLITUDE* PW IS THE PULSE WIDTH* T IS PERIOD ***
145 REM * TR IS THE RISE TIME**TF IS THE FALL TIME*
220 FOR J = 0 TO XXX :INPUTRE(J) = J*SLOPETR
240 NEXT J
250 FOR J = XXX TO YYY :INPUTRE(J) = A
270 NEXT J : K = 1
290 FOR  J = YYY TO ZZZ :INPUTRE(J) = A-(K*SLOPETF) :K = K + 1
310 NEXT J
320 FOR J = ZZZ+1 TO N-1 :INPUTRE(J) = 0
340 NEXT J
```

Figure 2.3 FFT for rectangular waveform generation with program resident data.

```
10 REM  AUTO SCALING AND PROMPTS FOR RECTANGULAR WAVE PARAMETERS**
130 INPUT "WHAT IS AMPLITUDE?";A
140 INPUT "WHAT IS PULSE WIDTH?";PW
150 INPUT "WHAT IS PERIOD OF WAVEFORM?";T
160 INPUT "WHAT IS RISE TIME ?";TR
170 INPUT "WHAT IS FALL TIME ?";TF
180 REM *ROUTINE TO CREATE RECTANGULAR*
190 RRR = 0 :RR = RRR :TS = WW/N TTT = T/TS :TT = TTT :NC = WW/T
250 XXX = TR/TS :XX = XXX :SLOPETR = A/XXX
260 YYY = PW/TS :YY = YYY
300 ZZZ = ((PW+TR/2+TF/2)/TS)-1 :ZZ = ZZZ :SLOPETF = A/(TF/TS)
340 FOR L = 0 TO NC - 1
350 FOR J = RRR TO XXX :INPUTRE(J) = (J-TTT+TT)*SLOPETR
370 NEXT J
380 FOR J = XXX TO YYY :INPUTRE(J) = A
400 NEXT J : K = 1
420 FOR  J = YYY TO ZZZ :INPUTRE(J) = A-(K*SLOPETF) :K = K + 1
450 NEXT J
460 FOR J = ZZZ+1 TO TTT-1
470 INPUTRE(J) = 0
480 NEXT J
490 RRR = RRR+TT:XXX=XXX+TT:YYY=YYY+TT:ZZZ = ZZZ+ TT:TTT = TTT+TT
540 NEXT L
```

Figure 2.4 FFT with prompts for rectangular waveform generation.

Prompts for Input Parameters

The input method in which the computer prompts for the input parameters is more elegant and excludes the possibility of the program being accidentally changed by the user. This method of input, shown in Figure 2.4, allows the program to be compiled so that it will run much faster in the computer.

Manual Data Input and Graphical Digitizing

Data can be manually digitized and inserted into the array INPUTRE. The READ-DATA commands may be used to put the input data in array INPUTRE as shown in Fig. 2.5. Manual digitizing can be accomplished by photographing the waveform with a scope

```
10 REM        MANUAL DATA INSERTION INTO ARRAY INPUTRE
90 REM   ****INPUT DATA FOR XFORM HERE--REAL DATA INTO ARRAY
100 FOR I = 0 TO 255 :READ INPUTRE(I)
120 NEXT I
130 DATA 0,.2,.7,1.5,2.5,3.8,5.8,7,14,17,19.5
140 DATA 20.7,23.1,25.3,27,28.3,29.4,30.5
150 DATA 31.6,32.7,33.8,35,38.4,38.4,31,36,34
160 DATA 32.9,33.3,33.6,33.4,33.3,33.1,32.9
480 REM  Lines 170 trough 480 not shown for brevity
490 DATA 2.5,2.2,1.98,1.77,1.56,1.34,1.13
500 DATA .91,.7,.47,.23,0
510 K = 0
520 FOR I = 0 TO 254 :TMP(K) = INPUTRE(I) :K = K+1
550 TMP(K) = INPUTRE(I) - ((INPUTRE(I) - INPUTRE(I+1))/2) :K = K+1
570 NEXT I
580 FOR K = 0 TO N-1 :I = K :INPUTRE(I) = TMP(K)
610 NEXT K
620 FOR J = 0 TO N-1 :INPUTRE(J) = (INPUTRE(J)*2.7)+65
640 NEXT J
650 FOR J = 0 TO N-1 STEP 8
660 LPRINT INPUTRE(J) INPUTRE(J+1) INPUTRE(J+2) INPUTRE(J+3) INPU-
TRE(J+4) INPUTRE(J+5) INPUTRE(J+6) INPUTRE(J+7)
670 NEXT J
```

Figure 2.5 Manual data insertion into array by hand digitizing scope digitized data.

camera. The photograph may need to be enlarged with a copy machine to provide better resolution. Transparent graph paper can be placed over the copy so that the data points can be recorded. Any convenient scale can be used to record the data. The data can then be scaled appropriately in the computer as shown in Fig. 2.5, line number 620. To increase the resolution of the data, a linear interpolator is included in the program to insert extra points between the recorded points. When the input waveform is displayed by the program, a comparison can be made to the original photograph to ensure that the data is representative of the actual waveform.

Digitized Data from a Storage Scope or a Waveform Recorder

A digital storage scope or a waveform recorder may be interfaced directly to the computer or the data may be recorded on a floppy disk and transferred to the computer. The data are loaded into array INPUTRE. To improve the quality of the data, many waveforms may be averaged to get better results. About five cycles is the optimum number so that the frequency domain resolution will be maximized. A detailed description of the best methods is presented in *Fundamentals of FFTs*.

2.6 AN ENHANCED VERSION OF FFT

An FFT version that includes prompts for easy generation of common switched waveforms is listed in the appendix. This FFT program will analyze rectangular waveforms such as power switching waveforms and computer clock signals. The input sections of Figs. 2.2 through 2.5 or any desired mathematical relation can be used in place of the prompted type of inputs.

Running the FFT Program

After loading the program in BASIC, press the run key F2. The program prompts for the number of input data points. The response must be a binary number such as 256, 512, or 1024. The program then prompts for the period of the window, period of the waveform, pulse width, rise and fall times, and amplitudes. The window period is considered to be repeating to infinity both forward and backward in time. The time domain data input prompts are shown in Fig. 2.6 with example user data inputs underlined.

```
Ok
RUN
RECORD LENGTH MUST BE A POWER OF 2.
INPUT RECORD LENGTH.? 512
WHAT IS THE TIME PERIOD OF THE WINDOW WIDTH?
PERIOD IS? 20e-6
WHAT IS AMPLITUDE A ?? 28
WHAT IS AMPLITUDE B ?? 28
WHAT IS PULSE WIDTH?? 2e-6
WHAT IS PERIOD OF WAVEFORM?? 10e-6
WHAT IS RISE TIME ?? 1e-9
WHAT IS FALL TIME ?? 1e-9
```

Figure 2.6 Input data prompts.

The time domain data will be displayed in the window with the vertical axis automatically scaled to fill the CRT monitor (Fig. 2.7). The program finds the maximum and minimum vertical values and scales the display accordingly.

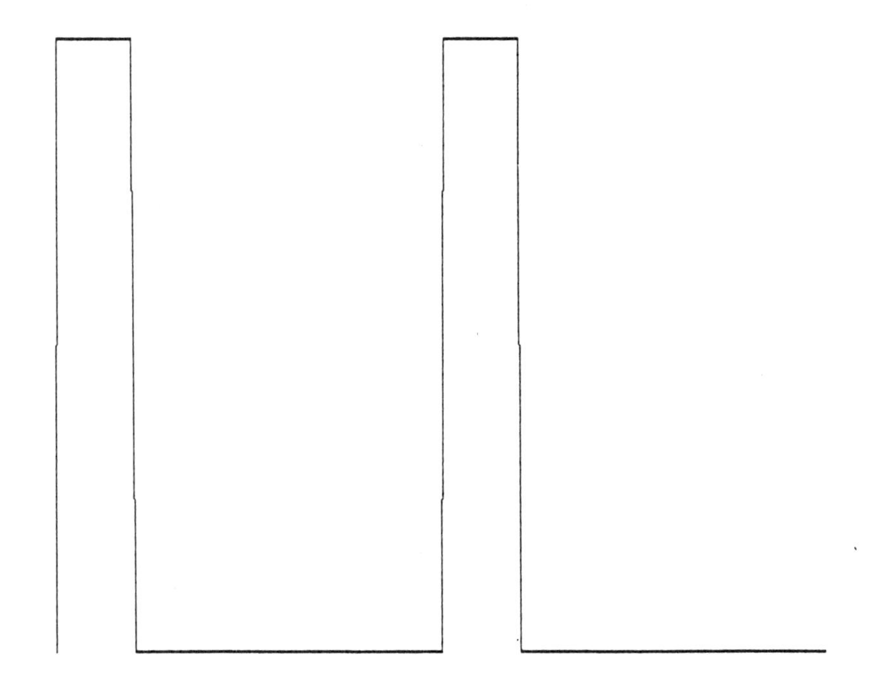

Figure 2.7 Plot of input data displayed on the monitor screen.

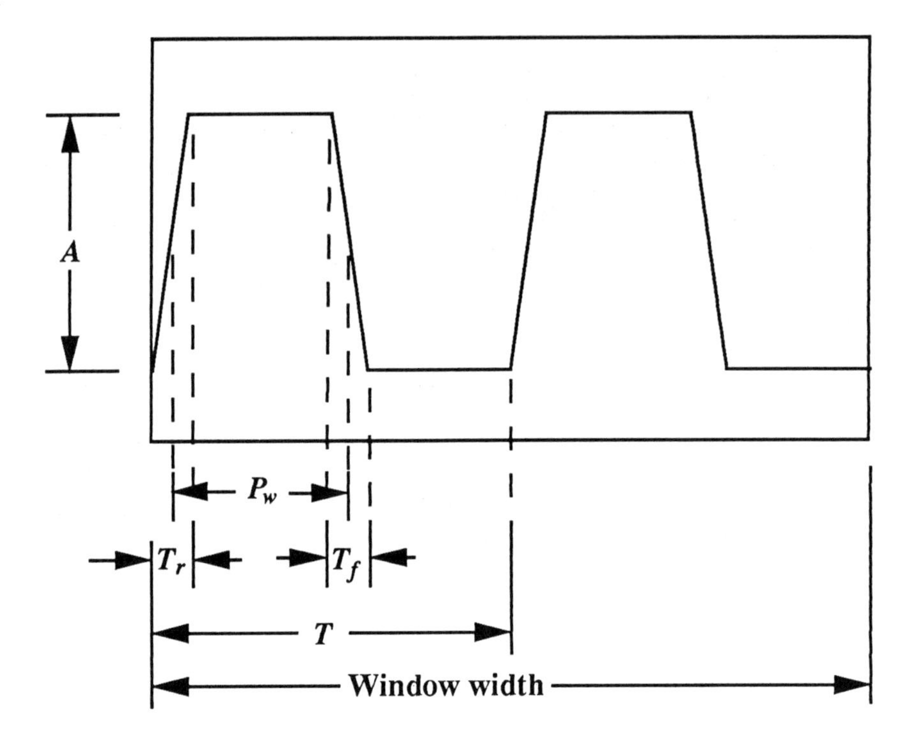

Figure 2.8 Diagram of input variables used for the prompted FFT version.

The program prompts for amplitude, pulse width, period, rise time, and fall time, which are graphically defined in Fig. 2.8. The maximum and minimum values of each plot are displayed and a question mark prompt is displayed. The return key is pressed to continue at each question mark prompt. These pauses can be used to "screen print" the graphs and data as desired.

At the prompt for forward or inverse FFT, the program is instructed to process the input arrays as time or frequency domain data. The forward FFT treats array INPUTRE(J) as time domain data and ignores INPUTIM(J). The real part of the frequency domain output data amplitudes is stored in array INPUTRE(J) and the imaginary part of the data is stored in array INPUTIM(J). The data are converted to time domain amplitudes by Eq. (2.1) and stored in the array MAG(L).

When the FFT algorithm is processed, the output data is out of order and requires manipulation to be presented in an orderly fashion. This data manipulation is a relatively significant portion of the program.

The plots for the FFT inverse time domain output and the frequency domain output phase data plots have not been developed. These developments are left to the reader.

The frequency domain output amplitude data are first plotted on a linear versus linear frequency scale (amplitude versus frequency) as shown in Fig. 2.9. Zero frequency (dc) is displayed in the center of the plot. Negative frequencies are plotted left of center and positive frequencies are plotted to the right of center. Because of the wide range of amplitudes of important data, the log versus log plot, shown in Fig. 2.10A, is a better way to visually observe the data. The scale is selected to be either decibels above 1μA or decibels above 1μV as needed. Horizontal graticules are added after the question mark prompt has been responded to. At this point in the program, the frequency components can be labeled with the screen cursor and a screen print can be performed to print out the scaled and labeled plot. A complete table of output data can

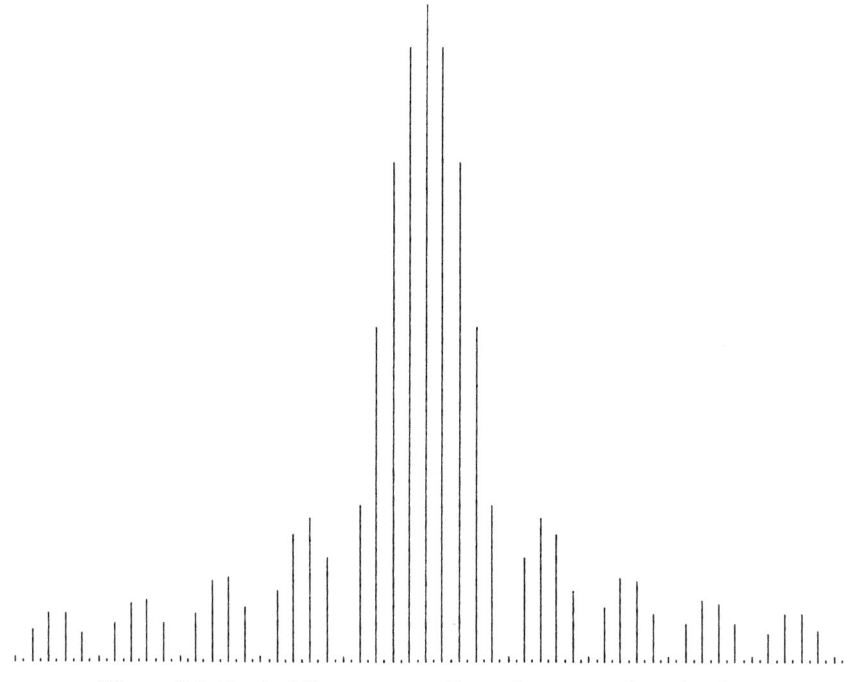

Figure 2.9 Typical linear versus linear frequency domain plot.

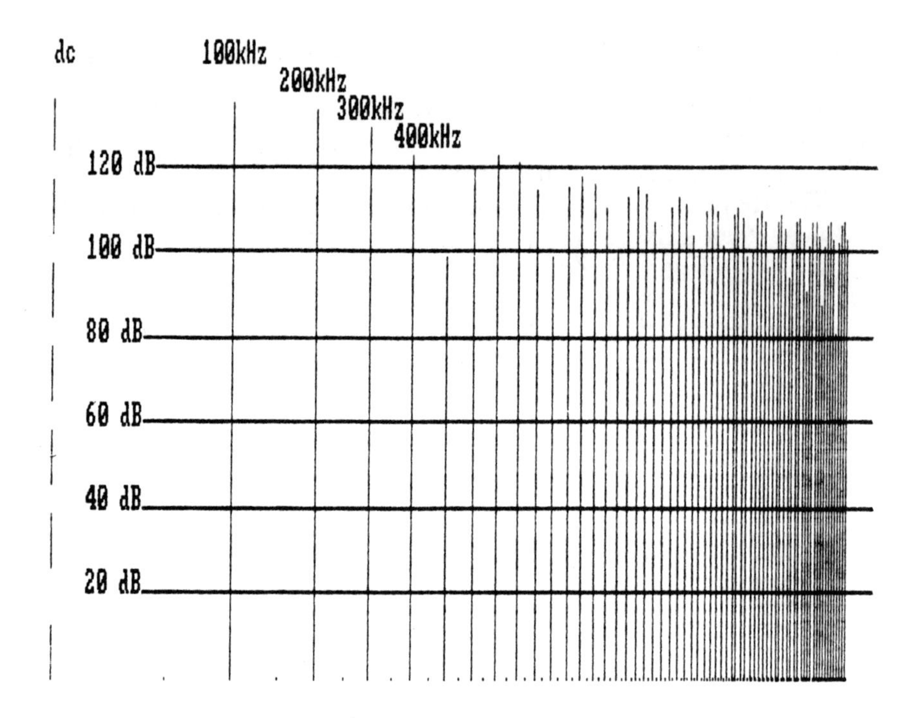

Current (A)	dB/micro Amp	Frequency
5.6875	135.0984	0
.000001	0	50000
5.311409	134.5042	100000
.000001	0	150000
4.271308	132.6112	200000
.000001	0	250000
2.807362	128.9660	300000
.000001	0	350000
1.245895	121.9096	400000
.000001	0	450000
8.834908E-02	98.9240	500000
.000001	0	550000
.9561202	119.6103	600000
.000001	0	650000
1.259725	122.0055	700000
.000001	0	750000
1.056218	120.4751	800000

Figure 2.10 Typical FFT log versus log plot and printed table of data.

be printed if desired. A table of the first 17 components is presented in Fig. 2.10B.

Frequency markers can be set to any harmonic desired so as to identify high-frequency features. High-frequency harmonics are quite closely spaced and are hard to discern. The lower-order harmonics can be removed from the CRT display so that the higher-frequency components can be inspected more closely. The number of lower harmonics to be removed can be selected at two consecutive prompts.

2.7 EXAMPLES OF FFT CONVERSIONS FROM TIME TO FREQUENCY DOMAINS

Our first example is conversion of dc to the frequency domain. The program input for a continuous 10 V dc is as follows:

```
FOR J = 0 TO N-1
INPUTRE(J) = 10
NEXT J
```

The frequency domain representation, shown in Fig. 2.11, is a single component at zero frequency or dc.

Program input for a sine wave is:

```
FOR J = 0 TO N-1
INPUTRE(J) = 10*SIN(J*2*3.14159/N)
NEXT J
```

A pure sine wave transformed into the frequency domain will have only one component and will be located at its frequency as shown in Fig. 2.12.

Multiple sinusoids can be created as shown in Fig. 2.13. This waveform is a fundamental with amplitude of 10, the second harmonic is at an octave above with amplitude of 1, and the fourth harmonic is two octaves above the fundamental, with an amplitude of 1/10. The program input is:

```
FOR J = 0 TO N-1
INPUTRE(J)=10*SIN(J*2*3.14159/N)
+SIN(J*4*3.14159/N)+.1*SIN(J*8*3.14159/N)
NEXT J
```

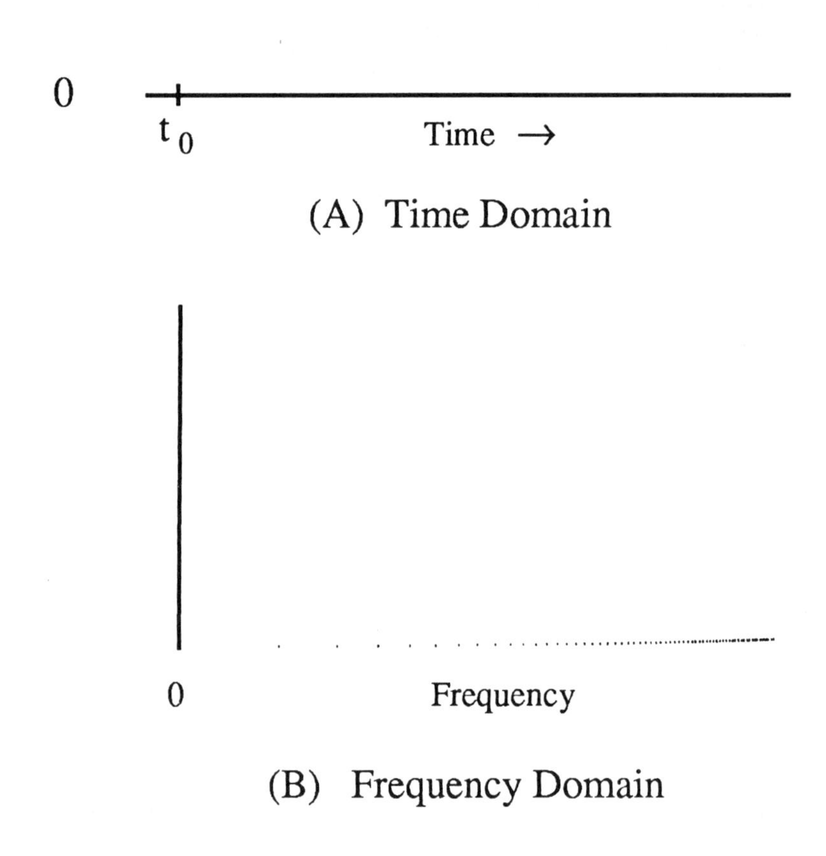

Figure 2.11 Time domain to frequency domain conversion for dc.

Rectangular waveforms with fast rise times have many high-frequency components as shown in Fig. 2.14. Rectangular waveforms with relatively slow rise times have very few high-frequency harmonics. The shape appears to be somewhat sinusoidal. There are far fewer high frequency components in Fig. 2.15B than in Fig. 2.14B.

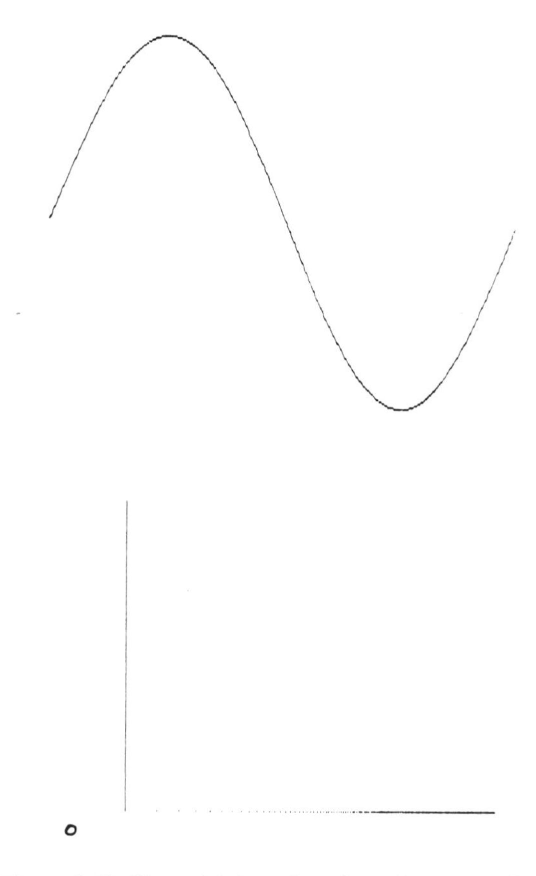

Figure 2.12 Sinsuoid time domain to frequency domain.

Rectangular waveforms generally have exponentially damped sinusoids that occur at the rise and fall times. The damped sinusoids are usually referred to as "ringing"; if they were audio frequency, the damped sinusoids would sound somewhat like a bell ringing. This ringing is caused by the resonant nature of the circuit layout capacitances and inductances responding to the voltage and/or current changes. The program input is shown in Fig. 2.16.

The trapezoidal current waveform in Fig. 2.17 is a typical waveform in switching power supplies' circuits. It is the input current waveform for buck and forward convertors. The typical damped exponential "rings" are included at the switch rise and fall times.

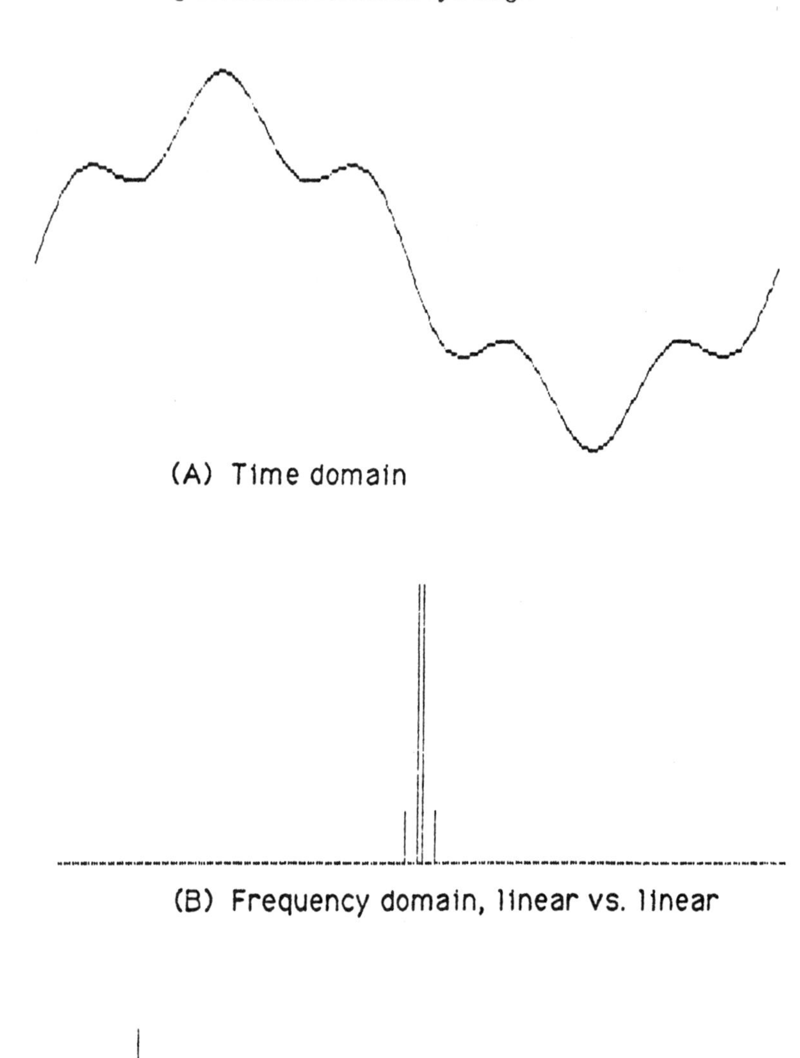

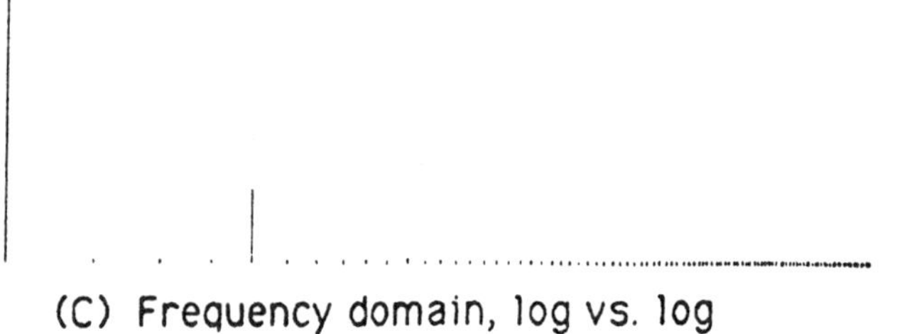

Figure 2.13 Multiple sinusoid time domain to frequency domain.

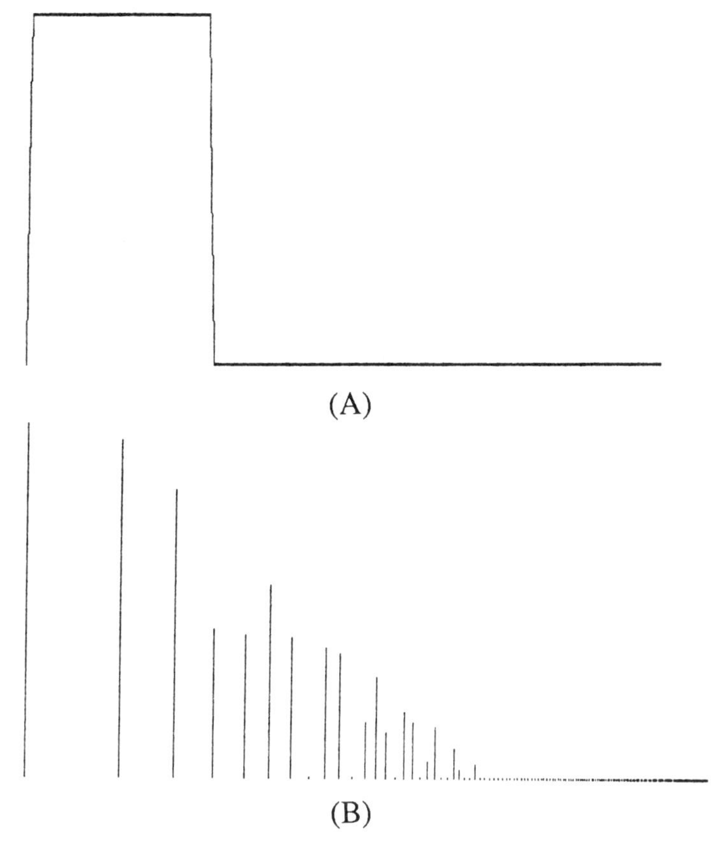

Figure 2.14 Rectangle fast rise time domain to frequency domain.

The damped exponentials have much less energy than the trapezoidal components. The frequency domain components of the damped exponential are corresponding small compared to the exponential components in the resulting frequency domain representation. Often what is referred to as a noisy waveform in the time domain is somewhat like not being able to see the forest because of all of the trees. The program input is shown in Fig. 2.18.

Phase angle control is a very common method of controlling ac power for motors, lamps, and many other applications. With this control method the EMI is maximum at a 90° conduction angle. The high frequency components of the EMI are caused by the fast rise or fall time at the time of switching the conduction state. At 90° the switching voltage amplitude is at a maximum. For large conduction angles the frequency domain representation has a large

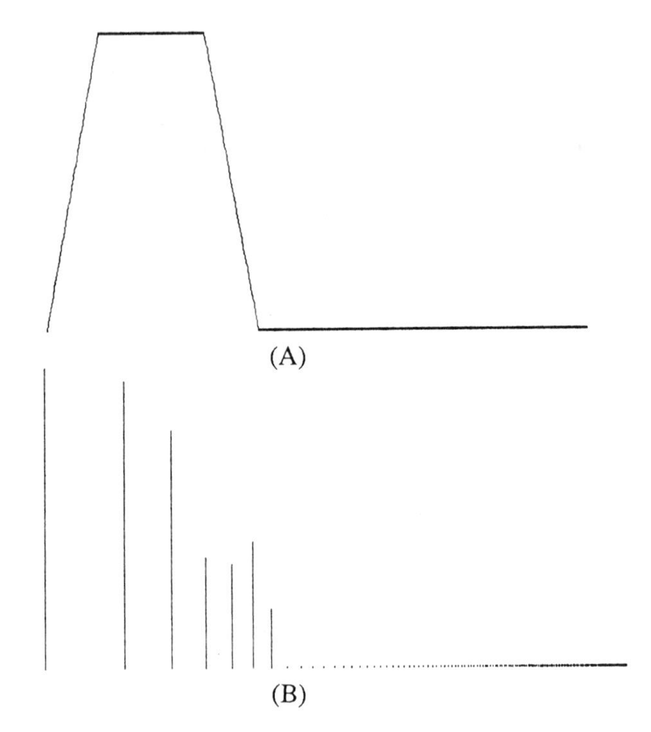

Figure 2.15 Rectangle slow rise time domain to frequency domain.

```
100 INPUTRE(0) = 0
102 INPUTRE(1) = 25
104 INPUTRE(2) = 50
106 INPUTRE(3) = 75
108 FOR J=4 TO 149
110 INPUTRE(J) = 100+(J/10)+20*SIN(J)*EXP(-.06*J)
112 NEXT J
114 INPUTRE(150) = 75
116 INPUTRE(151) = 50
118 INPUTRE(152) = 25
120 FOR J = 153 TO N-1
130 INPUTRE(J) = 20*SIN(J)*EXP(-.06*(J-152))
140 NEXT J
180 FOR K = 0 TO N-1
190 INPUTIM(K) = 0
200 NEXT K
```

Figure 2.16 Trapezoid with damped exponentials.

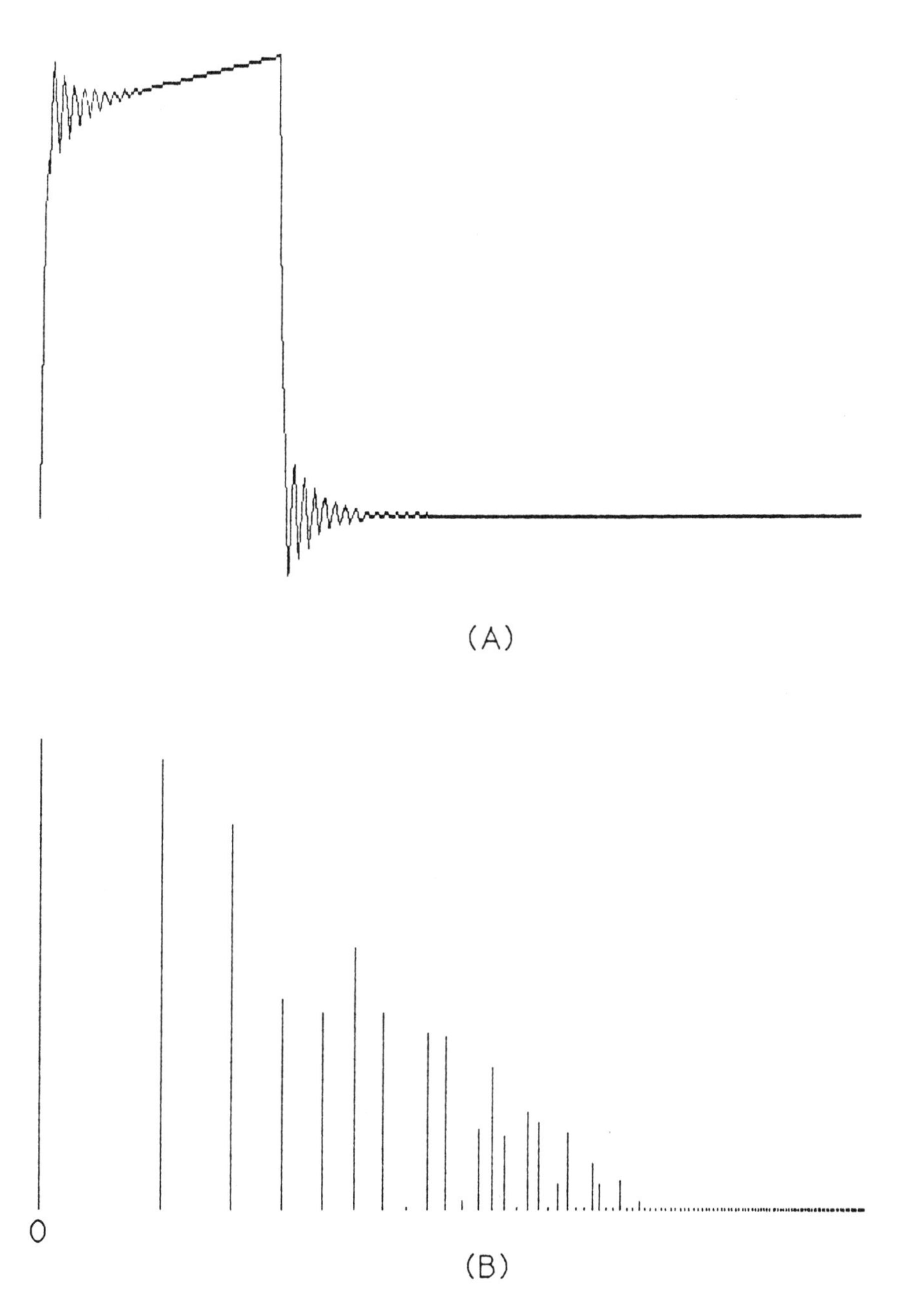

Figure 2.17 Trapezoid with exponential ringing time domain to frequency domain.

```
100 FOR J=0 TO ((N-1)/8)
101 INPUTRE(J)=0
102 NEXT J
103 FOR J=((N-1)/8) TO ((N-1)/2)
104 INPUTRE(J)=155*SIN((2*3.1416*J)/(N-1))
105 NEXT J
106 FOR J=((N-1)/2) TO ((5*(N-1))/8)
107 INPUTRE(J)=0
108 NEXT J
109 FOR J=((5*(N-1))/8) TO N-1
110 INPUTRE(J)=155*SIN((2*3.1416*J)/(N-1))
111 NEXT J
180 FOR K = 0 TO N-1
190 INPUTIM(K) = 0
200 NEXT K
```

Figure 2.18 Double damped exponential.

sinusoidal component plus a spectrum that is caused by the small triangular shape of the waveform that is the nonconducting portion of the waveform (see Figs. 2.19A and B). For very small conduction angles, the fundamental sinusoidal component is greatly reduced (see Figs. 2.19C and D), but the high-order harmonics are the same. The frequency domain representation components resulting from the small triangular shape during the "on time" of the waveform in Fig. 2.19C are hardly distinguishable from a triangle waveform of the same duty cycle and amplitude.

2.8 SOME POSSIBLE PITFALLS

The start and stop data points must be chosen so that the repeated time domain data in the window replicates the desired waveform. The data in Fig. 2.20A "starts and stops," therefore the repeated waveform is a sinusoid with no discontinuities. The corresponding frequency domain data is shown in Fig. 2.20B. The same sinusoidal waveform is shown in Fig. 2.20C, but the start and stop data points make a discontinuity at the edges of each window. The resulting frequency domain data with aliasing is shown in Fig. 2.20D. The repeated waveform of Fig. 2.20C is shown in Fig. 2.21. If this were the desired waveform, the results would not be described as aliased.

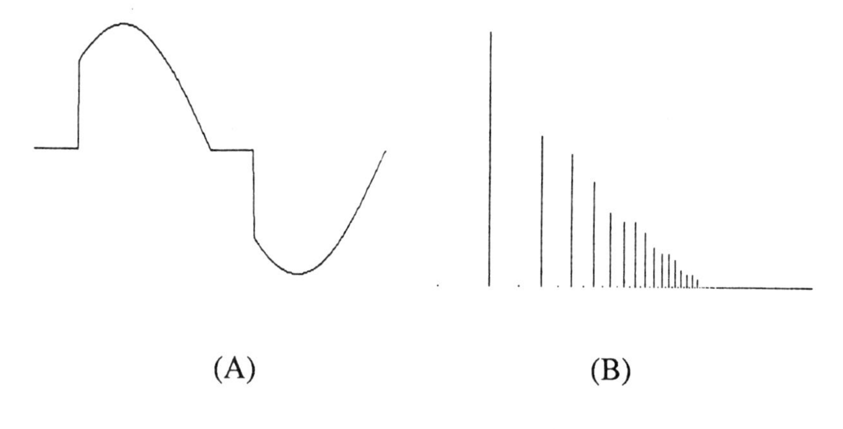

Conduction angle = 135 °

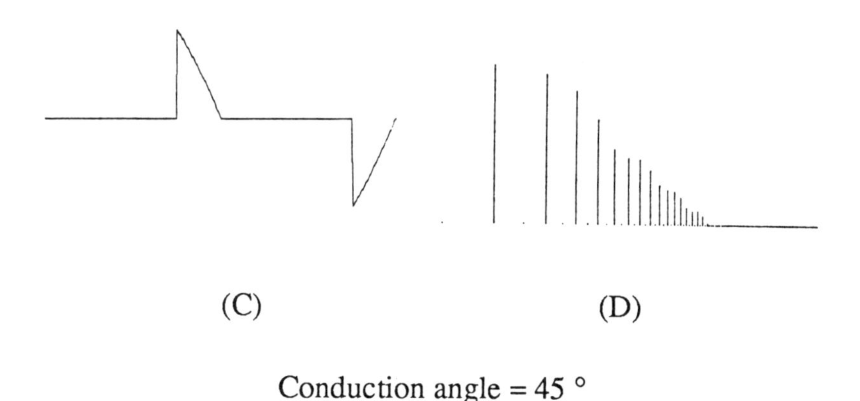

Figure 2.19 Phase control time domain to frequency domain.

2.9 SUBHARMONICS

A waveform that is very stable (exactly repeating) has no subharmonic frequency content. A waveform that varies (doesn't repeat exactly) will have subharmonic components at the frequency of variation of the waveform. If the frequency varies, the spectral components will spread correspondingly at each harmonic.

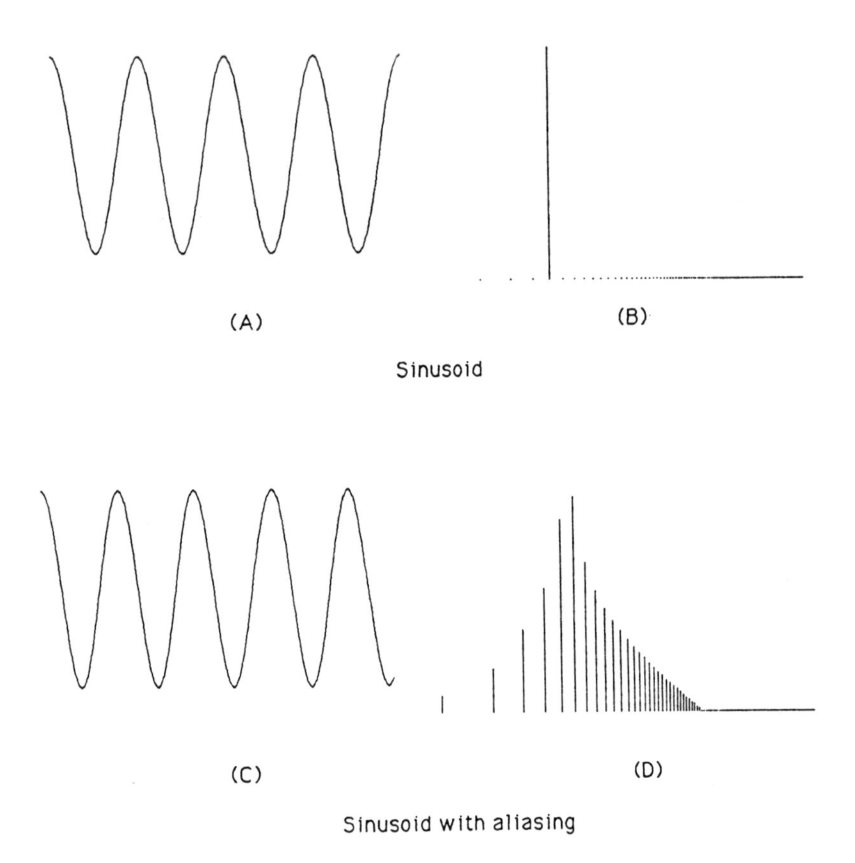

Figure 2.20 Aliasing.

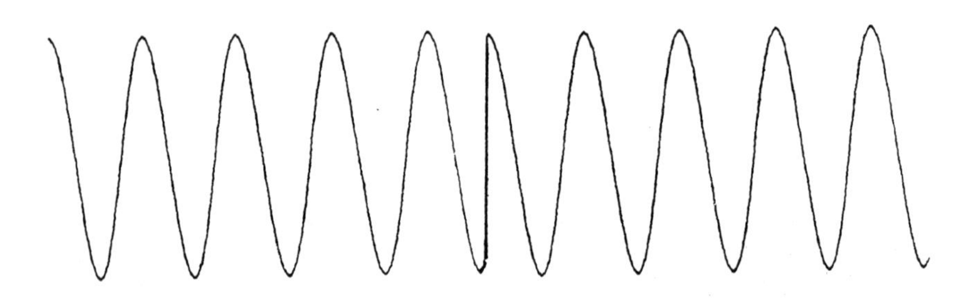

Figure 2.21 Aliased time domain waveform.

QUESTIONS

2.1 Why do we generally use a frequency domain of circuit waveforms in EMI work?

2.2 Contrast the Fourier transform and the Fourier series.

2.3 For what applications, in EMI simulation or analysis, could the FFT be used?

2.4 List the kinds or methods of data input to the FFT.

2.5 Why do we like to plot the frequency domain data logarithmically?

2.6 What time domain waveform feature affects the amplitude and bandwidth of the high-order harmonics in the frequency domain?

3

CAPACITOR MODELING

There are many types of capacitors available to the power supply designer for both commercial and military use. The choice of a particular type of capacitor for each specific application depends upon the requirements. Capacitor applications include such functions as:

1. Bypassing (local energy storage)
2. Filtering
3. Timing
4. Holdup

Often the criteria for choosing capacitors for particular applications are:

1. Effectiveness
2. Size
3. Reliability

The model in Fig. 3.1A is a model of an ideal capacitor. The capacitive reactance X_c is $1/(2\pi fC)$. An ideal capacitor has no losses so that the impedance is equal to the reactance. The ideal capacitor impedance reduces with frequency at the rate of 20 dB per decade. The reactance is proportional to $1/f$.

A capacitor can be constructed from two large plates separated by a small distance. The capacitance is equal to $\sigma A/d$, where σ is the permeability of free space ($8.855\ e^{-12}$), A is the area of the plates (square meters), and d is the distance between the plates (meters). A capacitor made with two plates in air would behave somewhat like an ideal capacitor. A capacitor made with plates of $1m^2$ separated by 0.1 mm. (approximately the thickness of a human hair) has a capacitance of about 0.09 µF. The physical size of this capacitor is too large to be very useful in most applications.

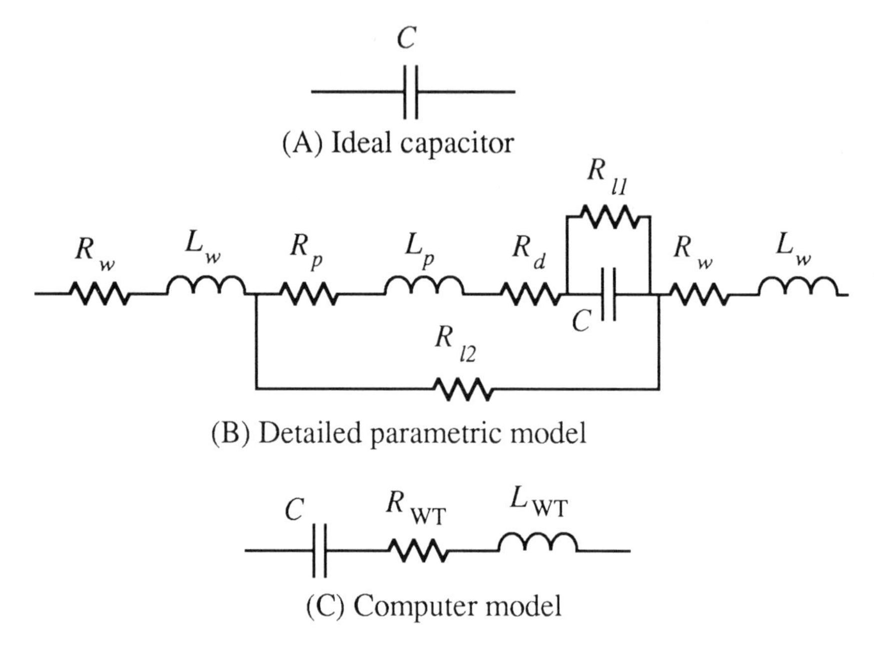

Figure 3.1 Capacitor model.

To reduce the size of the capacitor a dielectric substance of many times the permeability of free space is placed between the plates. The high-frequency performance is greatly affected by the added dielectric. We are willing to trade the dielectric losses and poor performance incurred to get a much larger capacitance for a given size. We must consider and allow for these accepted parasitics in the design process. At high frequencies the dielectric losses increase the equivalent series resistance and limit the minimum impedance at resonance. This series resistance is often a major design parameter in low voltage power conversion.

The capacitor model in Fig. 3.1B includes the detailed parasitic elements. These features are:

R_w	lead wire resistance
L_w	lead wire inductance
R_p	plate resistance
L_p	plate inductance
R_d	dielectric losses
C	capacitance
R_l	leakage resistance

Leakage Resistance

The leakage resistance is dependent upon the area of the plates of the capacitor. The capacitance is directly proportional to the plate area. Generally the equivalent leakage resistance is very large and does not affect the filter performance of a capacitor.

Resistance

The equivalent series resistance of a capacitor is the summation of the lead, plate, and equivalent resistances of the dielectric losses. These resistive losses dissipate energy in the form of heat. These losses also play a significant role in the performance of a capacitor when used in a filter. These losses are all frequency dependent.

At high frequencies, R_w increases because of the "skin effect." The impedance in the center portion of the wire increases so that current flows mostly in the outer "skin" of the conductor. With increasing frequency the effective conductor cross-section area decreases. The plate resistance suffers the same skin effect, but it has much less of an effect since the plate surface area is large.

Dielectrics

Dielectric materials used in capacitors are insulators in which the constituent molecules can be polarized so as to enhance the amount of capacitance possible in a given volume. To avoid confusion, two concepts should be differentiated:

1. Dielectric constant is a measure of the increase in stored charge in a capacitor with a dielectric material placed between the two electrodes of a capacitor.
2. Dielectric strength is a measure of the voltage stress a material can withstand without physical degradation or arcing.

The two most important dielectric dipole characteristics are mobility (speed) and amount of work required for polarization. For example, aluminum electrolytic dielectrics have relatively long molecules. They produce the most capacitance per unit volume but polarize slowly because of the length of the dipole molecules and their mobility. Tantalum electrolytic dipoles are much shorter in length than aluminum oxide dipoles and also are much more

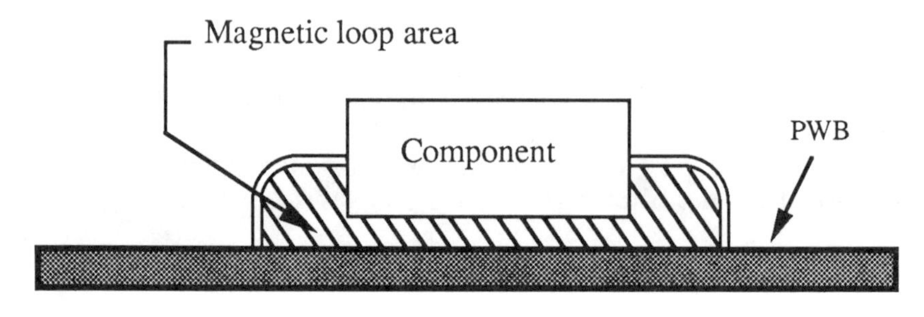

(A) Small loop area

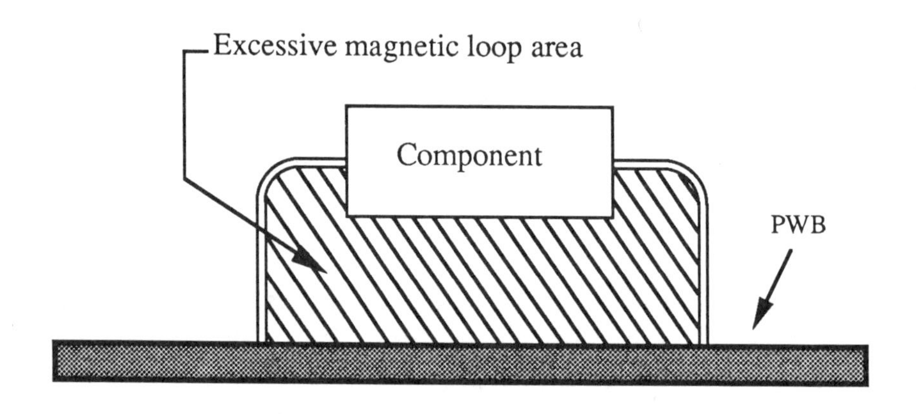

(B) Large loop area

Figure 3.2 Excessive loop area lowers resonant frequency.

mobile. This difference accounts for the better frequency performance of tantalum electrolytic capacitors compared to aluminum.

Capacitor Inductance

The inductance L_w associated with the lead wire typical of most capacitors is about an average of 20 nH per inch. The lead wire length contributes to the loop area of the capacitor conduction path as shown in Fig. 3.2. The larger loop area causes a larger induc-

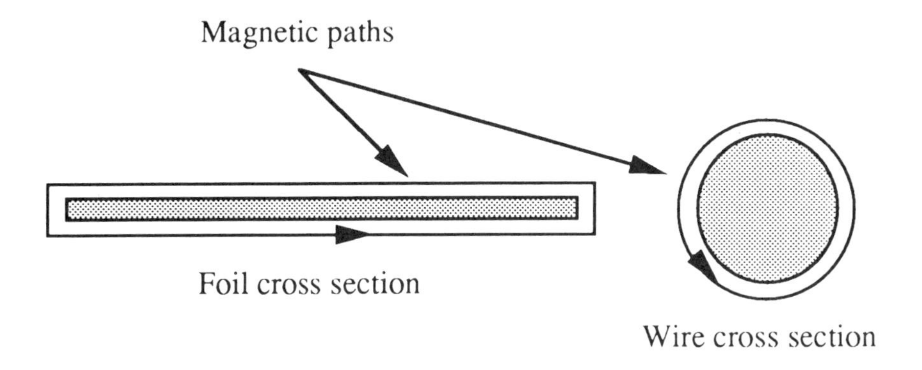

Figure 3.3 Inductance of wire versus foil.

tance. This parasitic inductance causes a lower "in circuit" resonant frequency and thus limits the effective bandwidth of the capacitor.

The inductance of the plate of a capacitor is very small. For a given amount of current flowing in a wire or plate (equal conductor cross-sectional area) the flux around the wire is greater than the flux around the plate (see Fig. 3.3). The permeability of air is 1, so the flux density B is equal to the magnemotive force H, which is equal to NI/L_m, the number of turns times the current divided by the magnetic path length. For a wire or plate the number of turns is one. The current in the wire and plate are given to be equal. It is also assumed that the loop areas of all the circuits are equal. The magnetic path length around the plate is much longer than the wire so that the flux density around the wire is much larger. The plates in a capacitor are closely spaced so that the loop area internally is very small. A plate has much less inductance than the wire for a given length and current. If this were not true, capacitors would be very inductive and would resonate at such low frequencies that they would not be usable.

3.1 THE CAPACITOR MODEL

The model in Fig. 3.1C represents the simple summation of the resistances and inductances in the model. The leakage resistance is removed since it does not affect filter operation appreciably. This model is useful for most EMI conducted emission analysis.

Capacitor Impedance Curve

The impedance of an ideal capacitor reduces at a rate of 20 dB per decade or $1/f$ as frequency increases to infinity. "Real world" capacitors resonate at the frequency of $1/(2\pi(LC)^{1/2})$ where L is the series inductance of the capacitor. At frequencies above resonance the capacitor is inductive and the impedance increases at the rate of 20 dB per decade. At frequencies above the resonant frequency the inductance is proportional to the slope of the impedance curve. In the frequency range around resonance the capacitance value is not very useful and is often referred to as the "apparent" capacitance. Capacitance or inductance measurements are difficult in the frequency range of resonance because the slope of the impedance curve is nearly zero. Measurements of the small difference between two large numbers are difficult. When doing analysis, the impedance is the most useful quantity as a measure of effectiveness throughout the frequency range.

The curves in Fig. 3.4 are impedance plots of the elements of a generic capacitor model. The inductive and capacitive reactances and the resistance are plotted separately. The plot labeled Z = square root of $(R^2 + X^2)$ is the impedance of the whole model. The resonant frequency is labeled W_0. At frequencies lower than W_0 the model is capacitive. For frequencies higher than W_0 the model is inductive. This simple model is very useful in EMI and filter analysis. The model is very accurate from dc through the RF frequency range.

3.2 PARASITIC ELEMENTS OF CAPACITORS

The elements R and L are parasitic to the performance of the filter, but they are not shown on schematics. These parasitics come about from the choice of component types and the circuit layout. The parasitics are crucial design parameters yet their importance is often underestimated. The control of these parasitics is a crucial part of the design effort for EMC.

The circuit designer cannot pass a schematic on to a packager or a layout technician and expect to achieve a good design. The circuit designer must be involved in the PWB layout so that the para-

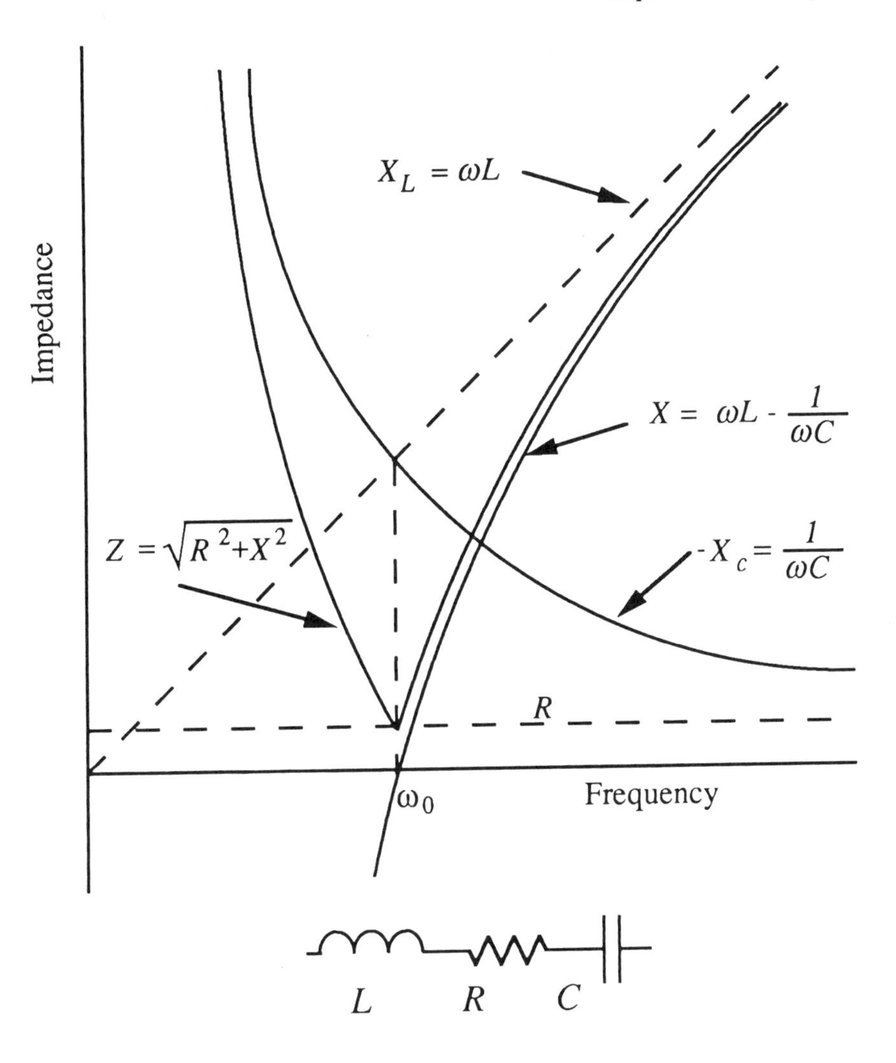

Figure 3.4 Impedance curves of capacitor model.

sitic elements do not degrade the filter to a degree that it will not meet the design requirements. Component types are so different that a parts list must be more specific than to just call out a 10–μF 50-V-dc capacitor. In many cases, the substitution of a capacitor of the wrong style or type but same capacitance and voltage rating may cause the circuit to not function as required. It is important that the designer chooses the appropriate capacitor type for each application.

Measured Capacitor Impedance Curve

The impedance of a capacitor can easily be measured with a Hewlet Packard 4274A or 4275A Multi-Frequency LCR Meter or a similar capacitance bridge. A form used for recording the data is shown as Fig. 3.5. A table of measured data for a typical tantalum capacitor is shown in Table 3.1. The data were measured with a Hewlet Packard 4275A Multi-Frequency LCR Meter.

The data are then keyed into the impedance curve program on a Mentor workstation. The measured impedance data are plotted in Fig. 3.6. The dotted line is the ESR, which is composed of the resistance in the wire leads and capacitor plates added to the equivalent resistance caused by the dielectric losses. The wire and plate resistance is a function of frequency because of the skin effect. The dielectric equivalent resistance is also a function of frequency because of the increase in work required at higher frequencies to polarize the electric dipoles in the dielectric between the capacitor plates. The capacitor impedance at resonance is resistive and is the ESR.

Capacitor Model Impedance Curve

Impedance is defined as Z = square root of $(R^2 + (X_c\text{-}X_l)^2)$, where R is the resistance, X_l is the inductive reactance, and X_c is the capacitive reactance. The value of X_l is $2\pi fL$. The value of X_c is $1/(2\pi fC)$. The capacitor model impedance measurement circuit is shown in Fig. 3.7. The voltage across the capacitor divided by the current through the capacitor is the impedance. At the resonant frequency the reactance of the capacitor and the inductance are equal and cancel each other so that measured impedance is an equivalent series resistance.

A plot of the data generated from the model of the test circuit in Fig. 3.7 is presented in Fig. 3.8. The modeled plot is much smoother than the measured data because there are many more points plotted. In the measured plot of Fig. 3.6 straight lines are plotted between points. Even though the ESR in the model is not a function of frequency, there is a very good correspondence between the measured data and the modeled data. At either higher or lower frequencies around the resonant frequency the reactance is large with respect to the ESR, therefore the frequency effects of the ESR are generally insignificant for modeling purposes.

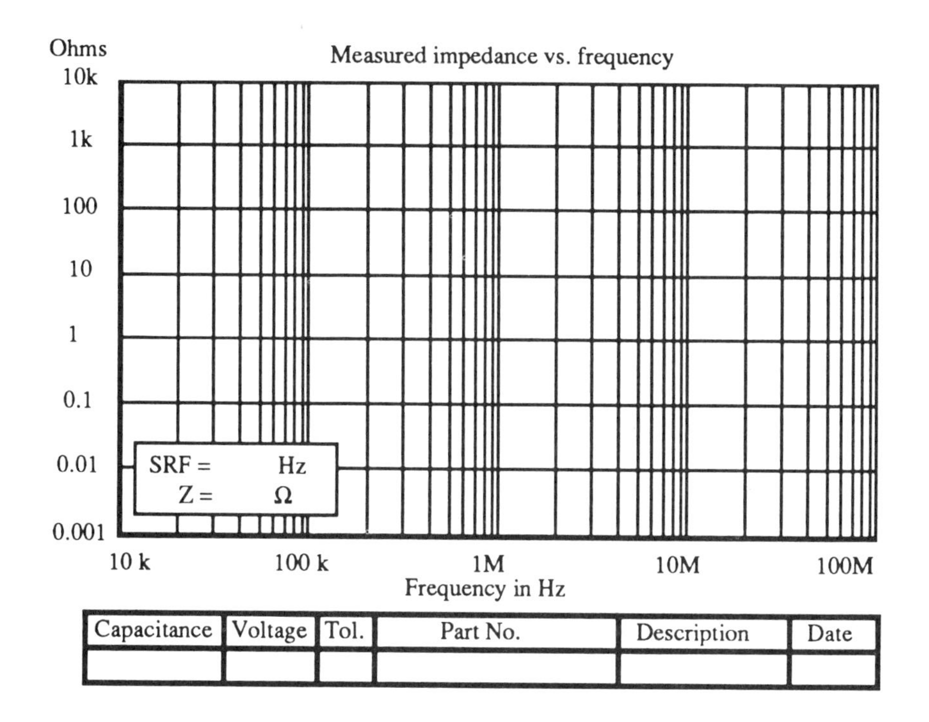

Table of Measured Data

Frequency	C (μF)	ESR (Ω)	Z (Ω)	Phase	EPC (pF)
10k 20k 40k					
100k 200k 400k					
1M 2M 4M					
10M 20M 40M					
100M					

Figure 3.5 Capacitance data worksheet.

Table 3.1 Measured capacitor data.

Frequency	C (μF)	ESR (Ω)	Z (Ω)	Phase	EPC (pF)
10k	0.905	0.100	175.000	-89.94	
20k	0.905	0.100	87.930	-89.94	
40k	0.905	0.100	43.970	-89.91	
100k	0.916	0.074	27.570	-89.81	
200k	0.948	0.074	8.755	-89.66	
400k	1.085	0.069	4.308	-89.34	
1M	0.903	0.069	1.740	-87.94	
2M	0.904	0.074	0.812	-84.98	
4M	0.905	0.081	0.375	-77.46	
10M			0.098	-8.80	
20M			0.238	63.90	17.00
40M			0.542	75.00	20.80
100M			1.383	81.34	21.70

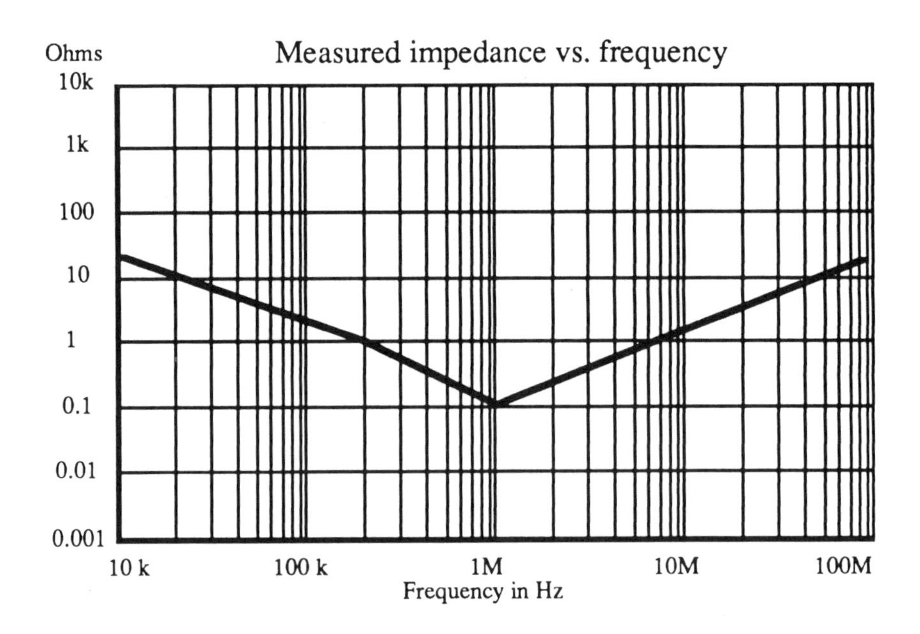

Fig 3.6 Plot of measured capacitor data.

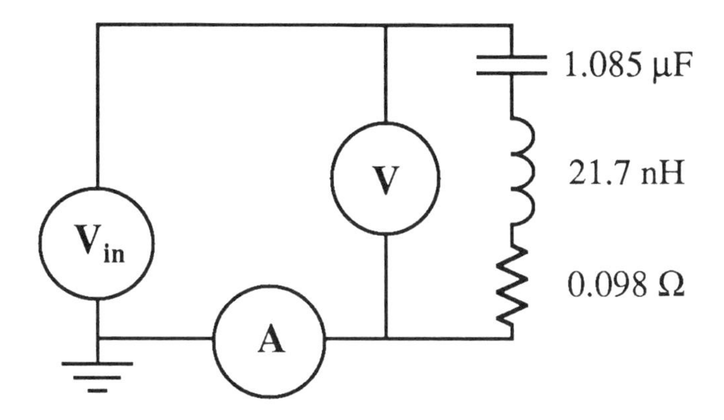

Figure 3.7 Capacitor model and test circuit.

3.3 CAPACITOR TYPES

A few of the most important capacitors in EMI applications have been modeled. The following is a brief description of the capacitor types analyzed along with their impedance curves, models, and the impedance curves of the models.

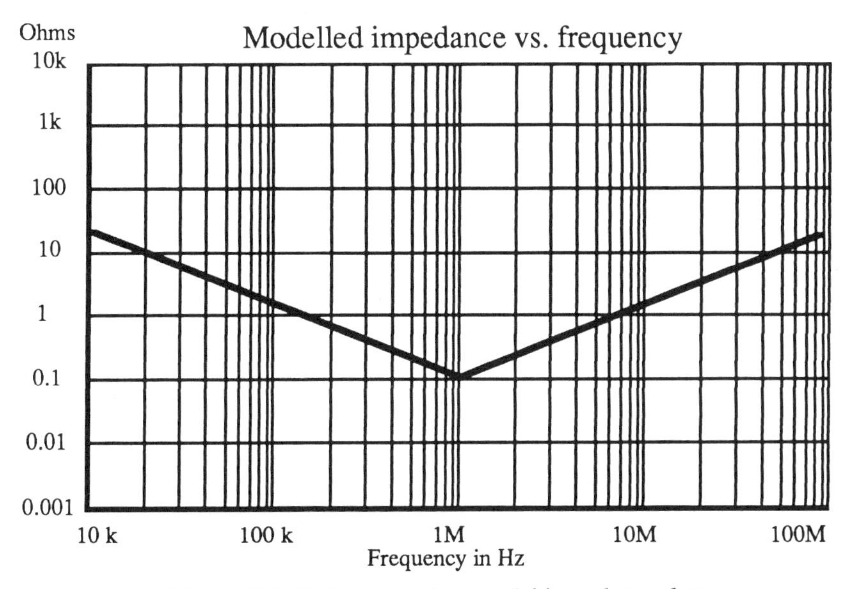

Figure 3.8 Plot of capacitor model impedance data.

Electrolytic Capacitors

Aluminum electrolytic capacitors are intended for use in filter and coupling applications where large capacitance values are required in small cases and where excesses of capacitance over the nominal value can be tolerated. In applications above 75 Vdc, aluminum electrolytics are usually chosen over tantalum electrolytics because of the relatively lower voltage rating of tantalums.

Tantalum electrolytic capacitors are the most stable and most reliable electrolytic capacitors available, having a longer life characteristic than any of the other electrolytic capacitors. Because their passive electrolyte is solid and dry, these capacitors are not temperature sensitive; they have a lower capacitance temperature characteristic than any of the other electrolytic capacitors.

The dielectric in CSR21 tantalum capacitors takes less work (force through a distance) to polarize the dipoles than the CSR13 tantalum capacitors, therefore the ESR is smaller in value. The CSR21 tantalum capacitors are also more stable than the CSR13 type. On the other hand, the dielectric strength of the CSR13 is higher than the CSR21. The CSR21 tantalum capacitors are only available with voltage ratings up to 50 Vdc. The CSR13 tantalum capacitors are rated up to 75 Vdc. Tantalum capacitors resonate in the frequency range of 100 kHz to 400 kHz. The military types M39003/01 and M39003/09 are the equivalents to the commercial types CSR13 and CSR21, respectively.

The wet slug tantalums (M39006/*xx* or CLR) are not recommended for new designs. Wet slug tantalums have poor reliability and performance when compared to the dry solid tantalums.

Ceramic Capacitors

Ceramic type dielectrics come in many varieties. Ceramic capacitors are not as stable as some other dielectrics and are not recommended for frequency determining circuits in general. There are some special temperature compensated types that can be used in some less critical frequency determining circuits. Ceramic dielectrics have a very good frequency response into the 50 MHz range. The CKR05 and CKR06 types are extremely compact and have inherent low series inductance because of their construction. The placement of the leads facilitates making close-coupled low-inductance connections. They are useful for bypass applications and for general purpose small capacitors. They are also very useful in some filter applications.

The new multilayer ceramic capacitors offer very low ESRs. These multilayer ceramic capacitors can handle larger ripple currents than the tantalums. As a result, the new ceramics are fast becoming an important component for use in filters. The new ceramic ESR is so low that the resulting high Q can cause problem oscillations at the resonant frequency of the capacitor and its inductive leads. Ceramic capacitors are not recommended for input filters for power supplies that switch at frequencies between 100 kHz and 300 kHz. They are excellent for output filters and applications where their low ESRs do not cause resonance problems.

Polycarbonate Capacitors

Polycarbonate capacitors of the CRH type (military version M83421/*xx*) use metallized plastic film for a dielectric. These are nonpolar capacitors that can be used in ac, dc, or a combination of ac and dc voltage applications. The dielectric absorption is relatively low. The capacitance is very stable with temperature changes. The polycarbonates have very high resonant frequencies and are very good capacitors. The dielectric strength is high; therefore higher voltage types (up to 600 Vdc) are available.

The relatively low dielectric constant of the polycarbonate causes these capacitors to be physically larger for a given capacitance than many other types of capacitors. Correspondingly the plate size is larger and so are the associated plate resistance losses. The ratio of plate resistance losses to dielectric absorption losses is relatively larger for the CRH capacitors than for the electrolytics or ceramics. The physical size of the CRH capacitors limits the ability to achieve low series inductance in a circuit layout. This is probably the most serious detriment to the high-frequency performance of the CRH type of capacitor.

3.4 CAPACITOR VOLTAGE RATINGS

The voltage rating for a dc capacitor is the maximum dielectric withstanding voltage; which is the voltage that the manufacturer deems will be safe from the dielectric breaking down. Arcs of current pass across the plates when the capacitor dielectric breaks down. For short periods of time the voltage may surge to a value 10 to 50 percent higher than the continuous maximum voltage rating without arcing or causing other damage to the dielectric.

Direct Current Capacitor Ripple Current Ratings

Direct current capacitors are generally used in applications where there is a relatively large dc voltage and a relatively small ac ripple voltage. The ripple current resulting from the ripple voltage is equal to $C(\delta v/\delta t)$. This ripple current through the lossy elements of the capacitor causes a power dissipation equal to the RMS value of the ripple current squared times the equivalent resistance of the lossy elements of the capacitor. This power is dissipated as heat. The maximum ripple current is based on the temperature that the capacitor can withstand without damage.

Alternating Current or Line Capacitor Voltage Ratings

Capacitors designed for ac applications are rated differently from capacitors designed for dc applications. Alternating current or line capacitors are designed for use directly across power lines or in applications where the voltage across the capacitor is virtually all ac. The working voltage rating for an ac capacitor is really a ripple current rating. This is a continuous operation rating. The manufacturer of an ac capacitor calculates (and measures) the RMS current for a given capacitance at the line frequency and voltage. The capacitor is designed to withstand the maximum temperature that the capacitor will be subjected to from its internal heating and the external environment. The dielectric withstanding voltage is generally much higher and determines the maximum surge voltage that can be applied. The thermal time constant of the capacitor is an important factor in rating the allowable time period of a surge.

QUESTIONS

3.1 Describe a design situation (not mentioned in this text) in which the parasitic elements of the components or a component are the "key players" in the performance of the circuit.

3.2 Is a capacitor reactive at its self-resonant frequency?

3.3 Describe all of the loss mechanisms of a capacitor at self-resonance that contribute to the equivalent resistance (or the ESR Equivalent Series Resistance).

3.4 What are the package layout advantages of SMDs (Surface Mount Devices) that contribute to the performance improvements over the radial or axial leaded capacitors?

3.5 What capacitor characteristics are improved by the use of dielectric materials? Why?

3.6 What capacitor characteristics are degraded by the use of dielectric materials? Why?

3.7 At what frequency (in low impedance filter applications) does the ESR of a capacitor limit the filter attenuation performance?

4

INDUCTOR MODELING

An ideal inductor is modeled in Fig. 4.1. The inductive reactance is:

$$X_l = 2\pi fL \tag{4.1}$$

An ideal inductor has no losses so that the impedance is equal to the reactance. The ideal inductor reactance (impedance) increases with increasing frequency at the rate of 20 dB per decade. The impedance (reactance) is directly proportional to the frequency.

Five Turn Air Core Inductor Model

The model in Fig. 4.2 represents a five turn air core inductor. The five turns are labeled L_{t1} through L_{t5}. The dc resistances are labeled R_{t1} through R_{t5}. Capacitors C_{tt1} through C_{tt5} represent the turn to turn capacitance. Resistors R_{tc1} through R_{tc5} provide the equivalent resistance at high frequencies. The insulation material coated on the inductor wire is dielectrically polarized. The associated absorption losses are modeled as the equivalent resistances of R_{a1} through R_{a5}. Air core inductors are used in receivers and for other high frequency filter applications.

Air core inductor performance is closer to the ideal inductor than any other device. A simple air core inductor can be constructed by coiling a single layer of 21 turns of #14 AWG wire, 2 in. in length and 1/2 in. in diameter. The inductance would be approximately 1 μH. For a 1 mH coil the physical size would be

Figure 4.1 Ideal inductor.

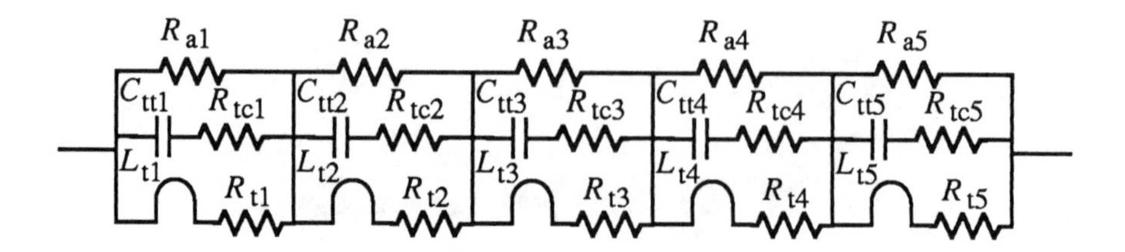

Figure 4.2 Air core inductor.

64 in. long by 1/2 in. dia. The applications for this 5 foot 4 in. inductor would be greatly limited because of its length.

To reduce the size of this 1 mH inductor to the size of the 2 in. by 1/2 in. dia. inductor, a core of 1000 relative permeability could be inserted in the coil. The high-frequency performance is greatly affected by the added permeable core. At high frequencies, the core losses limit the maximum impedance and the core capacitance reduces the self-resonant frequency. At frequencies beyond resonance the inductor's filter effectiveness is greatly diminished. Even with these detrimental effects, we are willing to trade the core losses and poor high-frequency performance incurred to get a much larger inductance for a given size. But we must consider and allow for these accepted parasitics in the design process.

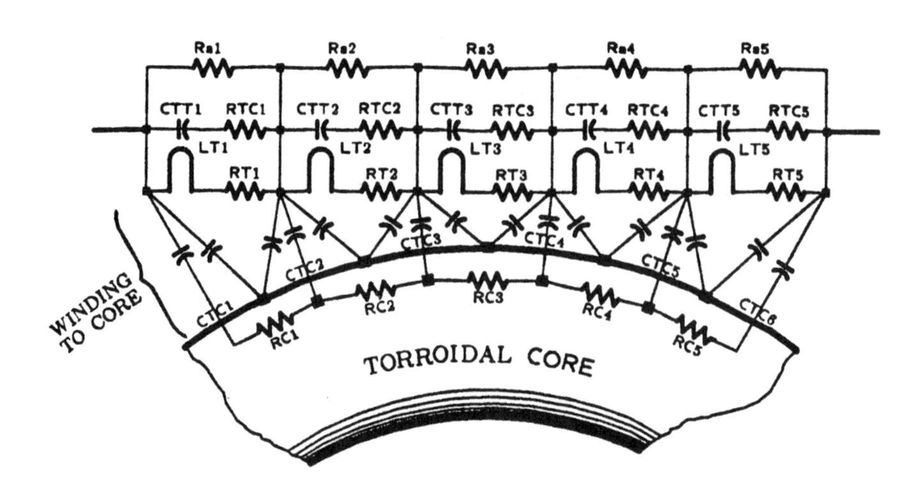

Figure 4.3 Detailed five turn torroid model.

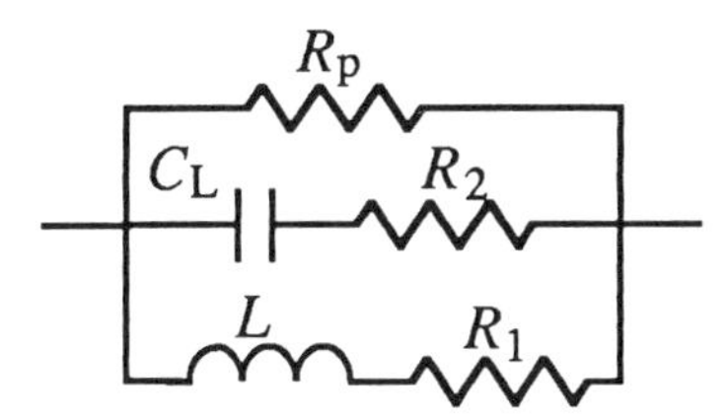

Figure 4.4 Computer simulation inductor model.

Five-Turn Inductor with a Permeable Core Model

The model in Fig. 4.3 represents five turns wound on a torroidal core. The capacitance effects in this model are common to inductors wound on all core geometries. Only a section of the torroid is drawn for the sake of simplicity. The five turns are labeled the same as in the air core model of Fig. 4.2 but with the addition of capacitances C_{tc1} through C_{tc6} to represent the turn to core capacitance. The core conductivity is modeled by resistors R_{c1} through R_{c5}. The winding to core capacitances are in parallel with the inductor. These capacitances lower the self-resonant frequency of the inductor. For inductors wound on cores of high resistivity, (low permeability distributed gap, for example) the turn to turn capacitance will be the largest factor in the parallel parasitic capacitance. For inductors wound on conductive cores, (ferrite, for example) the turn to core and core to turn capacitance may be the largest factor lowering the self-resonant frequency.

The model in Fig. 4.3 can be reduced down to the equivalent model in Fig. 4.4. The model in Fig. 4.4 is very useful in computer simulation. It is much less cumbersome than the model in Fig. 4.3 and performs as well in simulation. For most applications the differences are insignificant.

4.1 INDUCTOR LOSSES

Equivalent resistances are used to model the effects of the inductor losses, which are frequency dependent and can be classified into three types:

1. Core losses
2. Resistive losses
3. Dielectric losses

The two major core losses are hysteresis losses and eddy current losses. Hysteresis losses are created by the work of polarizing the magnetic domains within the core. The eddy currents are induced in the core, perpendicular to the flux path. The resistance of the core material causes a parasitic power loss from the eddy current. The energy is transformed to heat and is dissipated. The equivalent resistance R_p models these core losses.

The resistive losses in the wire are frequency dependent. At high frequencies, the impedance of the area in the center of the inductor wire is high so that the current flows on the outer "skin" of the wire. This resistive loss is modeled by the equivalent series resistances R_1 and R_2.

The dielectric losses of the insulation are insignificant in inductors wound on cores. The core losses are the principal loss mechanism. For air core inductors, there are no core losses; R_p (Fig. 4.4) is then used as the equivalent resistance for the dielectric losses.

4.2 INDUCTOR CAPACITANCE

The Effects of the Spacing of the Winding Turns

There are many methods for winding the turns of an inductor on a core. Performance, type of application, and manufacturing cost are some of the criteria in determining which is the best winding method. Generally the turns of a winding are as close together as possible to maximize magnetic coupling. Magnet wire used for inductors and transformers has a heavy synthetic coating for electrical insulation. This coating is usually only a few thousandths of an inch thick. The small distance between the turns sometimes creates a significant capacitance between turns.

Capacitances from turn to turn are formed as shown in Fig. 4.2. The turn to turn capacitances in a single layer air core are in series. The capacitance C_s of the series combination of n capacitors (C_{tt}) is equal to:

$$C_s = 1/(nC_{tt}) \tag{4.2}$$

where n is the number of turns. The fact that these capacitances are in series greatly reduces the negative effect they have on performance. This series combination of capacitors resonates with

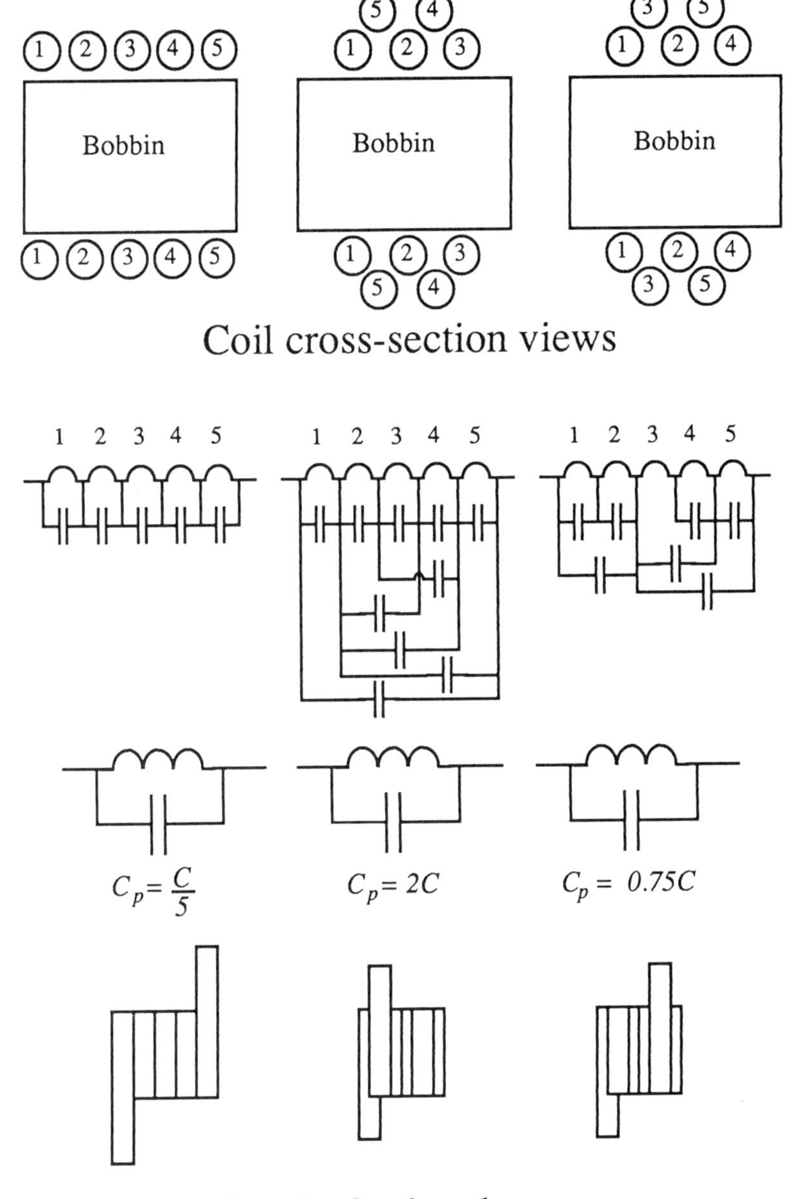

Figure 4.5 Winding methods.

the inductance of the coil. These parasitics affect the high frequency performance. The self-resonant frequency is:

$$F_{sr} = 1/(2\pi(LC_{tt}/n)^{1/2}) \quad (4.3)$$

The Effects of Winding Methods

Multilayer winding methods can exacerbate or reduce the inter-winding capacitance. Two methods of winding are compared in Fig. 4.5. The capacitance of turn 1 to turn 5 is directly paralleled across the inductor. This capacitance is a significant factor in lowering the self-resonant frequency of the inductor. Bank winding separates the input to output lead breakouts (which reduces the capacitance) so as to increase the self-resonant frequency of the inductor. The input and output leads in the circuit layout should also be separated since this capacitance is directly parallel to the inductor and can be very significant.

4.3 AIR CORE WITH CONDUCTOR NEAR EXPERIMENT

The Effects of Core Conductivity on Self-Resonance of Inductor

Conductors physically in the vicinity of inductors serve as capacitive paths around the inductor. This parallel capacitance resonates with the inductor and causes the "in circuit" self-resonance of the inductor to be lower in frequency. This effect can be observed by measuring the resonant frequency of an air core inductor connected to an impedance bridge. The impedance bridge is set to the resonant frequency of the air core inductor. When conductors are placed near the inductor, the resonant frequency and the impedance become lower.

4.4 INDUCTOR CORES FORM CAPACITIVE PATHS

To get more inductance from a given physical size, inductors are wound on permeable cores. Along with the increased inductance comes some detrimental effects. The core actually provides a capacitive path around the inductor that reduces the self-resonant frequency. The turn to core and core to turn capacitance can be larger than the turn to turn capacitance. High-permeability cores are generally more conductive than low-permeability cores. A conductive inductor core contributes more capacitance to an inductor than a less conductive core inductor with an identical winding. Core materials often used in inductor applications are made from powdered ferrite suspended in a filler and binder that provides a distributed air gap for energy storage. The conductance of this type of core material is much less than for solid ferrite core material.

An experiment was performed to compare the effect of core conductivity on the capacitance of inductors wound on permeable cores. The turn to turn spacing was also varied to compare the relative effects of spacing to the effects of core conductivity on the capacitance of inductors. Three inductors were wound with five turns of #20 AWG wire, one with an air core, one on a Moly-perm (distributed gap) core of 26-µ permeability, and the third using a ferrite 3E2A core (solid powdered ferrite) with approximately 4000-µ permeability (Fig. 4.6). The windings and cores of these three examples are the same physical size and configuration.

The five-turn air core inductor self-resonant frequency was unmeasurable with the available test equipment. From the phase angle measured at the highest frequency of the impedance bridge (112 MHz), it is estimated that the self-resonant frequency of the air core inductor is several hundred megahertz. The air core inductor capacitance is less than 1 pF.

The conductance of the Moly-perm core material is much less than that of solid ferrite core material. This type of core material is often used in inductor applications. With five turns on this core the inductance is 580 nH. When the five turns are closely spaced as in Fig. 4.6B, the self-resonant frequency is 101 MHz corresponding to 4.28 pF. When the five turns are spread evenly around the core as in Fig. 4.6C, the self-resonant frequency is 112 MHz, which corresponds to a capacitance of 3.48 pF. The turn to turn capacitance of the winding is only about 19 percent of the total capacitance of the inductor.

A high-permeability (4000 µ) solid ferrite torroid core of the same size as the previous examples was used for comparison. This high-permeability core is generally used for transformers and baluns. Five turns of #20 AWG wire on this core result in approximately 46.8 µH of inductance. With the turns closely spaced as in Fig. 4.6B, the self-resonant frequency is 3.3 MHz corresponding to a capacitance of 49 pF. When the five turns are spread evenly

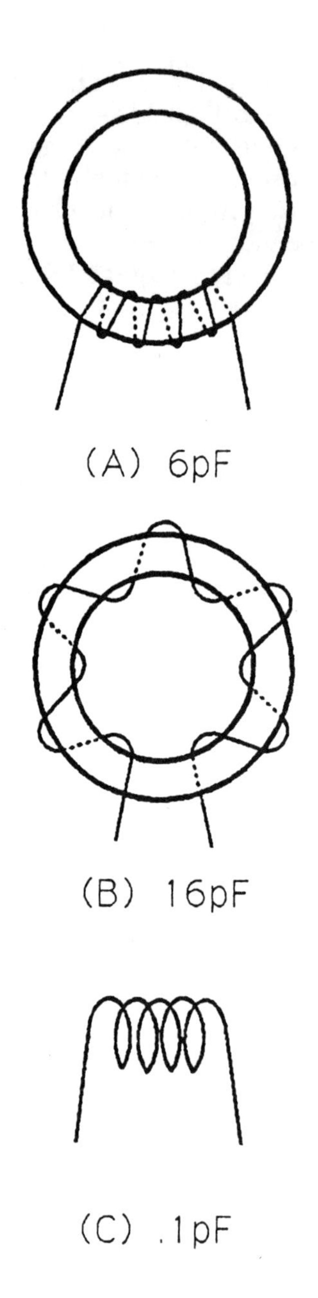

Figure 4.6 Inductor capacitance.

around the core as in Fig. 4.6C, the self-resonant frequency is 3.25 MHz, which corresponds to a capacitance of 51 pF. In this example, the turn to turn capacitance is only about 4 percent of the total inductor capacitance.

We can now compare the capacitance of the three inductors with different conductivities. The windings and cores of these three examples are the same physical size and configuration. The capacitance of the more conductive core inductor is about 50 pF. The capacitance of the less conductive core inductor is about 4 pF. The capacitance of the air core inductor (nonconductive) was less than 1 pF (unmeasurable with the available test equipment).

The capacitance of the winding on the higher conductivity core is more than 12 times larger than the winding on the lower-conductivitv core and more than 50 times the capacitance of the air core. The conductivity of the core material has a more significant effect on inductor capacitance than the capacitance from turn to turn. The core actually provides a capacitive path around the inductor and reduces the self-resonant frequency of the winding coil.

4.5 INDUCTOR IMPEDANCE CURVE

Inductance is the electrical "dual" of capacitance. A simple inductor model is a parallel connection of ideal components (Fig. 4.7).

The impedance of an ideal inductor increases at a rate of 20 dB per decade, proportional to frequency. The inductance is proportional to the slope of the impedance curve. Real world inductors resonate at the frequency of:

$$F_r = 1/(2\pi(LC)^{1/2}) \tag{4.4}$$

where C is the parallel capacitance of the inductor. At frequencies above resonance the inductor is capacitive and the impedance decreases at the rate of 20 dB per decade. At frequencies above the resonant frequency the capacitance is proportional to the slope of the impedance curve. In the frequency range around resonance the inductance value is not very useful and is often referred to as the "apparent" inductance. Capacitance or inductance measurements are difficult in the frequency range of resonance because the slope of the impedance curve is nearly zero. Measurements of the small

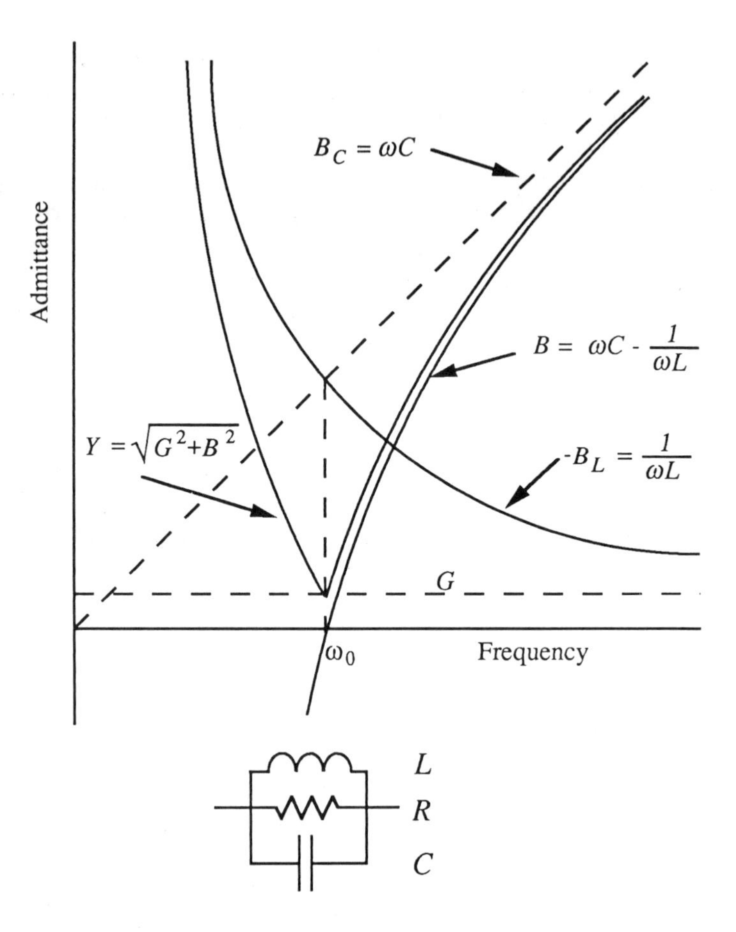

Figure 4.7 Impedance curves with elements plotted.

difference between two large numbers are difficult. When doing analysis, the impedance is the most useful quantity as a measure of effectiveness throughout the frequency range. The curves in Fig. 4.7 are impedance plots of the elements of a generic capacitor model. The inductive and capacitive reactances and the resistance are plotted separately. The plot labeled $Y = (G^2 + B^2)^{1/2}$ shows the admittance of the whole model. The impedance is the inverse of the admittance. The resonant frequency is labeled W_0. At frequencies lower than W_0 the model is capacitive. For frequencies higher than W_0 the model is inductive. This simple model is very useful in EMI and filter analysis. The model is very accurate.

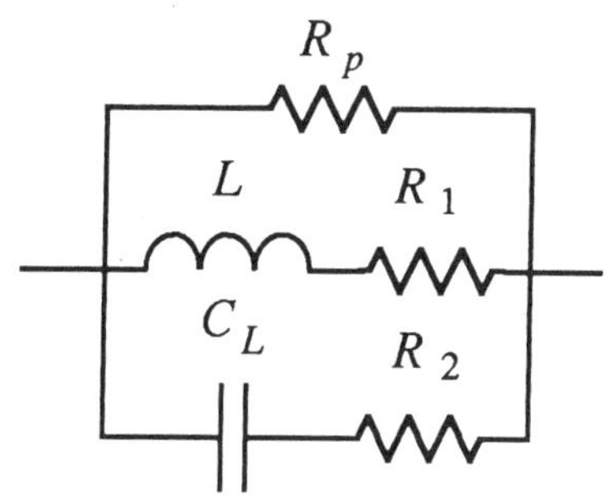

Figure 4.8 Computer simulation inductor model.

4.6 PARASITIC ELEMENTS OF INDUCTORS

The elements *R* and *C* are parasitic to the performance of the filter, but they are not shown on schematics. These parasitics come about in the choice of component types and the circuit layout. The parasitics are crucial design parameters yet are often underestimated in importance. The control of these parasitics is a crucial part of the design effort for EMC.

The circuit designer cannot pass a schematic on to a packager or a layout technician and expect to achieve a good design. The circuit designer must be involved in the PWB layout so that the parasitic elements do not degrade the filter to a degree that it will not meet the design requirements.

4.7 SIMULATION

The Simulation Inductor Model

Inductors used in filters are generally designed to operate in their linear region. For completeness of understanding and/or very special applications, a discussion of inductor saturation and the modeling of inductors in saturation is included in chapter 15.

Calculation of Inductor Model Element Values

The resistance R_1 is the measured dc resistance of the inductor (Fig. 4.8). Typically $R_p >> R_1$; therefore the dc resistance is set by R_1 and R_2 and is chosen to be about the same as R_1 but is larger in value by the percentage of the ratio R_{ac}/R_{dc} because of the skin effect resistance at high frequencies. The skin effect losses should be calculated at a frequency somewhat higher than the resonant frequency. The resistance R_2 affects the Q of the model and is the series resistance at high frequencies. The value of R_p is the measured impedance at the resonant frequency. For inductors wound on cores, R_p is an equivalent resistance that models the energy loss caused by the work of polarizing the magnetic domains within the core. For air core inductors, R_p is an equivalent resistance that is used to model the dielectric losses of the wire insulation.

The resistances in an inductor are all frequency dependent to some extent but generally can be modeled as frequency independent and still achieve good results. Most of the impact of the frequency effects on the resistances are reduced because they are used in conjunction with reactive elements.

The best way to choose a value for C_L is to measure the resonant frequency and calculate C_L from:

$$F_r = 1/(2\pi(LC_L)^{1/2}) \tag{4.5}$$

Inductor Impedance Curve Measurement Problem at Resonance

The measurement of inductor impedance at resonance can be difficult and deserves some special comment here. An ideal inductor would have infinite impedance at resonance. The impedance at resonance for an air core inductor is very high and the Q can be very large. The impedance of the measurement bridge may grossly affect the measurement accuracy. Even if the impedance of the bridge is high when compared to the impedance at resonance, a more difficult problem with impedance measurement exists. Impedance at a particular frequency is the voltage divided by the current. If the oscillator in the bridge has any harmonic distortion, current at harmonic frequencies will flow through the inductor, causing the measured impedance to be much smaller than the actual value. The losses in an air core inductor are so small that an accurate measurement of impedance at resonance is very difficult.

Measured Inductor Impedance Curve

The impedance of an inductor can easily be measured with a Hewlett Packard 4274A or 4275A Multi-Frequency LCR Meter or an impedance bridge. A form used for recording the data is shown in Fig. 4.9. A table of measured data for an inductor is shown in Table 4.1. This data was measured with a Hewlett Packard 4275A Multi-Frequency LCR Meter.

The data is then keyed into the impedance curve program on a Mentor workstation. The measured impedance data is plotted, as in Fig. 4.10, by the impedance curve program.

Inductor Model Impedance Curve

The inductor model impedance measurement circuit is shown in Fig. 4.11. The impedance of an inductor can be expressed as:

$$Z = (R^2 + (X_c - X_l)^2)^{1/2} \tag{4.6}$$

where R is the resistance, X_l is the inductive reactance, and X_c is the capacitive reactance. The value of X_l is:

$$X_l = 2\pi fL \tag{4.7}$$

The value of X_c is:

$$X_c = 1/(2\pi fC) \tag{4.8}$$

The inductor model impedance measurement circuit is shown in Fig. 4.11. The voltage across the inductor divided by the current through the inductor is the impedance. At the resonant frequency the reactance of the inductor and the inductor capacitance are equal and cancel each other so that measured impedance is an equivalent parallel resistance.

A plot of the data generated from the model of the test circuit in Fig. 4.11 is presented in Fig. 4.12. The modeled plot is much smoother than the measured data because there are many more points plotted. In the measured plot of Fig. 4.10, straight lines are plotted between points. Even though R_p in the model is not a function of frequency, there is a very good correspondence between the measured data and the modeled data. At either higher or lower frequencies around the resonant frequency the reactance is small with respect to R_p; therefore the frequency effects of R_p are generally insignificant for modeling purposes.

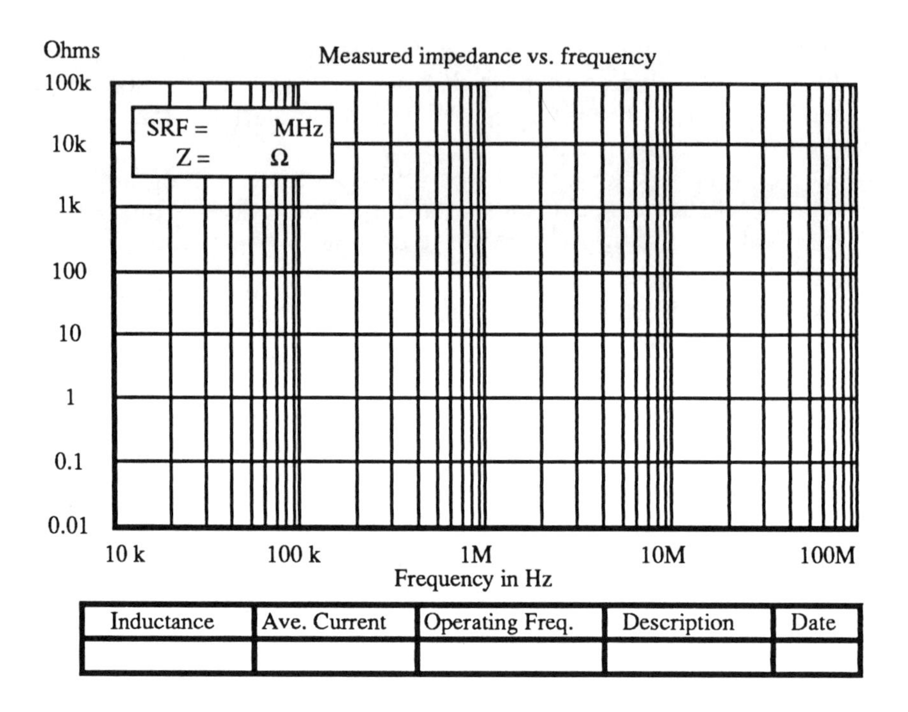

Inductance	Ave. Current	Operating Freq.	Description	Date

Table of Measured Data

Frequency	L (μH)	Z (Ω)	Phase	EPC (pF)
10k 20k 40k				
100k 200k 400k				
1M 2M 4M				
10M 20M 40M				
100M				

Figure 4.9 Inductor data worksheet.

Table 4.1 Measured data table.

Frequency	L (μH)	Z (Ω)	Phase	EPC (pF)
10k	302.8	19.0	89.50	
20k	302.6	39.0	89.60	
40k	302.4	76.0	89.60	
100k	302.1	190.0	89.40	
200k	303.5	381.4	89.09	
400k	312.3	785.0	88.47	
1M	405.7	2550	85.98	
2M		30500	-42.70	3.8
4M		2200	-87.00	17.9
10M		710	-88.90	22.4
20M				
40M				
100M				

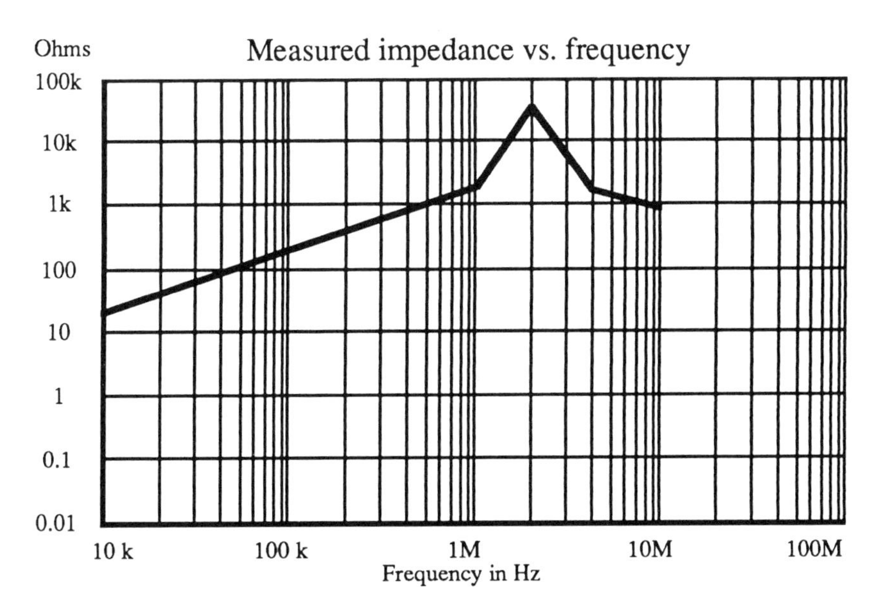

Figure 4.10 Measured data impedance curve.

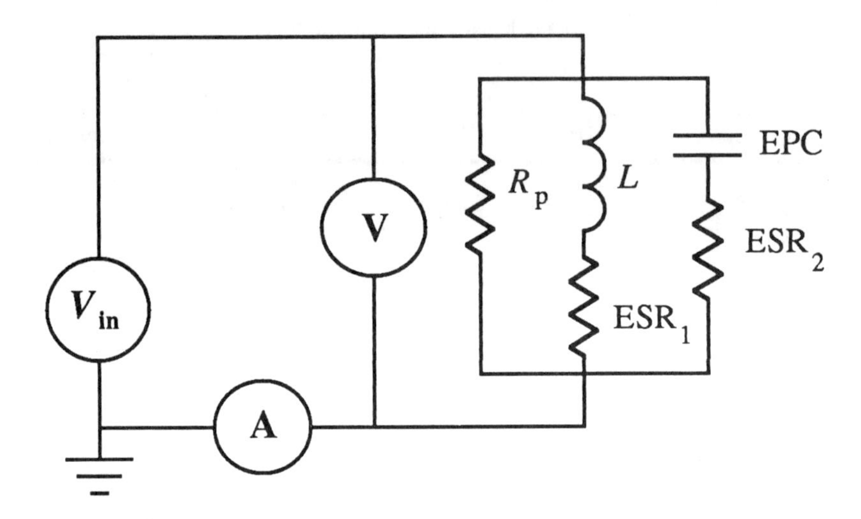

Figure 4.11 Inductor model and simulated test circuit.

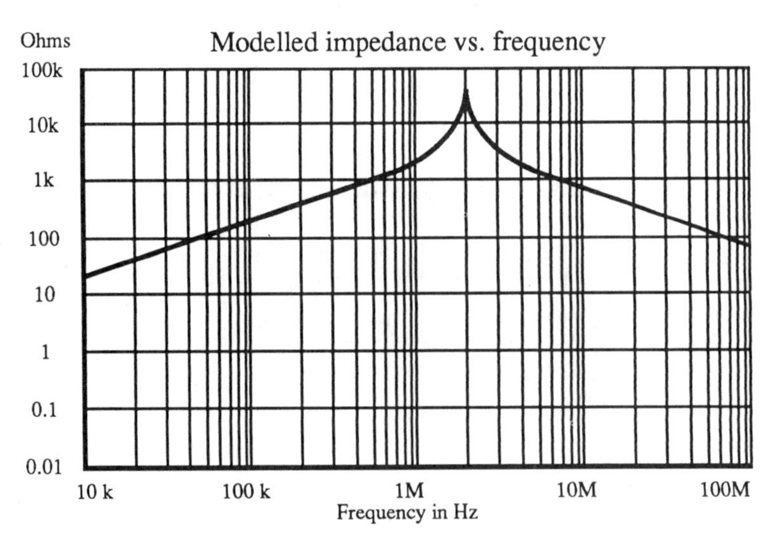

Figure 4.12 Simulated impedance curve data.

QUESTIONS

4.1 An air core inductor is connected to a 12-V-dc battery through a switch. After some time the inductor is at steady state. What will the voltage across the inductor be immediately after opening the switch? Can the current change instantaneously? Where will the current flow?

4.2 Is an inductor reactive at its self-resonant frequency?

4.3 Describe all of the loss mechanisms of an air core inductor at self-resonance that contribute to the equivalent parallel resistance.

4.4 What inductor characteristics are improved by the use of permeable materials? Why?

4.5 What inductor characteristics are degraded by the use of permeable materials? Why?

4.6 Why do high-permeability cores have a generally higher conductivity than low-permeability cores?

5

BALUN MODELING

Common mode inductors, sometimes called baluns, are useful in filtering common mode noise while passing differential currents virtually unfiltered. Another, and probably the first, use of baluns is to couple an unbalanced circuit or device (single ended, see Fig. 5.1) to a balanced circuit or device and vice versa. The term balun is derived from *bal*ance to *un*balance.

The balun, with two windings of equal number of turns, in Fig. 5.1 forces the currents i_{o1} and i_{o2} to be virtually equal. The equal currents in the loads R_{L1} and R_{L2} cause V_{o1} and V_{o2} to be equal and of opposite polarity. The outputs V_{o1} and V_{o2} are now balanced and are driven from the single-ended, unbalanced source V_s. Another balun circuit configuration is shown in Fig. 5.2. In this circuit the voltage V_{x1} is impressed across the load R_{L1} and is coupled to V_{x2}, which is impressed across R_{L2}. The voltages V_{o1} and V_{o2} are equal to V_{x1} and V_{x2}, respectively.

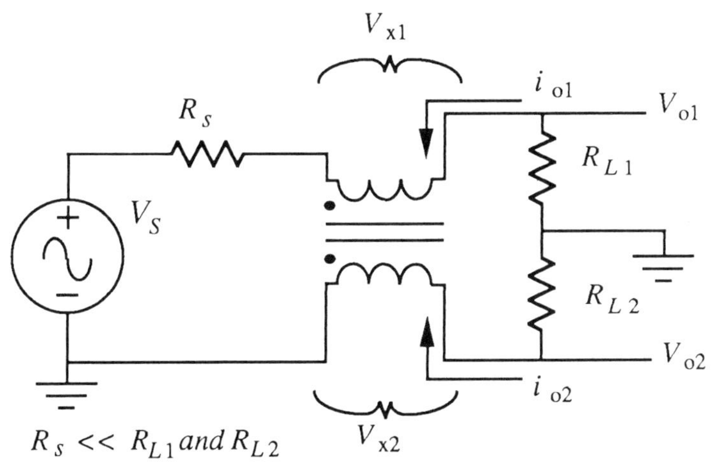

Figure 5.1 Unbalanced, single-ended conversion to balanced signal.

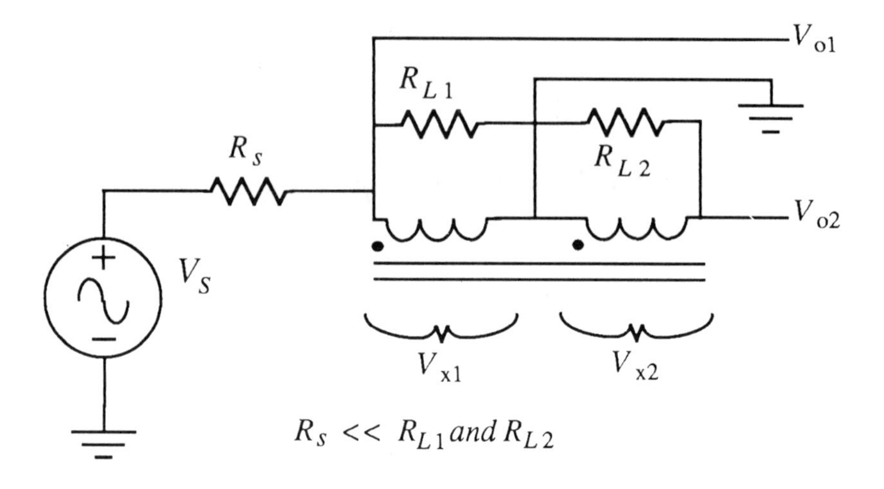

Figure 5.2 Another configuration of the circuit in Fig. 5.1.

In power conversion applications, baluns are useful for common mode filters. Any inductor or balun operated at a flux density beyond the saturation intensity has a greatly reduced inductance. Circuit components external to a saturated inductor or balun must limit the current levels to a safe value. To retain the designed amount of EMI attenuation, EMI filters must not saturate. Baluns can be used for common mode filters with large differential currents. Baluns only need to be designed to handle the expected worst case common mode EMI current.

5.1 DIFFERENTIAL MODE FLUX

Consider a voltage source V_{dm} (Fig. 5.3A) driving a load R_L. A balun is placed in the circuit path to operate as a filter. The output current i_{out} is equal to the return current i_{rtn}. The magnetic flux created in the balun by i_{out} is equal to the flux created by i_{rtn}. The magnetic flux caused by the differential currents i_{out} and i_{rtn} is in opposite directions. The net flux in the core is zero. The balun core will not be saturated by the equal differential currents regardless of the magnitude of the currents.

5.2 COMMON MODE FLUX

The balun in Fig. 5.3B is the same as the balun in Fig. 5.3A. The differential voltage source in Fig. 5.3B is set to 0 V. A common

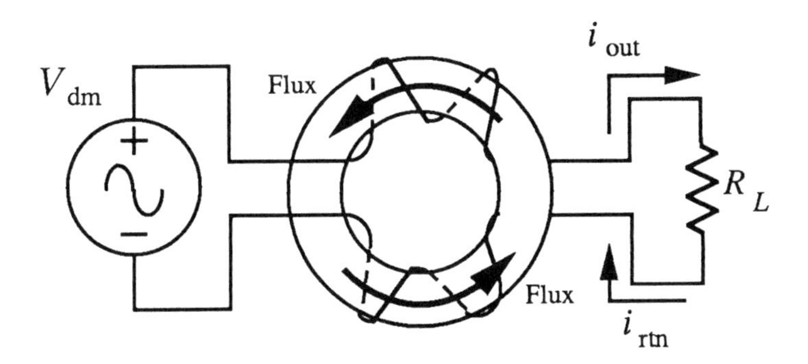

(A) Differential mode

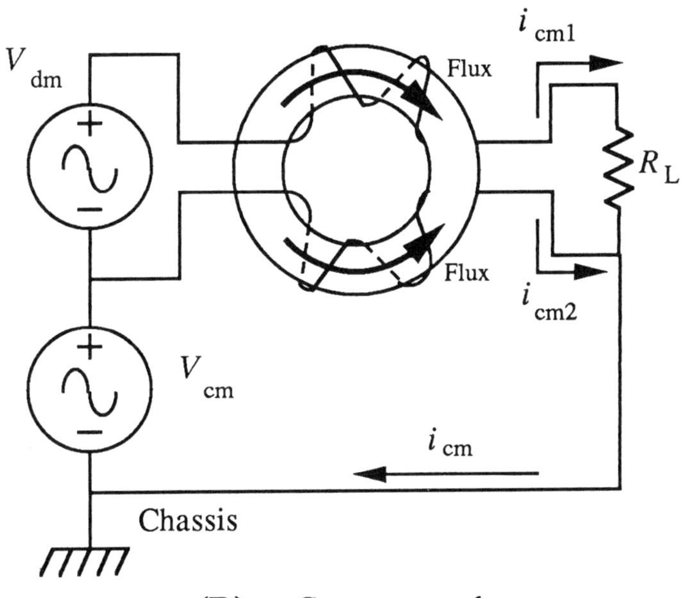

(B) Common mode

Figure 5.3 Flux in balun.

mode voltage source V_{cm} is applied to the balun, causing common mode currents i_{cm1} and i_{cm2} to flow. The flux caused by the common mode current is in the same direction in the core. The common mode flux adds and creates a net flux in the core. The balun acts as an inductor for the current in the common mode loop.

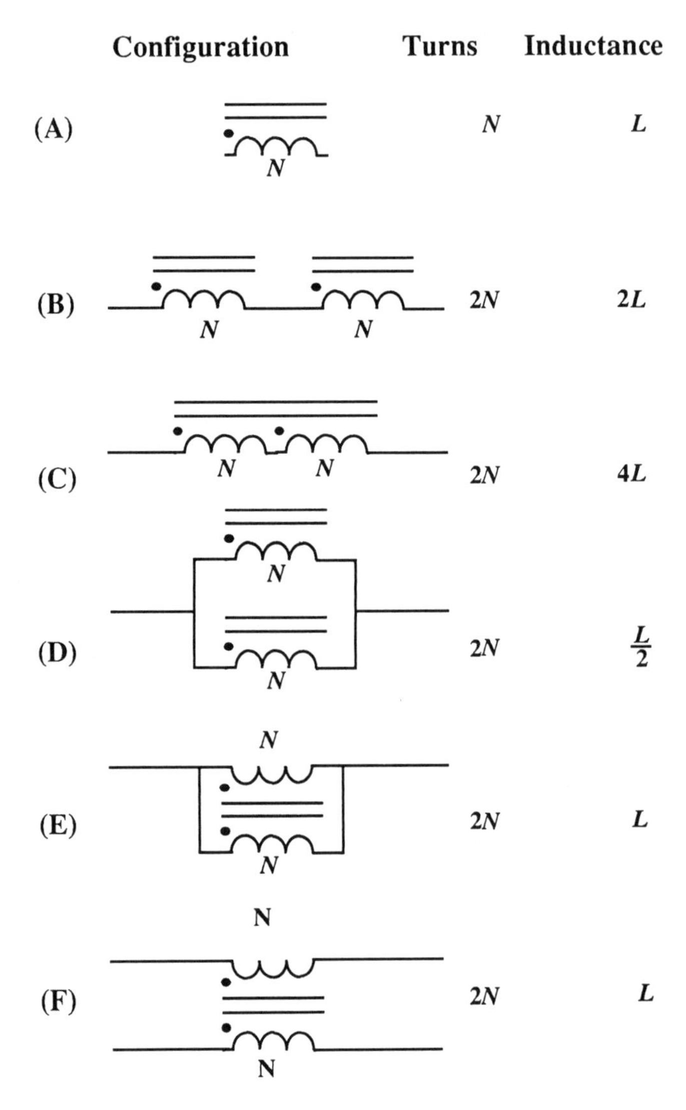

Figure 5.4 The balun inductance.

The designed value of the balun inductance can be much larger because only the common mode noise current can create flux and cause magnetic core saturation. The common mode balun is able to filter common mode noise without having flux in the core created by relatively large differential currents. If the common mode balun was uncoupled (windings X_{1a} and X_{1b} on separate cores), the differential currents would cause a high flux density in the core so that the core size necessary to alleviate saturation would be very large relative to the size of a coupled common mode inductor.

5.3 THE TRUTH ABOUT WINDINGS ON INDUCTOR CORES

All of the inductor cores in Fig. 5.4 are given to be identical. An inductor wound using one of these cores with N turns has an inductance equal to L. For two inductors in series with N turns wound separate cores the total inductance is $2L$. With the same number of turns $2N$, wound on a single core, the inductance will be $4L$ because the inductance is proportional to the square of the number of turns. When the same but separate cores are wired in parallel, the total inductance is 1/2 L. With two windings of N turns on the same core wired in parallel (as a balun) the total inductance is L.

5.4 COUPLING *K* FACTOR

The classic definition of coupling coefficient K for two winding transformers (or balun) is the ratio of the mutual inductance to the square root of the product of the self-inductances (open circuit inductance) of each of the two windings:

$$K = M/(L_1 L_2)^{1/2} \tag{5.1}$$

Each winding on a balun has the same number of turns so that Eq. (5.1) reduces to:

$$K = M/L \tag{5.2}$$

A simple way to measure the coupling coefficient is to apply a voltage source to the primary and measure the open circuit voltage

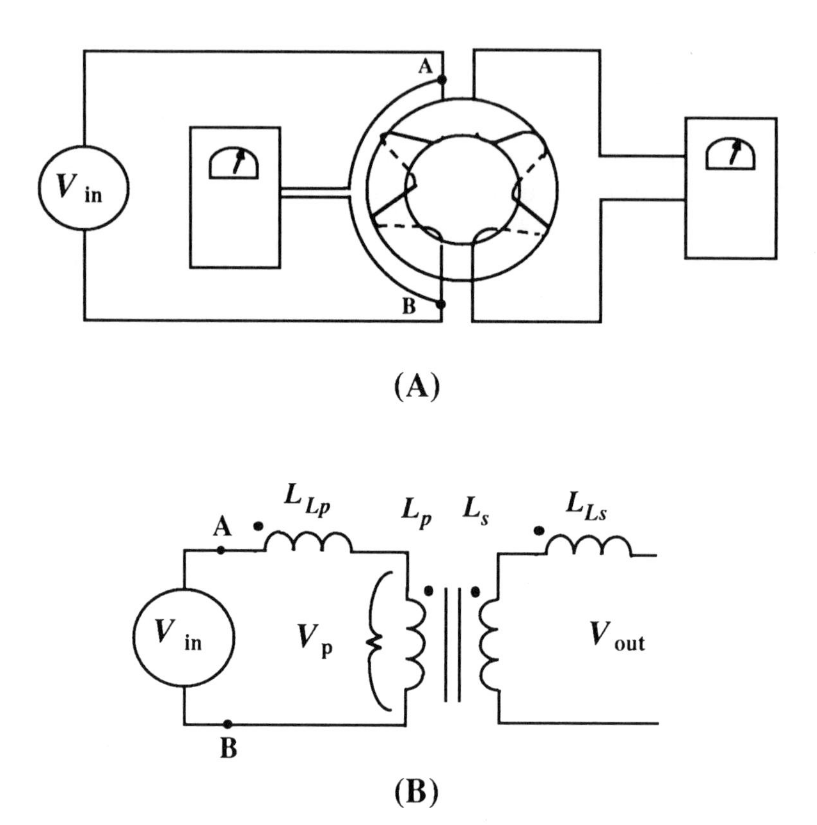

Figure 5.5 Coupling coefficient measurement, voltage method.

on the secondary, as shown in Fig. 5.5A. The voltage meter measurement on the primary must be made as close to the lead breakout points as possible (points A and B on the primary). The loop area of the meter leads should also be as small as possible. The voltage applied to the primary winding is:

$$V_p = V_{in}(L_p/(L_p+L_{Lp})) \tag{5.3}$$

For accurate measurements, the impedance of the voltmeter must be large compared to the leakage inductance reactance. With a high impedance meter, the voltage measurement on the secondary can be made with little regard to loop area of leads because there is no current flowing in the secondary and then:

$$V_p = V_s = V_{out} \tag{5.4}$$

The coupling coefficient K is:

$$K = V_{out}/V_m \tag{5.5}$$

The coupling coefficient K is also the ratio of the leakage induc-

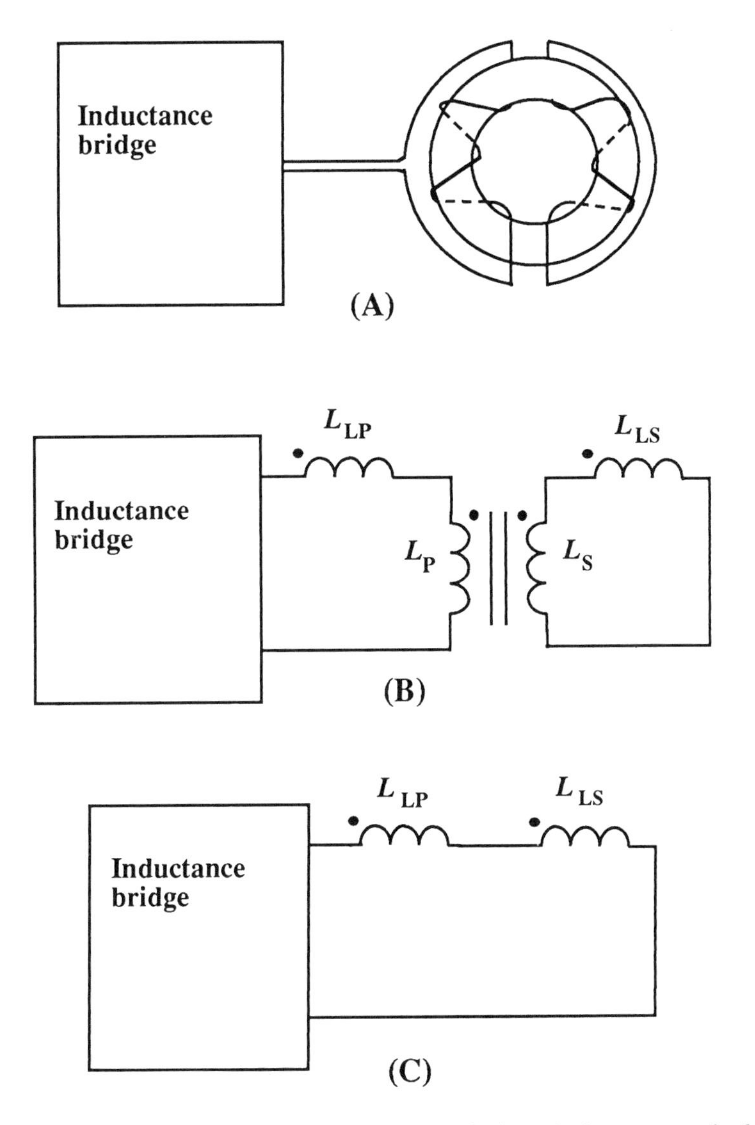

Figure 5.6 Coupling coefficient measurement, leakage inductance method.

tance associated with each winding divided by the open circuit inductance of the winding:

$$K = L_{Lp}/L_p \tag{5.6}$$

The inductance L_p and the leakage inductance L_{Lp} can be measured with an inductance bridge. The secondary winding leads are shorted with as small a loop as possible, as shown in Fig. 5.6A. The inductance bridge should have a special fixture for connection to the balun (transformer) primary side to minimize the lead loop area. Excess meter lead loop area will contribute leakage inductance to the inherent leakage inductance within the balun and cause

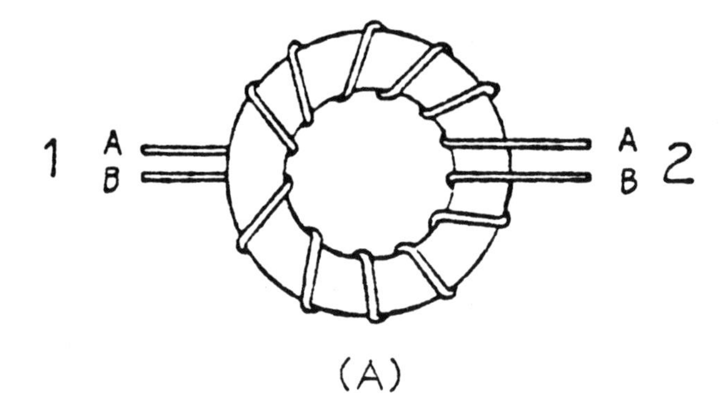

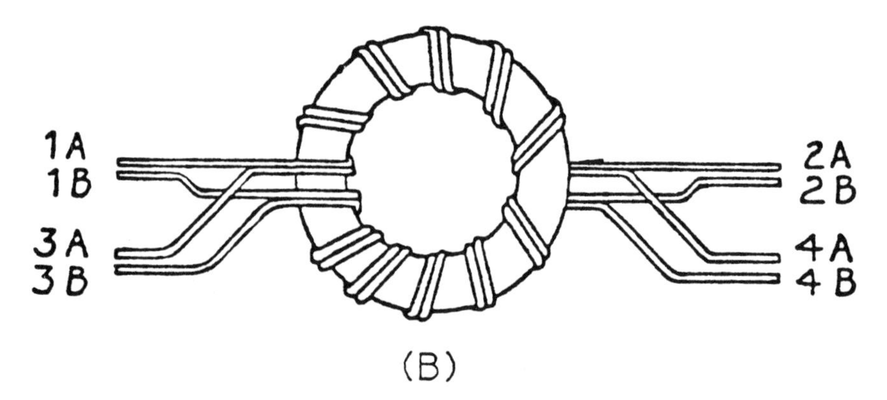

Figure 5.7 Balun winding methods.

a measurement error. A model of the leakage inductance measurement setup is shown in Fig. 5.6B. The leakage inductance is represented by L_{Lp} and L_{Ls}. Since L_p and L_s represent the perfectly coupled inductance portion and have a 1 to 1 ratio, the inductances L_p and L_s can be removed from the model. The simplified model is shown in Fig. 5.6C. The leakage inductance of the primary and secondary are in series so that the measured inductance L_{Lt} is:

$$L_{Lt} = L_{Lp} + L_{Ls} \tag{5.7}$$

The coupling coefficient can be increased by special winding methods. For simplicity and clarity, a special winding method for an inductor is shown in Fig. 5.7A. Two windings are used to create a single coil. Using the right-hand rule, we see that the flux created by each winding is in phase. The two windings are terminated together as shown and form a single coil inductor. The advantage of this type of winding method is the input and output lead breakout separation is the best possible configuration (opposite ends of the inductor). There are layout advantages to this configuration and the self-resonant frequency is improved.

To make a tightly coupled balun the same winding method can be used but each winding is done with "two in hand." The two coils (Fig. 5.7B) are terminated as they were for the single coil inductor. Each winding with this configuration is not only coupled through the flux in the core but also from coil to coil.

A comparison of coupling was made between the two winding configurations of Figs. 5.7A and 5.7B. Two five-turn baluns were wound on 846XT250-3C8 cores. The voltage ratio method was used to measure the coupling coefficients. The coupling coefficient of the balun in the configuration of Fig. 5.7A is 0.982. The coupling coefficient for the tightly coupled configuration of Fig. 5.7B is 0.998. The manufacturing costs of the tightly coupled balun are higher, but it is sometimes used when tight coupling is necessary.

5.5 DIFFERENTIAL BALUN INDUCTANCE

The simple circuit in Fig. 5.8 includes a voltage source, a model of a balun, and a load. The circuit is driven differentially by the voltage source and the resulting differential current is measured by an

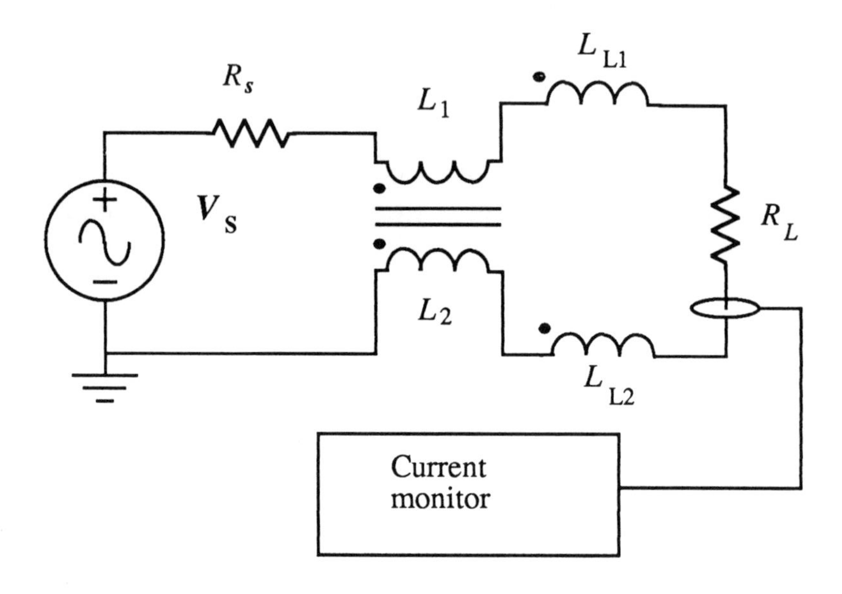

Figure 5.8 Simulation of differential mode balun operation.

ammeter labeled I_{cm}. The currents in coupled windings L_1 and L_2 are in opposite directions so that there is no resulting flux in the core. The uncoupled flux of the balun windings is the leakage inductance. The coupling coefficient gives the ratio of the leakage inductance to the common mode balun inductance. The coupling coefficient is typically 0.98. The leakage inductance for each winding is 40 μH (0.02 times 2 mH). The leakage inductances L_{L2} and L_{L1} add together to give a differential inductance:

$$L_{L1} + L_{L2} = 80\ \mu\text{H} \tag{5.8}$$

At low frequencies the current flow is governed by the load resistance. At frequencies above:

$$F_0 = R_L/(2\pi L_d) \tag{5.9}$$

the amplitude of current flow is attenuated by the reactance of inductance L_d, which reduces the current at a rate of $1/f$. For this case, as shown in Fig. 5.9, the turnover frequency is:

$$F_0 = R_L/2\pi L_d = 10/(2\pi 80\ \mu\text{H}) = 20\ \text{kHz} \tag{5.10}$$

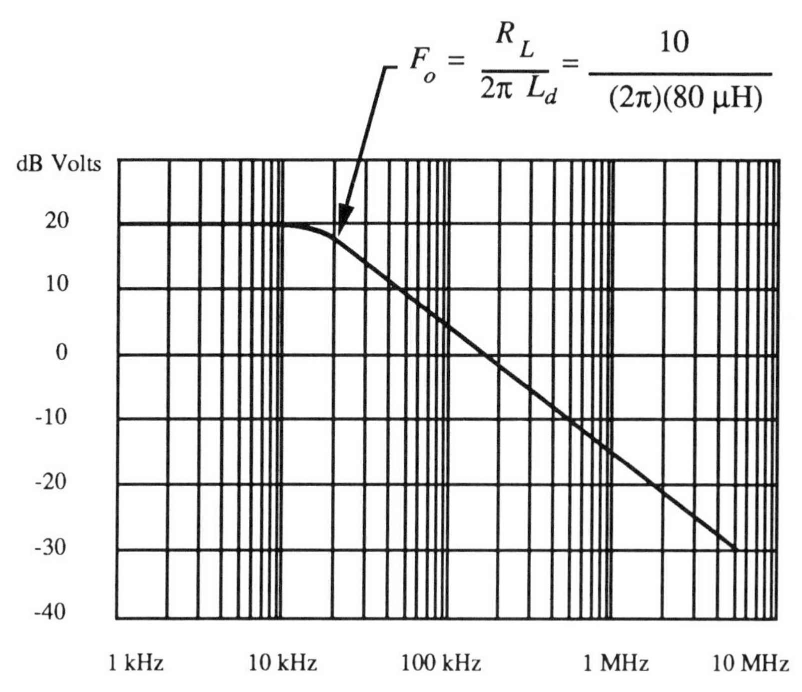

Figure 5.9 Balun differential mode simulation data.

5.6 COMMON MODE BALUN INDUCTANCE

The circuit in Fig. 5.10 is driven by the common mode voltage source V_{cm}. The common mode currents flow in the same direction through the balun windings and cause a flux in the balun core.

The load resistance is set to 1 $\mu\Omega$. At low frequencies the total current flow I_{cm} depends upon R_S, the source resistance (see Fig. 5.11). For frequencies above:

$$F = R_S/(2\pi L) = (0.1)/(2\pi \bullet 2\text{mH}) = 7.95 \text{ Hz} \tag{5.11}$$

the total current I_{cm} depends upon the reactance of the balun and reduces at a rate of $1/f$. The current in the load is one-half the total current, as shown by the 6 dB difference in the two curves.

The same circuit as in Fig. 5.10 was analyzed but with the load resistance set to 10 Ω. The resulting data is presented in Fig. 5.12. The voltage from node A to node B is the voltage across L_2 and L_{L2} and is equal to the voltages across L_1, L_{L1}, and R_L. The volt-

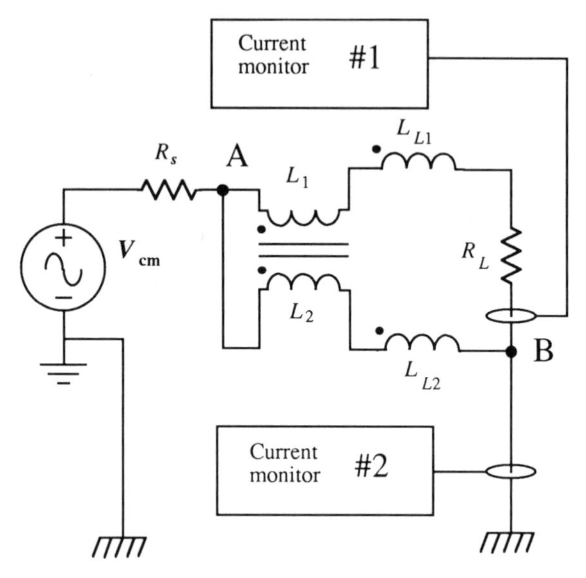

Figure 5.10 Simulation of common mode balun operation.

age across L_1 equals the voltage across L_2 so that the voltage across L_{L2} equals the voltage across L_{L1} and R_L.

At mid frequencies, and since $R_L >> R_S$, the current flow in winding L_1 depends upon R_L. The currents in windings L_1 and L_2 are equal at frequencies above where the reactance of L_1 is greater than R_L:

$$f > 10/(2\pi 40\ \mu\text{H}) = 39.8\ \text{kHz} \tag{5.12}$$

The current in the load is one-half the total current (6 dB difference in the two curves) at frequencies above 39.8 kHz.

5.7 EFFECTS OF LOAD AND SOURCE RESISTANCES ON ATTENUATION

Key Idea

If R_{load} is small compared to X_c and/or $(L/C)^{1/2}$, then L_{balun} and R_{load} are players at 20 dB per decade.

If R_{load} is small compared to X_c and $(L/C)^{1/2}$, L_{balun} and C_{cm} are players at 40 dB per decade.

A balun in a balanced circuit acts as an inductor in the common mode loops of the circuit. A model of a balanced common mode filter circuit is shown in Fig. 5.13. This model is analyzed with various load resistances.

A family of data curves is presented in Fig. 5.14 to illustrate the effects of relative impedances and *LC* resonances on the common mode filter noise attenuation. The bottom curve with 5000 Ω balanced loads is the typical characteristic curve for an unloaded *LC* filter because 5000 Ω is large compared to all of the other impedances in this circuit. The resonant frequency is:

$$f_r = 1/(2\pi(LC)^{1/2}) \tag{5.13}$$

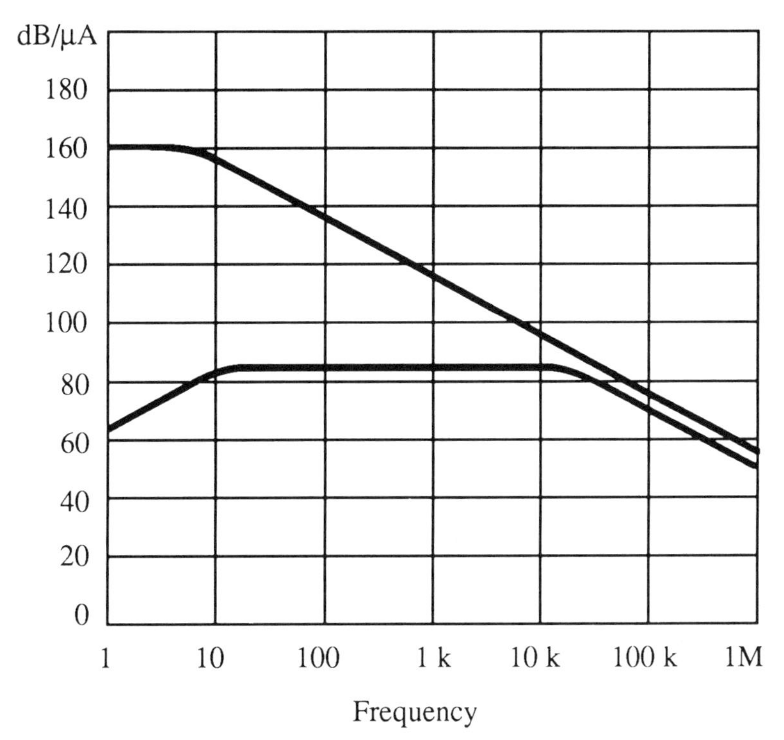

Figure 5.11 Balun common mode simulation data.

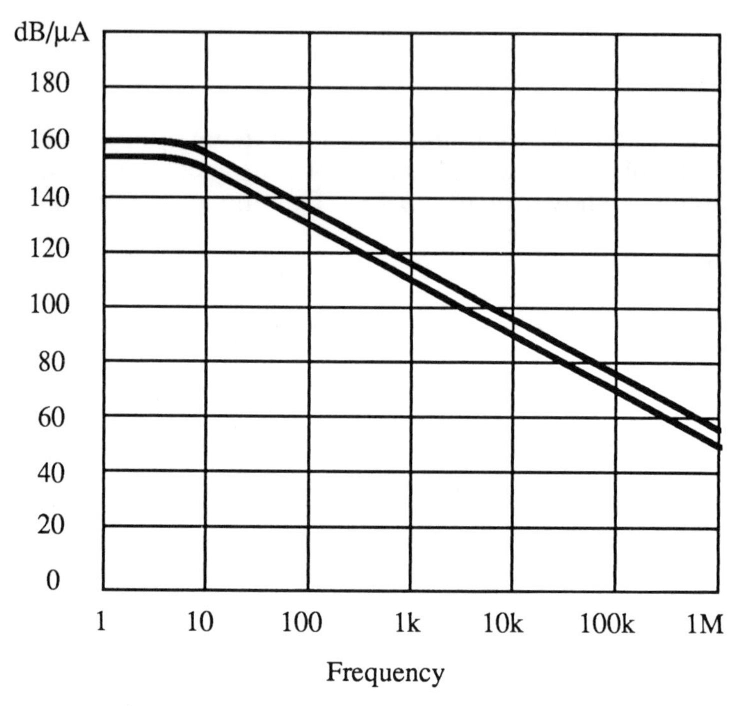

Figure 5.12 Balun circuit operation.

where L is the inductance of the balun and C is C_{cm1} and C_{cm2} in parallel. The two poles are at 35.6 kHz, and above this resonant frequency the attenuation increases with increasing frequency at the rate of $1/(f^2)$.

5.8 BALUN DRIVING IMPEDANCE

The driving (characteristic) impedance of this filter is:

$$Z = (L/C)^{1/2} \tag{5.14}$$

where L is the balun inductance and capacitance C is the parallel combination of C_{cm1} and C_{cm2}. For $L = 100$ μH and $C = 0.2$ μF the driving impedance is 22 Ω. From observation of the family of curves, it can be seen that as the driving impedance approaches R, the load plus the source resistance $(R_S + R_L)$, the resonance of the balun inductance L and C_{cm1} and C_{cm2} is damped out. For values

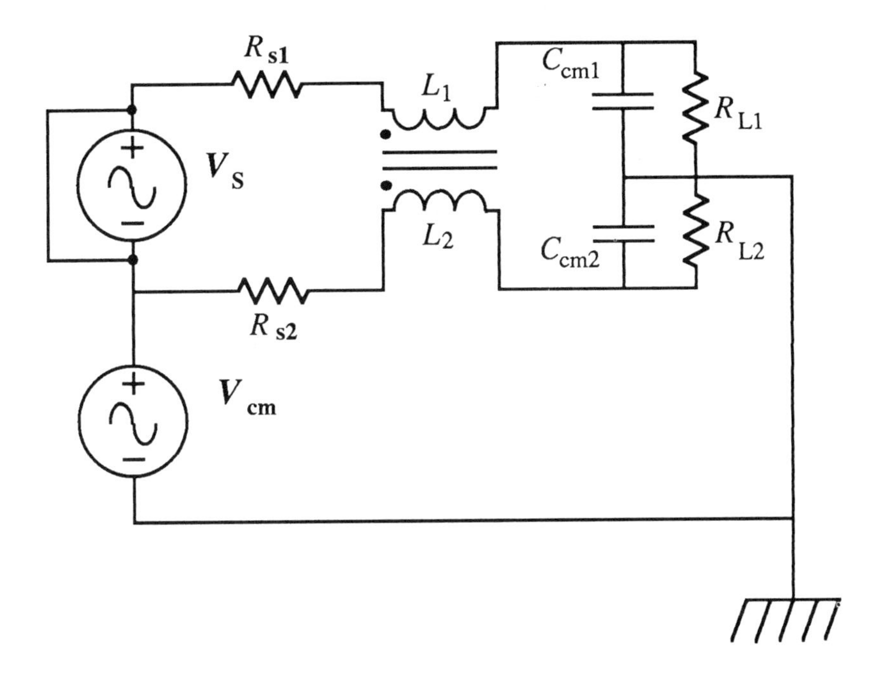

Figure 5.13 Model of balanced common mode filter circuit.

of R (load resistance plus source resistance) that are smaller than the driving impedance, the filter poles separate in frequency. The 0.5 Ω curve has a pole at:

$$F_p = R/(2\pi L) = 800 \text{ Hz} \tag{5.15}$$

As frequency increases, the filter is first order and attenuation increases at a rate of $1/f$. Another pole occurs when the reactance X_c of C_{cm1} plus C_{cm2} is equal to R (load resistance plus the source resistance). The frequency of this pole is:

$$F_p = 1/(2\pi RC) = 3.2 \text{ MHz} \tag{5.16}$$

5.9 BALANCED CIRCUITS

A balanced system of resistors or impedances gives the minimum common mode current flow. With balanced circuits and loads

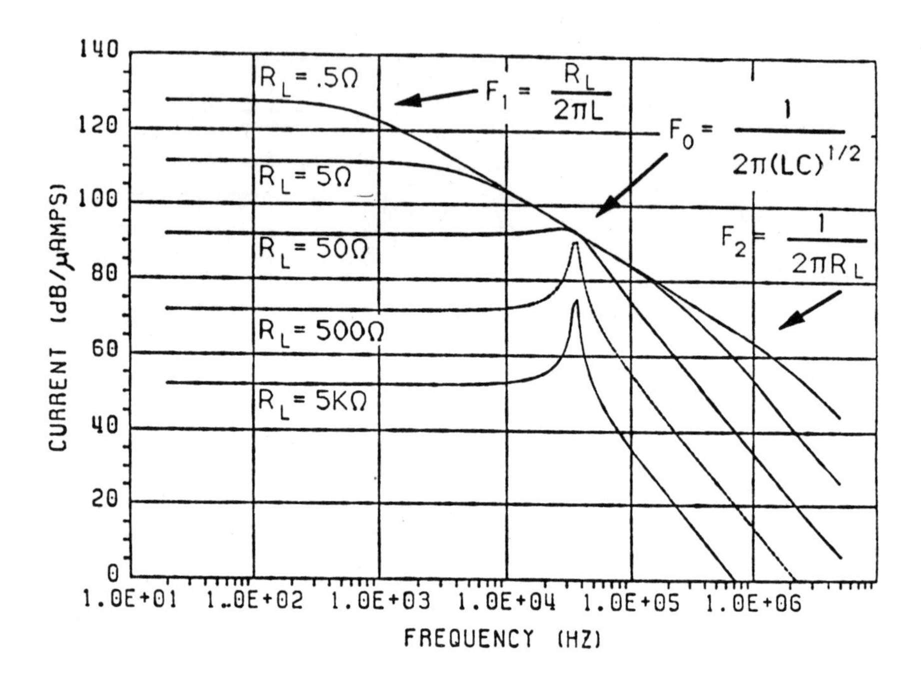

Figure 5.14 Effects of relative impedances.

there is significant impedance in both common mode current paths. When a differential circuit is balanced, the common mode currents are equal, and the voltages across balun windings are equal. There is no net flux in the core and the leakage inductances do not affect the differential performance. In Fig. 5.15A R_{pwr} and R_{rtn} represent the signal (output power) and return lines and components of a typical circuit. The voltage source V_{cm} represents the typical equivalent common mode sources that occur in electronic circuits. For example, the voltage source V_{cm} is set to 10 V ac. In Figs. 5.15A and 5.15B, the total resistance is 100 Ω. In Fig. 5.15A the total current flow is 400 mA. The circuit in Fig. 5.15B is unbalanced; R_{pwr} is 99 Ω and R_{rtn} is 1 µΩ. The total current in this circuit is 10.1 A, more than 25 times the current in the balanced circuit with the same total impedance. For a given amount of impedance, a balanced circuit results in less current flow than any other ratio of impedances. In most circuits, the current returns and/or grounds are of very low impedance and the signal or

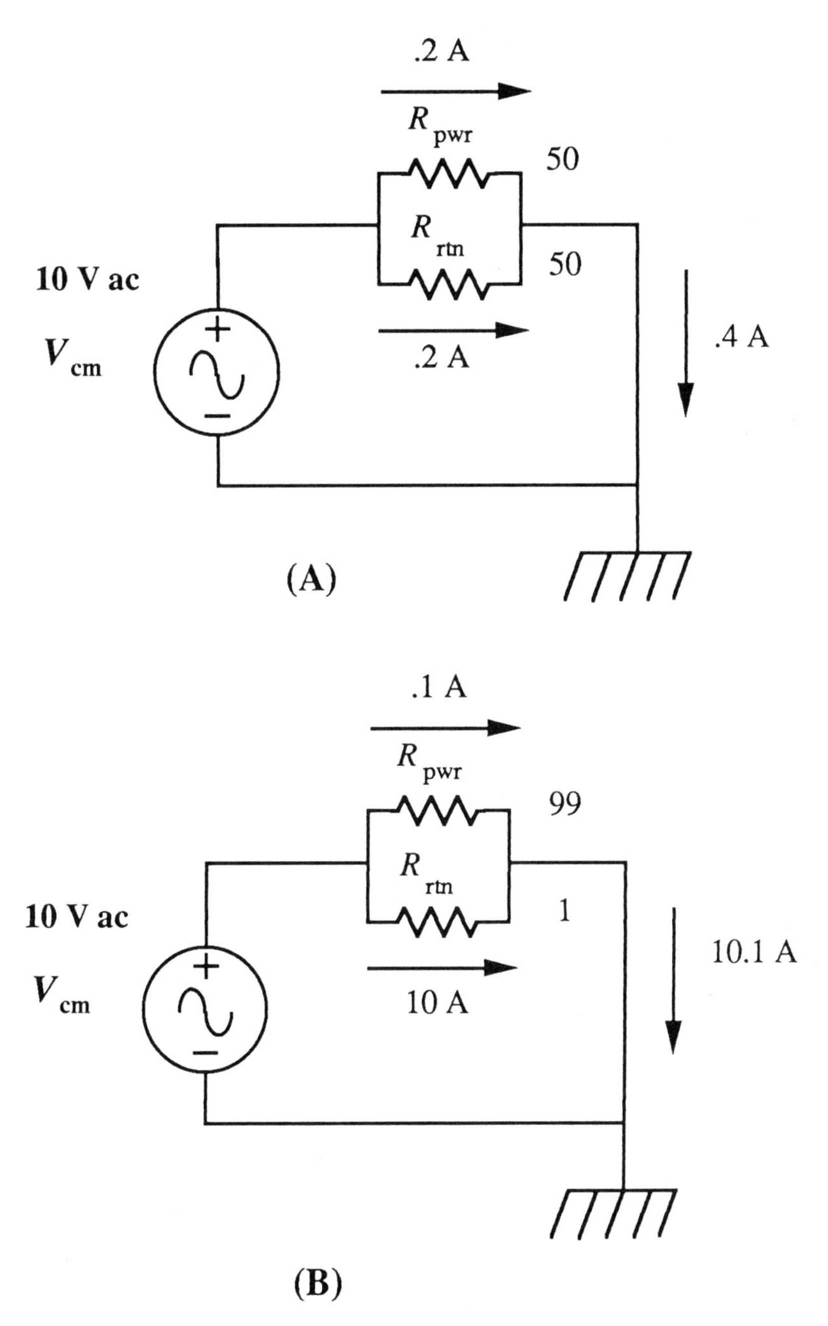

Figure 5.15 Balanced circuits result in minimum cm current flow.

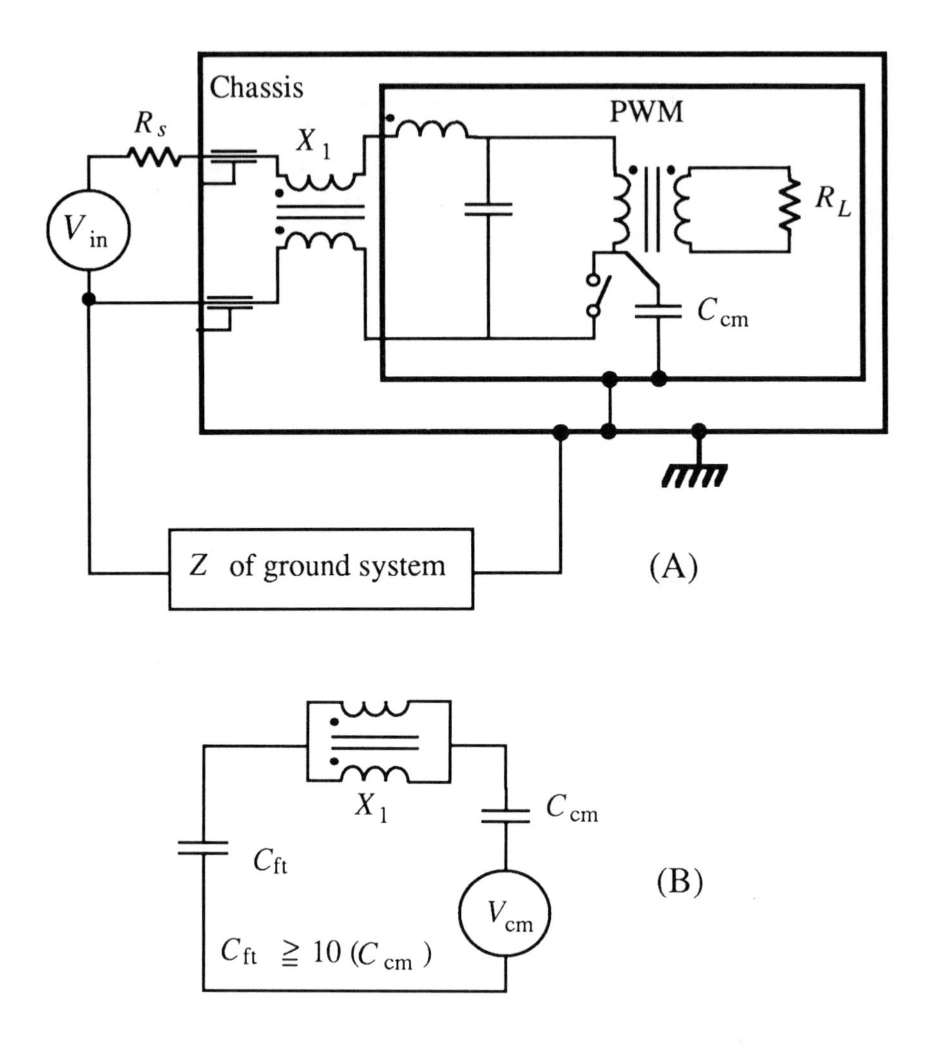

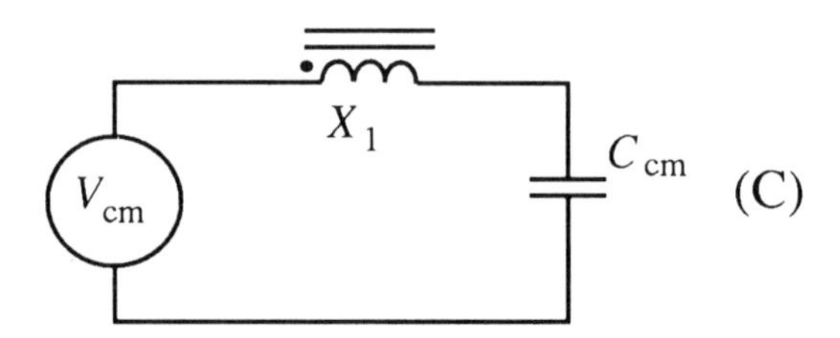

Figure 5.16 Common mode loop equivalent model.

power lines generally have relatively higher impedances (filter inductors, for instance).

Balanced circuits are often used in digital communications, but power supplies and computers are virtually always proportioned as in Fig. 5.15B. To balance these kinds of circuitry, a large increase in component count is required. The economical choice is to use good layout practices to reduce the common mode noise sources.

5.10 DESIGN CRITERIA

A simplified diagram of a typical generic switching power supply input is illustrated in Fig. 5.16A. The section labeled PWM (pulse width modulator) contains both a common mode noise source and a capacitive coupling path to the chassis. The circuit in Fig. 5.16A can be reduced to the common mode circuit of Fig. 5.16B. The feed-through capacitance is generally much larger than the common mode capacitance C_{cm}. The capacitors are in series, so the effective capacitance is approximately equal to C_{cm}. The common mode loop is resonant at:

$$F_r = 1/(2\pi(L_{x1}C_{cm})^{1/2}) \tag{5.17}$$

The common mode capacitance should be reduced to as low a value as possible by using good layout practices. The common mode capacitance can be calculated from the geometry of the printed circuit board and component electrodes that change voltage rapidly. Or alternatively, if a brassboard is constructed, the common mode capacitance can be measured. Then the balun inductance should be designed so that the resonant frequency F_r is lower than the frequency of the circuit being designed. Typical aspects for consideration are switching frequency of a power supply or the clock frequency of a computer.

Layout practices that minimize common mode capacitance (C_{cm}) are important for all design. For switching power supply design, L_{x1} should be chosen so that F_r is less than the switching frequency of the power supply. The same discussion can be made for computer or digital circuitry by replacing the section labeled PWM with a computer clock circuit, for example. The resonant frequency in this case should be designed to be less than the clock frequency.

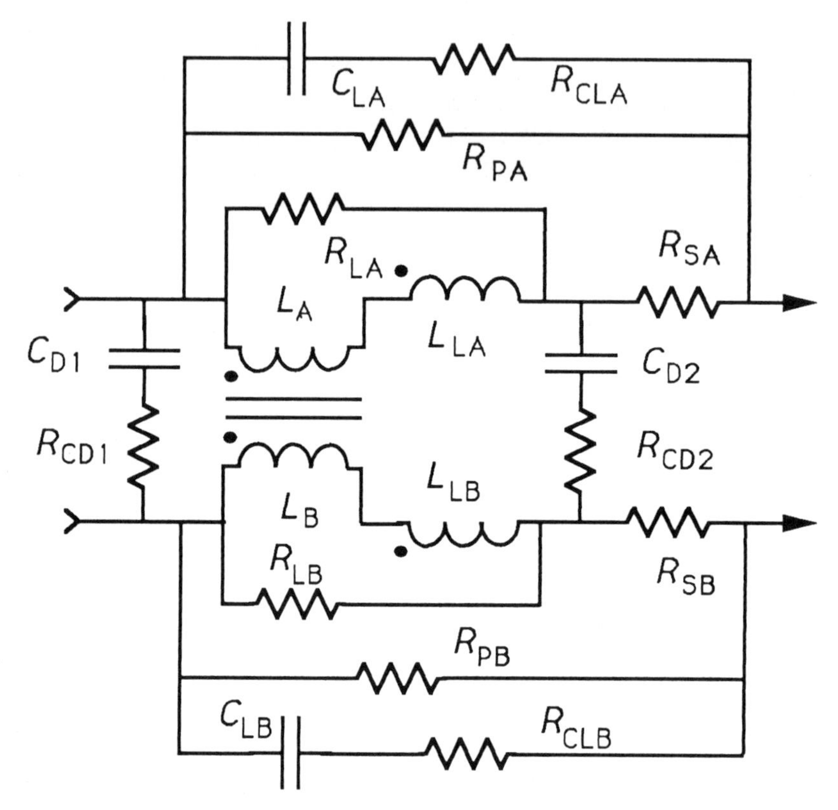

Figure 5.17 Balun model.

5.11 MODEL

A very useful balun model for use in SPICE simulations is shown in Fig. 5.17. The standard SPICE coupled inductor is the heart of this model. With the exception of C_{tt1} and C_{tt2}, the other elements of this model are the same as the inductor model (Fig. 4.4) of chapter 4 (for each of the two coupled windings). The capacitances C_{tt1} and C_{tt2} represent the total capacitance between the two windings of the balun. The coupling coefficient, of course, determines the magnitude of the leakage inductance.

QUESTIONS

5.1 What is the effect of the leakage inductance in a balun?

5.2 Is a balun an inductor or a transformer? Why?

5.3 Why don't we always design balanced circuits?

5.4 What are the differential effects of common mode currents in balanced circuits?

5.5 What are the applications of baluns in EMI filters?

5.6 Discuss the effects of coupling coefficient on balun performance as a common mode filter inductor.

5.7 Name two major applications of baluns in general.

5.8 Why does a balanced circuit have the least amount of total common mode current flow?

6

FILTERS

Passive filters are generally used for filtering unwanted EMI from power and signal lines in electronic equipment. Some filters do more than one function in a circuit. For example, an input filter that converts pulsating dc to smooth dc attenuates unwanted EMI and provides hold-up for momentary loss of input. Inductors and capacitors are the principle elements in passive filters. Filters are placed in the EMI path to attenuate unwanted EMI. Filters can be designed to bypass, absorb, and/or reflect the EMI energy. Impedance is the principle measure of filters and filter components. Relative impedances determine how much EMI current flows in each conducted path. Impedance curves provide a visual figure of merit of capacitors and inductors. Insertion loss curves are often used for characterizing filters.

Real inductors and capacitors fall very short in regard to performance when compared to the typical academic first order models. Knowledge of the performance advantages and disadvantages of each type of capacitor and inductor is important in the process of choosing the appropriate component for each particular application. EMC is often the most demanding electronic design requirement.

6.1 PARASITIC INDUCTANCES AND CAPACITANCES

The term "parasitic" is used by many engineers as an adjective to describe inductances and capacitances that don't appear on their schematics and cause the desired waveforms to have all sorts of unexpected features, usually called "noise." The term "stray capacitance" is often used for unspecified capacitances of an electronic assembly. "Stray," in Webster's dictionary, is defined as an adjective meaning scattered. Since any object, especially a conductor, has a capacitance to all other objects, stray is probably a fair description. These unspecified inductances and capacitances are crucial parameters in meeting EMI standards and often affect the ability of the electronic circuit to function.

These parasitics come about in the choice of component types and the methods of circuit layout. They are crucial design parameters yet their importance is often underestimated. The circuit designer cannot pass a schematic on to a packager or a layout technician and expect to achieve a good design. The circuit designer must be involved in the PWB layout so that the parasitic elements do not degrade the filter to a degree that it will not meet the design requirements. Parasitic elements are crucial design parameters in today's high-speed electronics. Component choice and methods of circuit layout will become more and more important as we go into the future.

6.2 ACADEMIC *LC* FILTER

A simple *LC* filter and insertion loss curve is presented in Fig. 6.1. The load R_L is large compared to the impedance of L and C, and the source resistance is small compared to the impedance of L and C; therefore the associated effects do not complicate the example. The resonant frequency of this *LC* filter is:

$$F_0 = 1/(2\pi(LC)^{1/2}) \tag{6.1}$$

At frequencies above F_0 the attenuation increases with increasing frequency at a rate of $1/f^2$ or 40 dB per decade. For an ideal filter, the attenuation would increase infinitely with increasing frequency. This is an academic model without parasitics.

6.3 SIMPLE REAL WORLD *LC* FILTER

A simplified real world model of the *LC* filter and insertion loss curve is presented in Fig. 6.2. The real world model of the inductor and capacitor includes the parasitic elements along with the schematic representations of the components. The insertion loss curve in Fig. 6.2 is a family of three curves that show the effect of the parasitic capacitance and resistance in parallel with the inductor (L) and the parasitic inductance and resistance in series with the

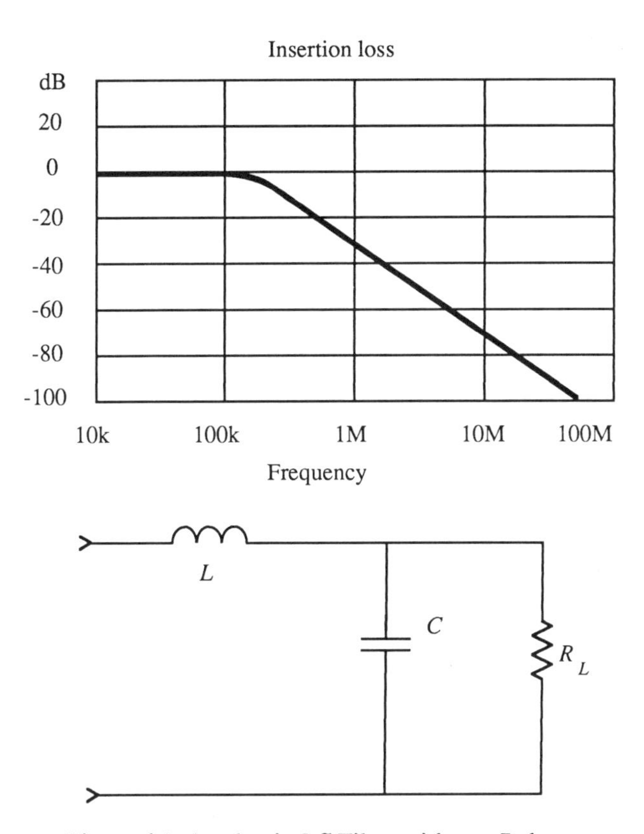

Figure 6.1 Academic *LC* Filter with two Poles.

capacitor (*C*). The zeros are picked to be coincident for purposes of illustration. The three frequency response curves are from three different possible combinations of layout parasitics resulting from three different layouts. At frequencies higher than the resonances of the parasitics (L and C_L, C and L_C) the insertion loss decreases. The filter attenuation is reduced at high frequencies where attenuation is still needed to pass EMI tests. The filter is not effective.

6.4 CONTROL PARASITICS BY DESIGN

The parasitic elements can be separated into two types. Some of the parasitics are elements inherent in inductors and capacitors. Other parasitics are caused by the circuit layout or packaging.

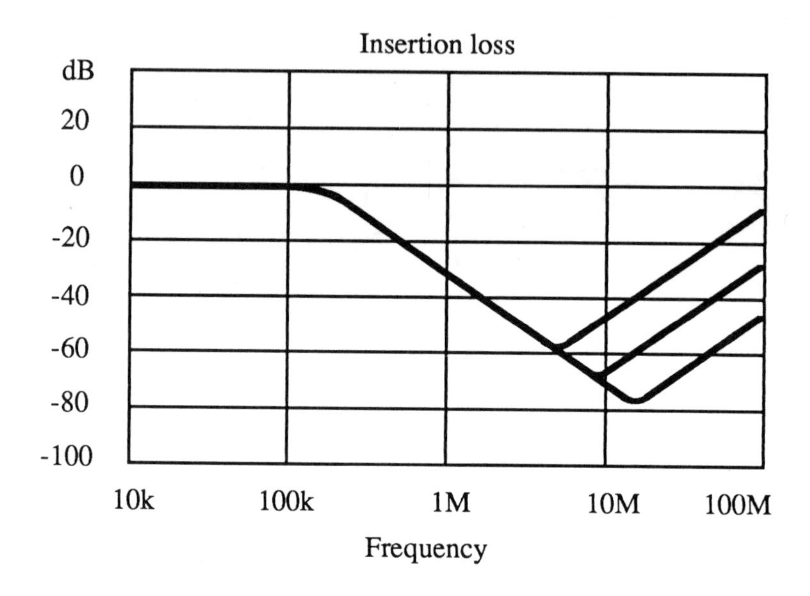

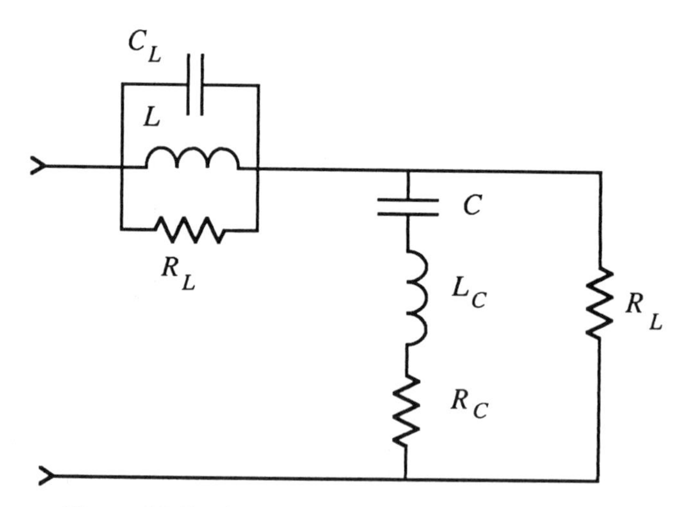

Figure 6.2 Real world *LC* filter with poles and four zeros.

Inherent Component Parasitics

The electronic design engineer is responsible for choosing the appropriate components for each application, which can reduce the detrimental parasitic effects of the components. When the parasitics inherent in filter components are taken into account early in the design process, the problems that arise later in the process are greatly reduced. The overall project costs are reduced and the project has a better chance of staying on schedule.

Inherent Capacitor Parasitics

The resistive parasitic effects (modeled by the equivalent resistance R_C) inherent in capacitors are:

1. Plate and lead resistance
2. Dielectric losses (modeled as equivalent resistance)
3. Skin effect resistance

Parasitic inductive effects modeled by equivalent inductance L_C are:

1. Plate inductance
2. Lead inductance

All capacitors should have lead lengths minimized to reduce series inductance. Even the best capacitors without leads have self-resonant frequencies in the range of 100 kHz to 20 MHz. The FCC and VDE impose conducted emissions limits up to 30 MHz and Mil-Std-461 CE03 imposes limits up to 50 MHz; therefore the self-resonant frequency of the capacitors must be considered in the design process.

Inherent Inductor Parasitics

Inductors are generally manufactured in house so that we have some control of the inherent parasitics. Even with the best design possible, there are still inherent inductor parasitics that are unavoidable.

Inherent inductor resistive parasitics (modeled by equivalent resistance R_l) are:

1. Lead and winding resistance
2. Hysteresis losses
3. Eddy current losses
4. Dielectric losses of the insulation
5. Skin effect losses

Parasitic winding capacitances (modeled by equivalent capacitance C_L) are:

1. Turn to turn capacitance
2. Turn to core and core to turn capacitance

Capacitance inherent in inductors should be minimized to increase the self-resonant frequency of the inductor. At frequencies higher than the self-resonant frequency, inductors are capacitive. Inductors wound on permeable cores generally resonate between 2 to 100 MHz. The FCC and VDE impose conducted emissions limits up to 30 MHz and Mil-Std-461 CE03 imposes limits up to 50 MHz, therefore the self-resonant frequency of the inductors must be considered in the design process.

6.5 PARASITICS CAUSED BY CIRCUIT LAYOUT

The electronic design engineer is responsible for passing layout information to the packaging engineer so that chosen components will perform as needed to meet the design requirements of the equipment specification. The "over the fence" approach will not work for packaging switching power supplies and most state of the art high-speed circuitry. The crucial layout concern for capacitors is the lead length, which contributes to the inductance of the parasitic inductance modeled as L_C. Excess lead length generally increases the area of the loop of the conductive path of the component. Inductance is directly proportional to the area of the loop. To reduce this parasitic inductance, the leads should be as short as possible and we must wire to and away from the cap as in Fig. 6.3. The crucial layout concern for inductors is the input to output capacitance, which is modeled as a portion of C_L in Fig. 6.2.

I/O Layout

The input and output leads should be physically spaced as far from each other as possible (see Fig. 6.4A). Conductors should not be physically close to the inductors that are not chassis or circuit grounded. A Faraday shield with feed-throughs on input and output is the best layout for an inductor used in an EMI filter. A Faraday shield (as in Fig. 6.4C) can be made from an aluminum box that totally encloses the inductor. The loop area of the com-

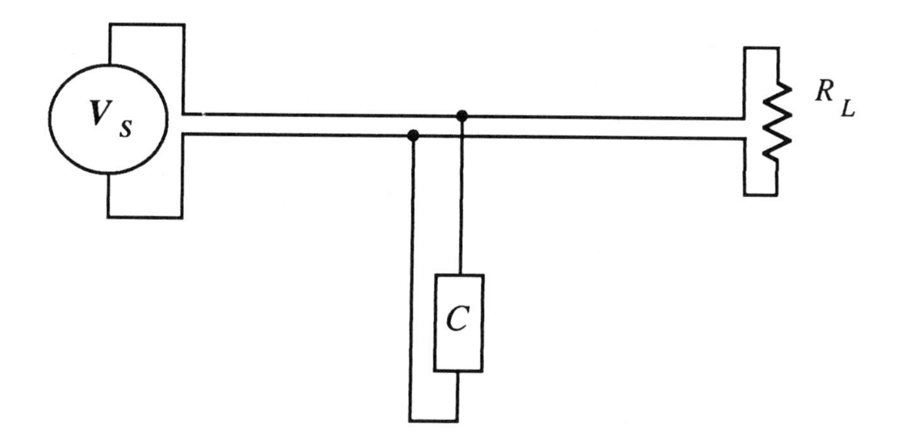

(A) Unacceptable

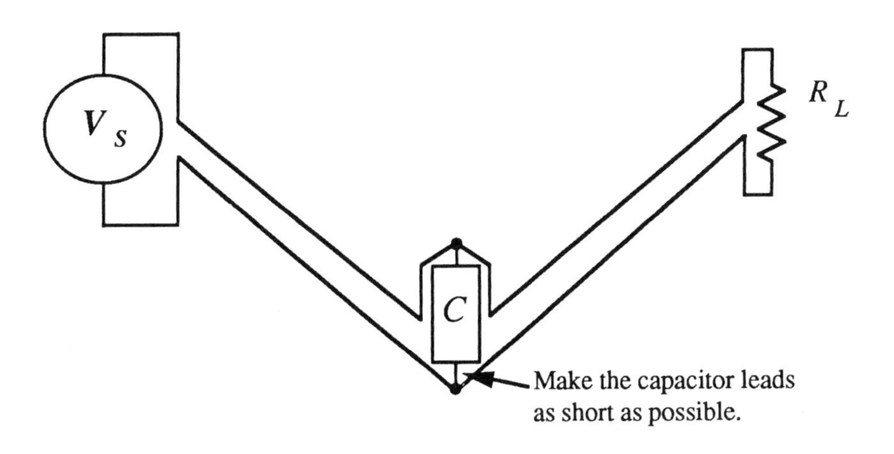

(B) Satisfactory

Figure 6.3 Capacitor layout.

ponent conductive path should also be minimized to reduce the magnetic radiated emissions and susceptibility. The best method of *LC* filter packaging is diagrammed in Fig. 6.5.

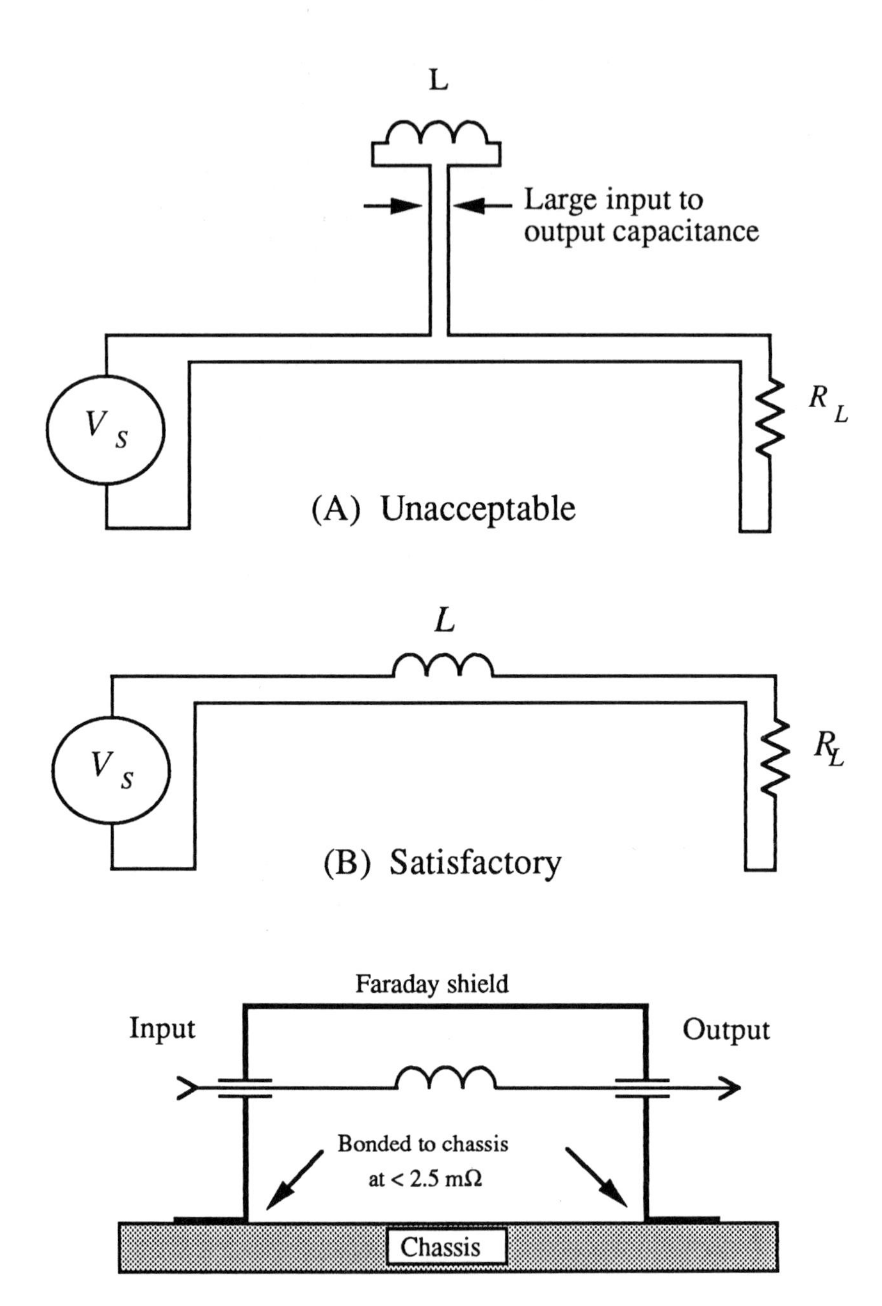

Figure 6.4 Inductor layout.

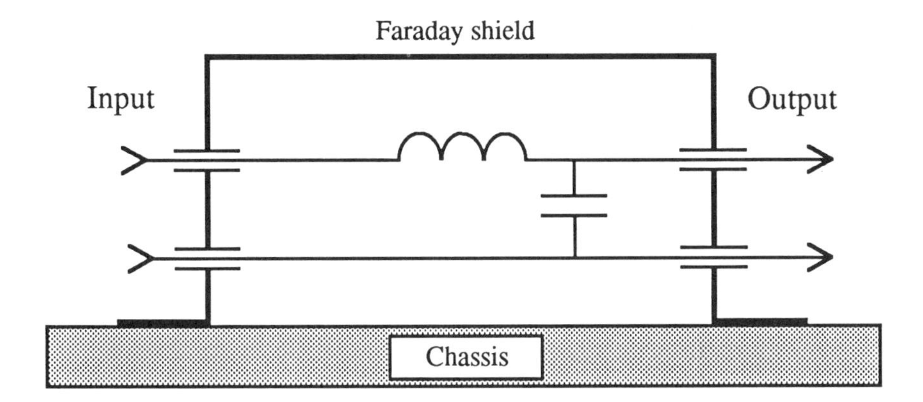

Figure 6.5 *LC* filter layout and packaging.

Three Different Results from Layouts

To get the desired performance from filter components, the layout in the packaging phase of the design depends upon good communication between the electronic designer and the packaging designer. The components are specified by the schematic and parts list.

An electronic designer may design a filter expecting the performance as shown in Fig. 6.2 but will get the performance in Fig. 6.1. A layout of the same filter and schematic by three different package designers could produce three different levels of performance as in the three curves in Fig. 6.2.

6.6 FILTER CIRCUIT DESIGN

Time Domain *LC* Filter Operation

To obtain overall well-rounded knowledge it is important to understand the time domain operation of *LC* filters. A simple example of an *LC* filter driven by a square wave generator is presented in Fig. 6.6. The characteristic (or driving) impedance of the *LC* filter is chosen to be much less than the load resistance:

$$(L/C)^{1/2} << .1R_L \qquad (6.2)$$

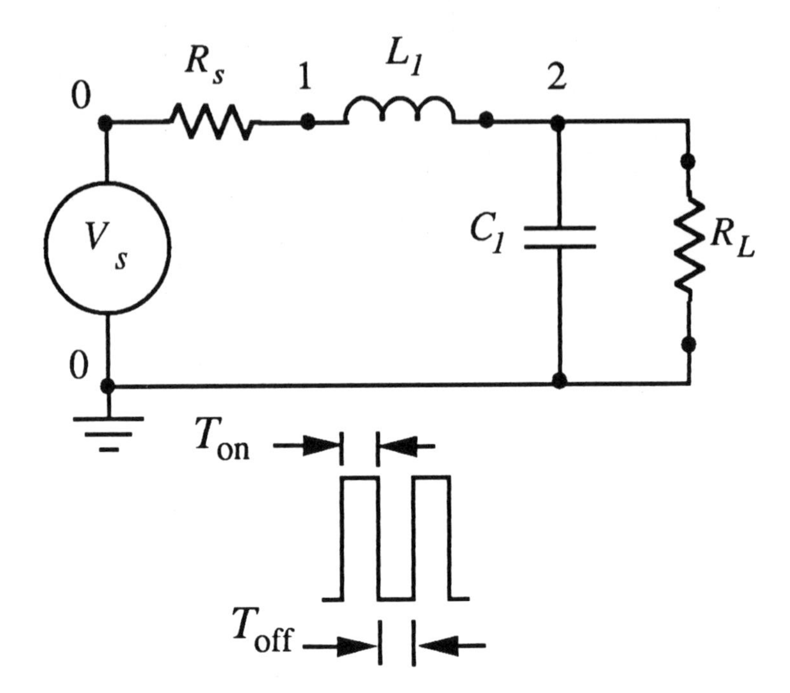

Figure 6.6 *LC* time domain operation.

The initial conditions are:

$$V_{in} = V_c = I_L = 0 \tag{6.3}$$

The capacitor voltage charges as in Fig. 6.7. The inductor voltage is:

$$V_L = V_{in} - V_c \tag{6.4}$$

The inductor voltage during the ON time is:

$$V_L = V_s - V_c \tag{6.5}$$

The inductor voltage (Fig. 6.8) during the OFF time is:

$$V_L = 0 - V_c \tag{6.6}$$

The voltage on a capacitor cannot change instantaneously nor can the current in an inductor. The current charging the capacitor is

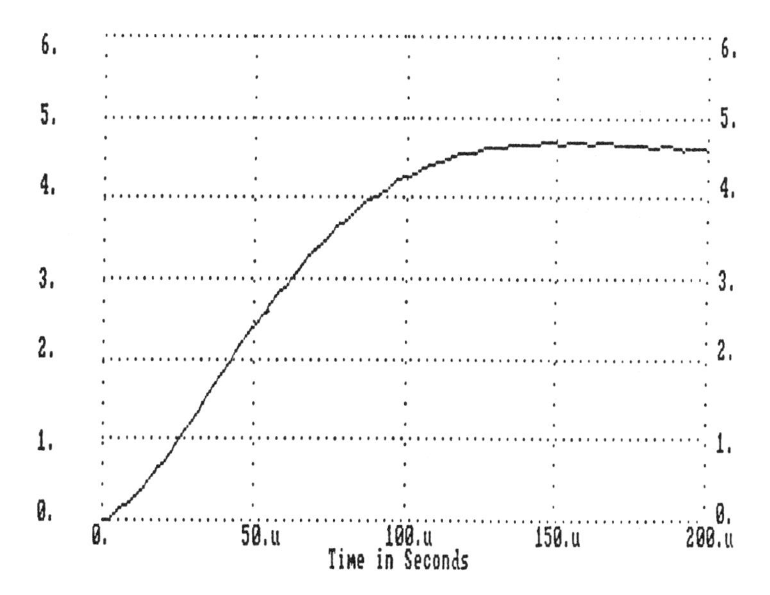

Figure 6.7 Capacitor voltage.

limited by the inductor. The voltage on the capacitor rises at a rate determined by the ratio of the charge to the capacitance:

$$V_c = Q/C \tag{6.7}$$

where Q is the charge on the capacitor. The inductor and capacitor act together to produce an output voltage (V_c) that is the average value of the voltage input. In the conducted emissions frequency range of 15 kHz to 50 MHz, generally inductors are a high impedance and capacitors are a low impedance. Fast input voltage changes are impressed across the inductor, and the capacitor voltage does not change quickly.

6.7 CHARACTERISTIC IMPEDANCE OF *LC* FILTERS

$$\text{We want } Z = (LC)^{1/2} \ll 0.1\, R_L \tag{6.8}$$

Key Idea 1

We want L small and C big for transient response.
We want L big and C big for EMI.

Power supply filters are designed to pass power at low frequencies and to attenuate higher frequency EMI. We want the output impedance of a filter to be small compared to the load it is driving. The frequencies of interest are well below 50 MHz. The wave length of a 50 MHz sinusoid is more than 18 ft. Transmission line reflections will not be possible under these circumstances.

The characteristic impedance (sometimes called "driving impedance") is defined as:

$$Z_0 = (L/C)^{1/2} \tag{6.9}$$

Increasing the inductance in an *LC* filter increases the characteristic impedance. Increasing the capacitance lowers the characteristic impedance.

The resonant frequency is proportional to the inverse of the square root of the product of *L* and *C*. Three different *LC* filters could have the same resonant frequency but entirely different characteristic impedances. When X_c is much larger than X_l, the *LC* filter has a low characteristic impedance. The transient response is better than an *LC* filter where X_l is much larger than X_c. When X_l

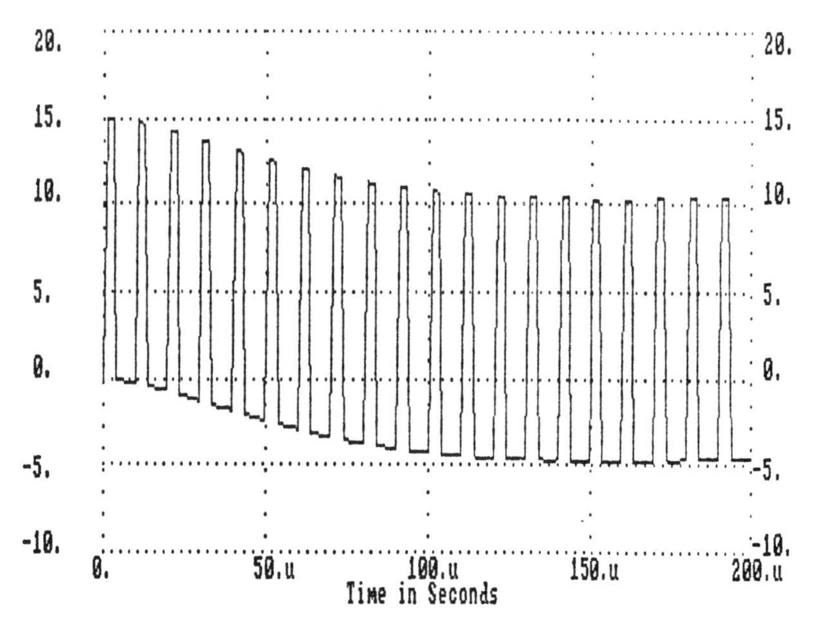

Figure 6.8 Inductor voltage.

is larger than X_c the *LC* filter will be a relatively higher impedance and the larger inductance will provide more input to output isolation.

For a given load the response of the two *LC* filters could be radically different from what was expected. For proper (non-interdependent operation) the characteristic impedance of a filter should be smaller than the load connected to the output and the reactance of the capacitor should be much less than the load resistance:

$$Z = (L/C)^{1/2} < 0.1R_L \qquad (6.10)$$

where

$$X_c < 0.1R_L$$

6.8 PARALLEL CAPACITORS TO LOWER THE ESR

The insertion loss and transient response of low-voltage (less than 300 V) power supply filters largely depend upon and are dominated by the ESR of the filter capacitors. Low-voltage switching power supply filter capacitors should be chosen such that the capacitive reactance X_c is small compared to the ESR of the capacitor at the switching frequency of the power supply. The ESR should also be small compared to the load resistance. To lower the equivalent series resistance of a filter, a number of capacitors can be placed in parallel to achieve the desired ESR. This method of design maximizes the effectiveness of the filter for a given size of volume.

The low-voltage filter effectiveness for EMI attenuation depends upon the series inductors and capacitor ESR ladder. The value of the capacitor makes little or no difference in the EMI attenuation as long as the reactance is small compared to the ESR. The two models, shown in Fig. 6.9A and B, are of a typical fourth-order (double *LC*) filter. The two capacitors in Fig. 6.9B have been replaced by short circuits. Insertion loss simulations were performed on both models. The insertion loss plots are presented in Fig. 6.9C and D, respectively. The two insertion loss curves are virtually identical.

For high-voltage (greater than 300 volts) power supply filters, the load resistance is generally much larger than the ESR or reactance of the filter capacitor. High-voltage circuit design implies

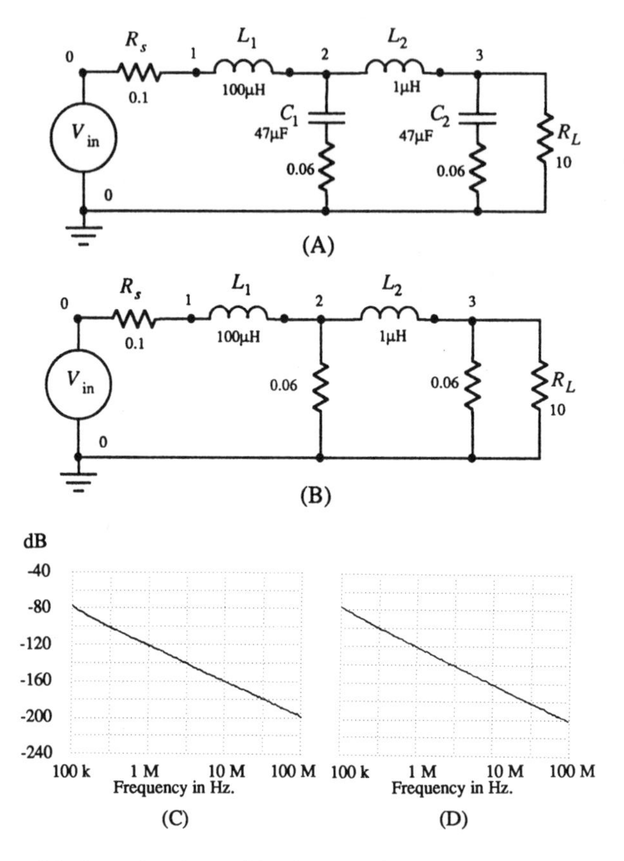

Figure 6.9 Insertion loss of *LR* ladder with and without capacitors.

high impedance. ESRs are so small compared to typical circuit impedances that they are generally insignificant. The reactance of filter capacitors in high impedance circuits is significant (X_c << R_L) so that paralleling capacitors to lower the ESR is ineffective.

To quantify a filter's ability to attenuate EMI, insertion loss analysis may be performed on the filter by hand or on the computer as was done in Fig. 6.9. It is useful to perform insertion loss tests on the filter to verify actual performance in the laboratory.

For a two-section LC the approximate attenuation Attn is:

$$\text{Attn} \approx \frac{R_{c1}(R_{c2} \text{ parallel } R_L)}{(2\pi f)^2 L_1 L_2} \tag{6.11}$$

For an n-section LC filter the approximate attenuation Attn is:

$$\text{Attn} \approx \frac{R_{c1}R_{c2}(R_{cn}\text{ parallel } R_L)}{(2\pi f)^n L_1 L_2 \cdots L_n} \quad (6.12)$$

The insertion loss in decibels is 20log(Attn). We will now calculate the insertion loss of the filter in Fig. 6.9 by hand (using a calculator of course).

The attenuation of the filter in Fig. 6.9 is approximately:

$$\approx \frac{0.06(0.06\text{ parallel } 10)}{((2\pi 100e^{3})^{2})1e^{-6}100e^{-6}}$$

$$\approx \frac{0.0036}{(3.9e^{11})(100e^{-12})}$$

$$\approx 92.3e^{-6} \quad (6.13)$$

Insertion loss at 100 kHz =

$$20(\log(92.3e^{-6}) = -80.7\text{ dB} \quad (6.14)$$

We now look at Figs. 6.9C and 6.9D and find that the approximation is extremely accurate. This approximation is very useful for quick calculations at specific frequencies. Equation 6.12 also provides insight into the parameters affecting the attenuation of LC filters.

Key Idea 2

The parallel element with the lowest impedance will dominate a parallel combination of elements.

A series element with highest impedance will dominate a series combination of elements.

The concepts expressed in Key Idea 2 are important to keep in mind when trying to identify the dominating "player" in a circuit. In a circuit of series elements, the element of maximum impedance is the dominant player and sets the minimum impedance of the series combination of elements. If the impedances of the other elements are small compared to the dominant player element, the value of the other series elements is insignificant. For parallel combinations of elements, the element of minimum impedance is the dominant player and sets the maximum impedance of the parallel combination of elements. If the impedances of the other elements are large compared to the dominant player element, the value of the other parallel elements is insignificant.

The identification of the dominant players as described is the same process as removing the insignificant mathematical terms when writing circuit node or loop equations. The reduced equivalent circuits (or equations) provide a better insight into the actual operation of circuits.

The first step in identifying the dominant players in a circuit is to choose the frequency of interest. Then label the reactance of each of the inductors and capacitors on the model schematic, which includes the parasitic elements. Then identify the key players by using the concepts in Key Idea 2. The insignificant elements can be removed and the result is a model of the significant players (at the frequency of concern) in the circuit under investigation. Keep in mind that the resistances are mathematically perpendicular to the reactances, but resistances or reactances can be neglected if they are insignificant as given by the concepts in Key Idea 2.

Key Idea 3

> Do not indiscriminately add capacitors without checking the "playing" inductance to avoid making tanks that resonate at frequencies that can cause problems.

There is a great temptation (apparently) for electronic engineers to put a capacitor anywhere where there might be some high-frequency noise in a circuit. The addition of a Band-aid capacitor often creates more problems than are cured. Capacitors can resonate with their own leads, circuit board traces, leakage inductances, circuit inductors, or transformer inductances. Care must be taken to make sure that there is no stimulation at frequencies close to the resonant frequency of an added capacitor and any inductance that may work with it to form a resonant tank.

6.9 *LC* FILTER

A single stage *LC* filter model and four insertion loss simulation data plots are shown in Fig. 6.10. The inductor and capacitor modeled are chosen to be the best possible components available. Optimistic values were chosen for the parasitic parameters of these models to show that even in the best circumstances possible, the real world detrimental aspects of components are serious design concerns. The *LC* filter is "lightly" loaded with a 10 kΩ resistor. The source resistance is set to 0.1 Ω. Neither the load or source resistance will affect the filter significantly in this circuit.

The plot in Fig 6.10A is of the *L* and *C* only with no parasitic elements. There are two poles at:

$$F_0 = 1/(2\pi(LC)^{1/2}) \tag{6.15}$$

$$F_0 = 1/(2\pi(10e^{-6}10e^{-6})^{1/2}) = 16 \text{ kHz}$$

The capacitor equivalent series inductance (ESL) is added to the model. The insertion loss curve for this configuration is shown in Fig. 6.10B. There are now two poles and two zeros. The zeros are located at:

$$F_1 = 1/(2\pi(\text{ESL}C)^{1/2}) \tag{6.16}$$

$$F_1 = 1/(2\pi(10e^{-6}20e^{-9})^{1/2}) = 356 \text{ kHz}$$

We now add the capacitance C_s in parallel to the inductor L. There are four poles and four zeros as shown in Fig. 6.10C. Two zeros are located at:

$$F_2 = 1/(2\pi(LC_s)^{1/2}) \tag{6.17}$$

$$F_2 = 1/(2\pi(10e^{-6}20e^{-12})^{1/2}) = 11.3 \text{ MHz}$$

And two poles are located at:

$$F_3 = 1/(2\pi(\text{ESL}C_s)^{1/2}) \tag{6.18}$$

$$F_3 = 1/(2\pi(20e^{-9}20e^{-12})^{1/2}) = 252 \text{ MHz}$$

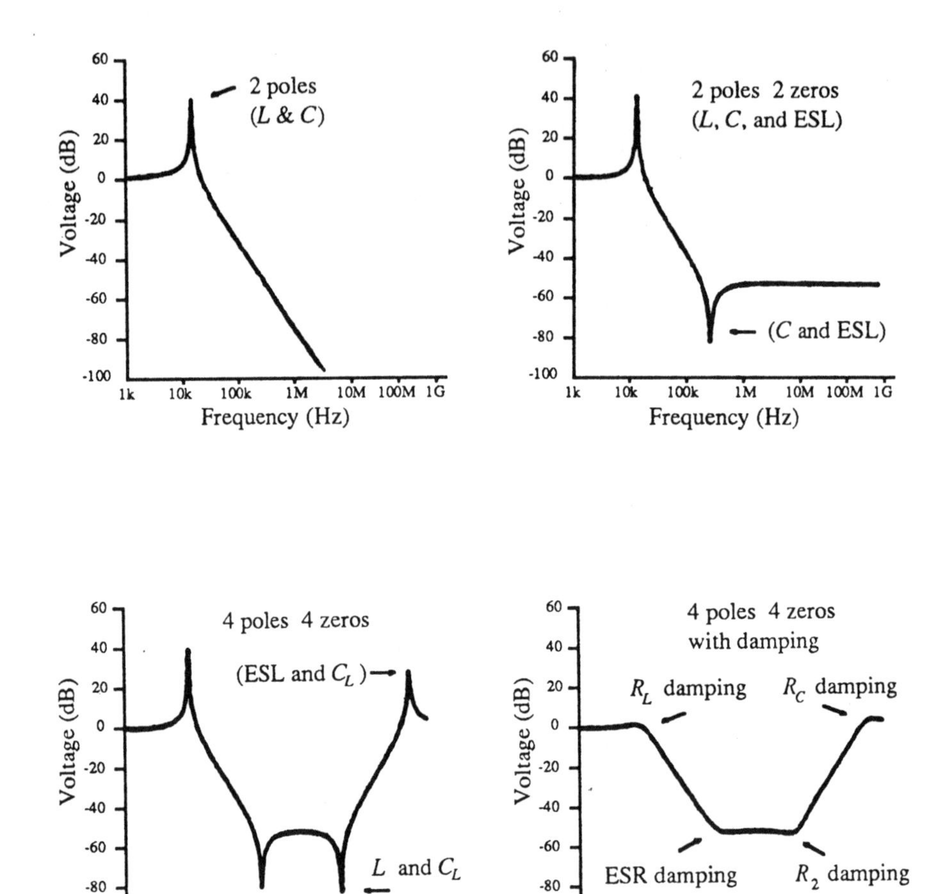

Figure 6.10 *LC* insertion loss curves.

Finally the resistances are added to the circuit. The damping effects of the resistances (losses) are shown in the insertion loss plot of Fig. 6.10D. A simple *LC* filter is actually a notch filter when the parasitic elements are considered. There is virtually no insertion loss above 252 MHz.

6.10 LINE IMPEDANCE STABILIZATION NETWORKS (LISN)

The purpose of discussing the LISN here is two-fold:

1. Line impedance stabilization networks are special filters used in testing when a constant line impedance is required. This an interesting type of filter.
2. The modeling of an LISN presents interesting challenges that require the use of accurate models. In order to model the LISN, the author was forced to make some detailed improvements to the inductor model that was previously in use.

A typical LISN will provide a constant 50 Ω impedance (± 1 dB) from 10 kHz to 1 GHz (see Fig. 6.11). Power transfer at frequencies below 10 kHz can occur because the LISN impedance is much less than 50 Ω. The 50 Ω impedance is chosen to be a standard for comparison. Line impedances of actual systems vary a great deal. A compromise was made between the desire for equal comparison to a standard and the desire to test the filter performance in actual operating conditions of the equipment under test.

An LISN is essentially an inductor with a capacitively coupled output for monitoring. An LISN can be constructed from a series of five inductors. The inductors are tuned to be resonant at five different frequencies. Each inductor provides 50 Ω impedance in its frequency range as is done in a three-way speaker system in a stereo high-fidelity audio system.

The five inductors are mounted on a 1/2-in-diameter wooden stick. Three of the inductors are wound on torroid cores. Two of the inductors are wound on the 1/2-in-diameter stick and function as air core inductors. The losses of the torroid cores play a significant role in making the impedance 50 Ω for the wide frequency ranges. The core losses in parallel with the inductors limit the maximum impedance at resonance. The capacitively coupled output is not shown in the model. The capacitor and 10 kΩ resistor isolate the monitor output from affecting the performance.

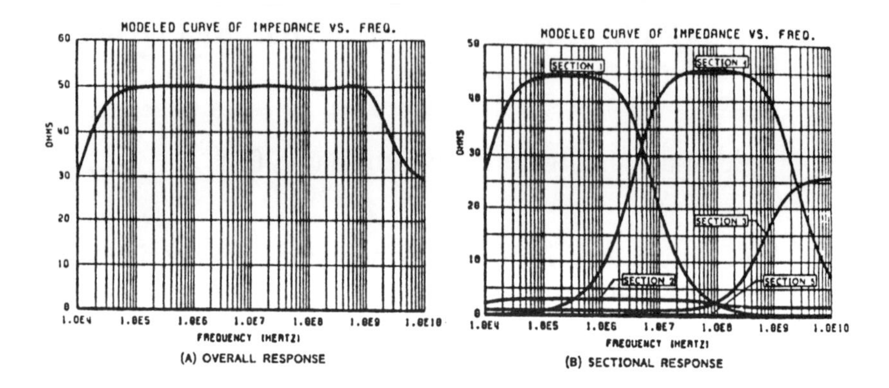

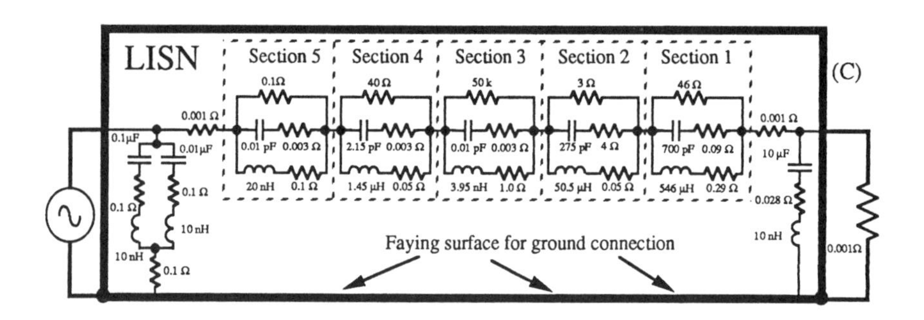

Figure 6.11 LISN.

6.11 FILTER LAYOUT AND PACKAGING DESIGN

Key Idea 4

> Inductor parasitics are parallel capacitances and resistive losses.
>
> Capacitor parasitics are series inductances and resistive losses.

The parasitics in capacitors and inductors lower their self-resonant frequency. In filter applications, this results in a reduced frequency range in which they will attenuate EMI. There are two important considerations to observe when laying out inductors in filters:

1. All adjacent metal (conductors) should be chassis grounded.
2. Maximize the spatial separation of inputs and outputs.

Two important considerations to observe when laying out capacitors in filters are:

1. Make the lead length as short as possible.
2. Wire to and away from caps.

Circuit Routing and PWB Layout

Key Idea 5

Current always flows in path of lowest impedance.

The use of parallel traces on adjacent layers for power and return lines minimizes the area and increases the power to return line capacitance. Closely spaced parallel traces greatly reduce the circuit wiring impedance and also reduce radiated emissions and susceptibility (see Fig. 6.12). Assemblies such as power transistor switches and transformers should be designed with lead routing as an important design parameter.

The 20-nH-per-inch Rule of Thumb

A rule of thumb commonly used for wire inductance estimation is 20 nH per inch. This is a very handy approximation but can be misleading if not understood. A piece of wire has no inductance. Only loops of wire have inductance. The 20-nH-per-inch approximation is an average value that occurs when an inch of wire is added to a typical capacitor or other component on a printed circuit board. The inductance is increased because of the larger loop area (see Fig. 6.13), not because of the longer length of wire. The two loops of wire in Fig. 6.14 can be used in an experiment to show the loop area effect on inductance and layouts. Both loops are made with 18 inches of wire. The loop on the left has an inductance of 1 nH. The loop on the right has an inductance of 800 nH. The area of the loop is the key parameter in the amount of inductance created.

Capacitive Coupling Paths around Inductors

Key Idea 6

Reduce filter input to output capacitance.

Figure 6.15 shows a generalized case of the capacitive coupling paths around an inductor. The same discussion can be applied to capacitive coupling paths around whole filters and electronic circuits. This is a model of a five-turn torroid mounted on a circuit card. The rectangle labeled "conductor" represents any conductors near the inductor. These conductors could be circuit traces, other components, or any metal structure (except chassis ground).

The capacitance labeled C_{cpLL} represents the capacitance directly from one lead to the other lead of the inductor. Spatial separation is a method to reduce C_{cpLL}.

The often overlooked and usually the most troublesome capacitive coupling path is from a component to another conductor and back to the component again. The capacitive coupling path from the inductor, through C_{cp2} to the conductor, and back through C_{cp3} to the inductor is often a lower impedance than any of the other external paths. The capacitances C_{cp2} and C_{cp3} are distributed with the connection node across the surface of the conductor. The sum total of all of these distributed series pairs of capacitances in parallel add to the capacitive path around the inductor.

The capacitances in Fig. 6.15 model the capacitive coupling paths, caused by the circuit layout, that allow high-frequency currents to flow around the filter elements that are intended to attenuate noise emissions.

All high-frequency inductors must have good input to output isolation to reduce capacitive paths around the inductors. A solution is to use a chassis-grounded aluminum case to totally enclose the inductors with feed-through connectors on input and output, which should be on opposite sides of the inductor. A better solution is to totally enclose the whole EMI filter (inductors and capacitors).

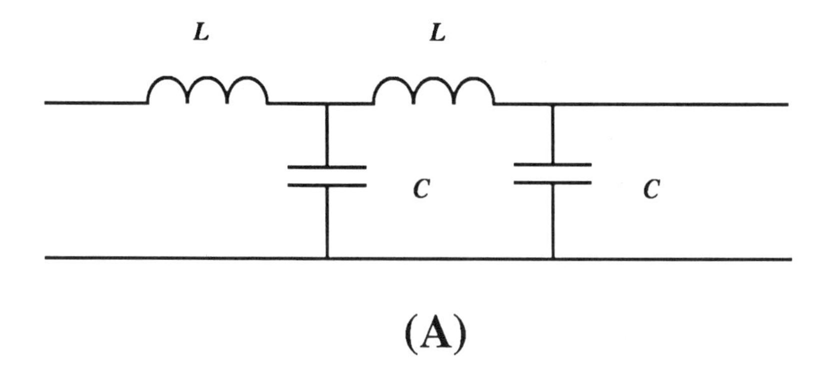

(A)

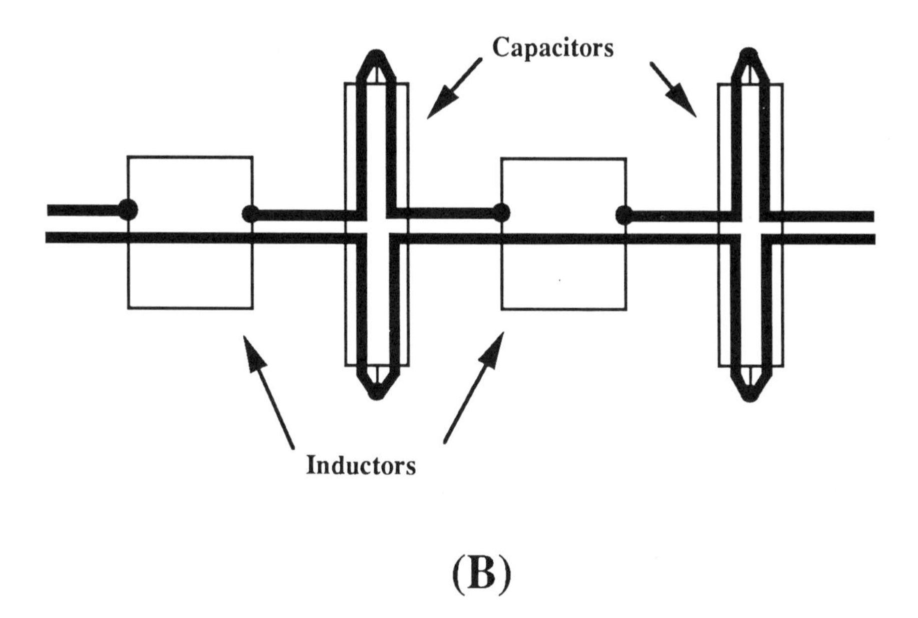

(B)

Figure 6.12 Trace routing.

All metallic conductors should be attached to the circuit or tied to the chassis through a low-impedance path. Ungrounded conductors provide capacitive coupling paths around inductors (see Fig. 6.15) and filters to the extent that in some cases the inductor or filter is totally ineffective.

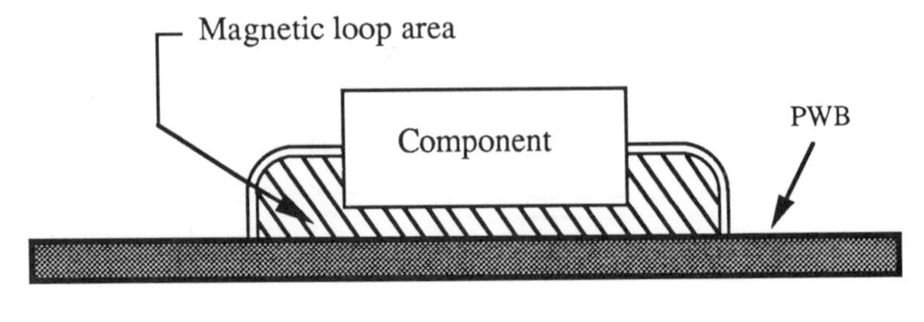

(A) Small loop area

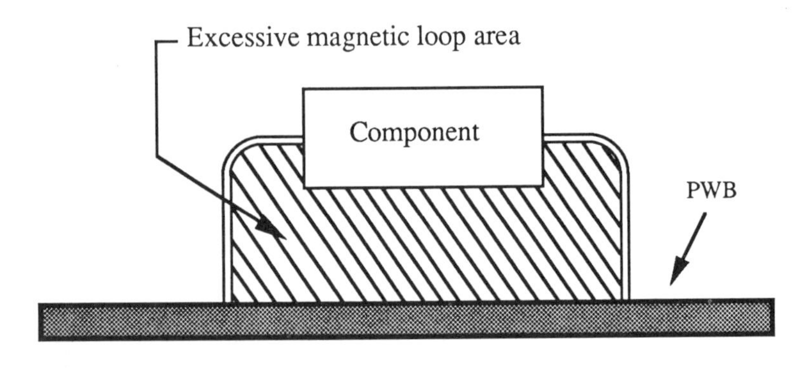

(B) Large loop area

Figure 6.13 Magnetic field caused by increased component circuit loop area.

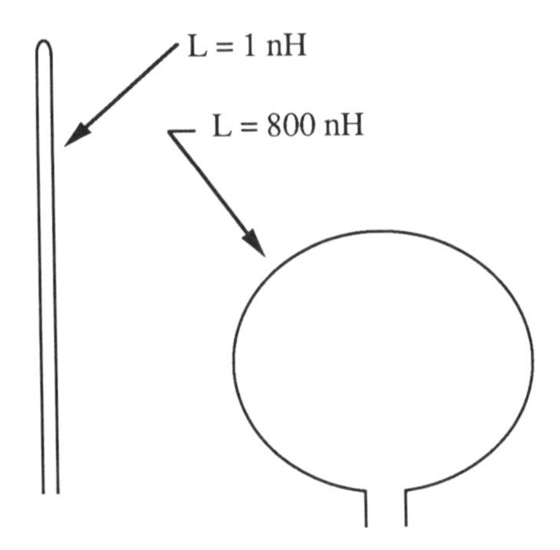

Figure 6.14 Inductance of 20 in of wire, large area versus small area.

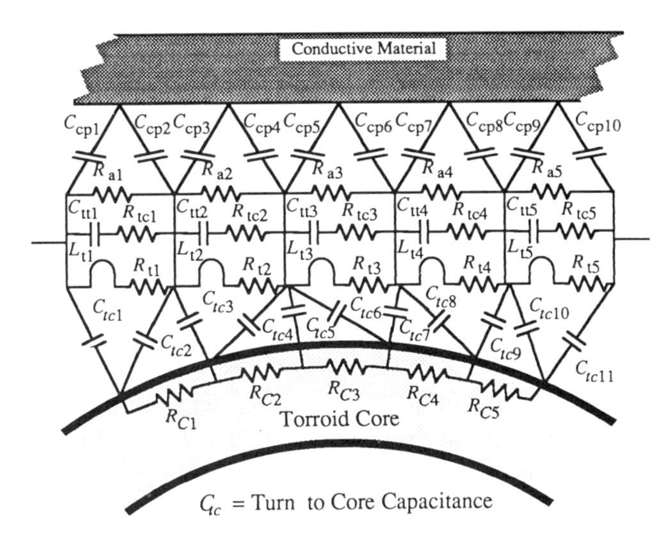

Figure 6.15 Capacitive coupling path around inductors.

Key Idea 7

Electronic ringing requires inductance and capacitance.

Exponentially damped sinusoids are very common in nature and always occur to some extent in electronic circuits. Damped sinusoids usually occur immediately following fast transient waveforms (Fig. 6.16).

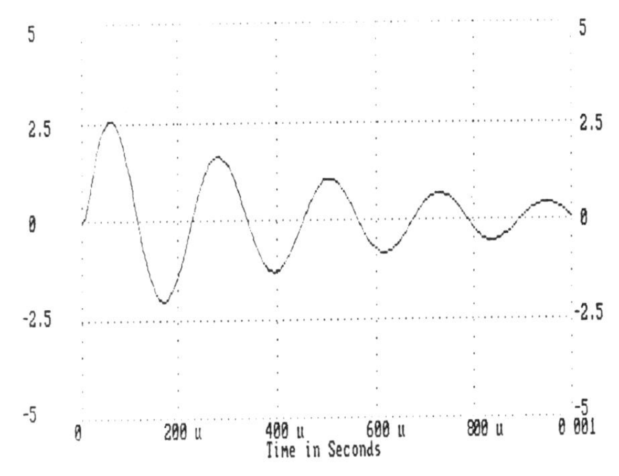

Figure 6.16 Single resonant damped sinusoid.

The damped sinusoid indicates the presence of an inductance and capacitance operating together (ringing) in a resonant "tank." Resonance can only happen if both inductance and capacitance are present. The resonant frequency will be:

$$F_r = 1/(2\pi(LC)^{1/2}) \tag{6.19}$$

where L and C are the operating inductance and capacitance. The length of the ring is determined by the Q of the resonant LC. The Q is calculated as:

$$Q = \omega L/R \tag{6.20}$$

where R is the equivalent resistance of the losses in the resonant LC tank. The Q is defined many ways, but the best concept to remember is in Key Idea 8.

Key Idea 8

> Q is the ratio of the stored energy in a circuit to the energy lost in each cycle of the waveform.

Often when a circuit node is subjected to a fast transient, more than one LC tank is stimulated into ringing. The waveform is more complex when multiple LC tanks operate together. Figure 6.17 is the waveform of two LC tanks ringing together.

Both component and circuit layout reactances can act together to form resonant LC tanks. Multiple tank circuits are often a combination of component and layout reactances. Printed circuit trace loop inductances and trace to trace capacitances affect even the best layouts in the 2- to 30-MHz frequency region. A capacitor's lead inductance resonates with its capacitance. An inductor winding capacitance resonates with its inductance. There are the self-resonances of inductors, capacitors, and printed circuit boards. Trace inductance adds to lead inductance to lower the in circuit self-resonance of capacitors. Input to output trace capacitance adds to the winding capacitance to lower the in circuit self-resonance of an inductor. Good circuit layouts can reduce the natural resonances of the interconnecting wiring. The natural response causes ringing (noise) when any significant voltage or current transition occurs. This ringing can cause both functional and emissions problems.

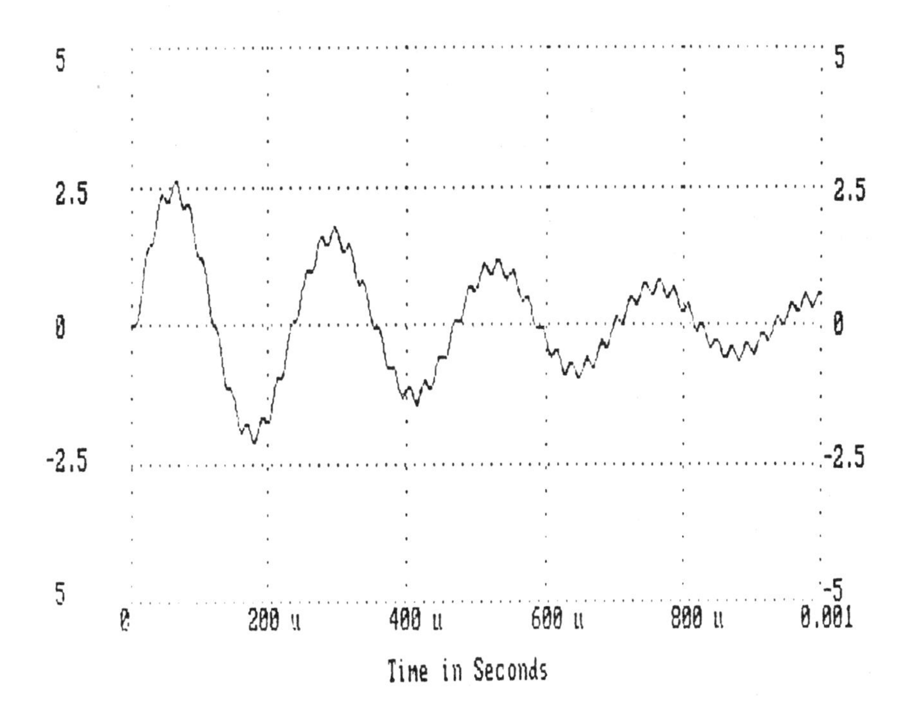

Figure 6.17 Two *LC* tanks ringing.

Interface EMI Filters

The use of a filter connector or an internal bulkhead with feed-through filters is recommended. Lossy pi filters provide insertion loss at high frequencies. The filter connector is mounted in a physical position that takes advantage of the chassis wall in which it is mounted. The chassis wall provides a barrier to reduce the capacitive coupling around the filter so as to perform well at high frequencies in the range of a few megahertz to beyond 50 MHz, the upper limit for CE03. This filter connector can also provide attenuation of unwanted common mode currents.

The EMI interface filter is the classic EMI filter. Generally the filters on printed circuit boards using standard inductors and capacitors resonate at frequencies well below the high frequency range of military or commercial conducted emission limits. The attenuation of frequencies above 10 MHz is done best at the input and output interfaces. Small components with special construction geometries for high-frequency applications are generally used.

They are totally enclosed in a Faraday shield that is bonded to the chassis. The bonding resistance should be less than 2.5 mΩ to be effective as a low-impedance path to the chassis. These classic EMI filters attenuate both outgoing and incoming noise. Some of the special types of inductors and capacitors used in interface EMI filters are baluns, beads, and feed-through capacitors.

QUESTIONS

6.1 Why are damped sinusoids so commonly found in electronic circuits?

6.2 How does the mechanical housing (package) affect the attenuation performance of an EMI filter?

6.3 What parameters must be considered when choosing components for an EMI filter?

6.4 Is an LC filter ladder a notch filter or a low pass filter? Why?

6.5 Under what circumstances is filter attenuation improved by paralleling many smaller capacitors with the same total capacitance in place of one larger capacitor?

6.6 What is a Faraday shield?

6.7 How do load and source impedance and the characteristic impedance of the filter affect filter attenuation performance?

6.8 Can an *LC* filter perform two circuit functions such as hold-up and filter attenuation? Give an example.

7

GROUNDING ELECTRONIC CIRCUITS

7.1 GROUNDING

Ground design must be done at the system, equipment, and circuit levels. Four different functions generally called grounding are:

1. Chassis grounding
2. Safety grounding
3. Connecting reference potentials
4. Current return

It is important to understand these functions and use them appropriately in the electronic design process. There are misconceptions about grounding that plague many electronic design engineers. The importance of proper grounding techniques grows continuously with the increase of the operating speed of electronic circuits. Poor ground designs cost manufacturers and users millions of dollars every year. Some human lives have been lost by the use of poor grounding techniques. In the medical electronics industry, grounding and human safety are the most important design criteria. Proper grounding techniques are important for all electronics endeavors.

Poor grounding can be detrimental to human safety in two ways:

1. Direct human shock
2. EMI causing upset in critical circuitry

Human shock occurs when a voltage potential is applied to the human body directly. With poor (high-resistance) contact to the skin, the current will hopefully conduct on the skin surface. Only a few milliamps are required to kill a human being if the current passes through the body. To get current to flow through a human body a very low-impedance contact is required. An example of conductive contact to the body is when a human stands in water and accidentally touches a high potential with another part of the body. Another dangerous situation for human safety also happens when body electrodes are attached to a patient in a hospital. Extreme precautions are necessary to prevent the accidental electrocution of the patient. There have been cases of accidental electrocution in hospitals because of ground faults.

EMI that causes an upset in critical circuitry can be indirectly detrimental to human safety also. For instance, an electrical upset in flight-critical electronic circuitry in an aircraft could cause the aircraft to crash. Or, the accidental ejection of a pilot in a military fighter aircraft could be caused by an EMI upset of an onboard control computer.

Definition of a Ground

A ground is a low-impedance connection that minimizes the voltage difference between parts of a system for the purpose of safety or signal voltage referencing.

The first three items in the list of four functions generally called grounding fit the definition of grounding. The fourth in the list, current return, does not fit and is technically not grounding. Calling a current return a ground is probably the most ubiquitous misconception in the electronics industry. This misconception has lead to a great deal of confusion for engineers and has complicated the very simple concept of grounding.

The probable reason for pervasiveness of the current return and ground confusion is the use of schematics. Here we must regress a little bit to sort out any confusion between wiring diagrams and schematics. The term "schematic" comes from the word "scheme."

A pure schematic is a diagram that shows, by means of graphic symbols, the electrical connections and functions of a specific circuit arrangement. The schematic diagram facilitates tracing the circuit and its functions without regard to the actual physical size, shape, or location of the component device or parts.

A pure wiring diagram, on the other hand, shows the physical connections of wires and components without concern for grouping or ordering in regard to their functions. The mechanical aspects are the important features. All ground wires are shown, and no ground symbols are used.

Typically the electrical diagrams used are a combination of schematic and wiring diagrams to fit individual needs. As a matter of convenience and clarity, current returns are generally depicted by ground symbols to simplify the diagram. This is where the confusion between grounds and current returns arises. Current returns are usually connected to grounds to set common references.

A schematic of three power supplies and loads is shown in Fig. 7.1. The power supplies are shown in block diagram form, but let's consider the power supplies as purchased assemblies; therefore Fig. 7.1 is a schematic by definition. The six ground symbols indicate that the low-potential side of the power supplies

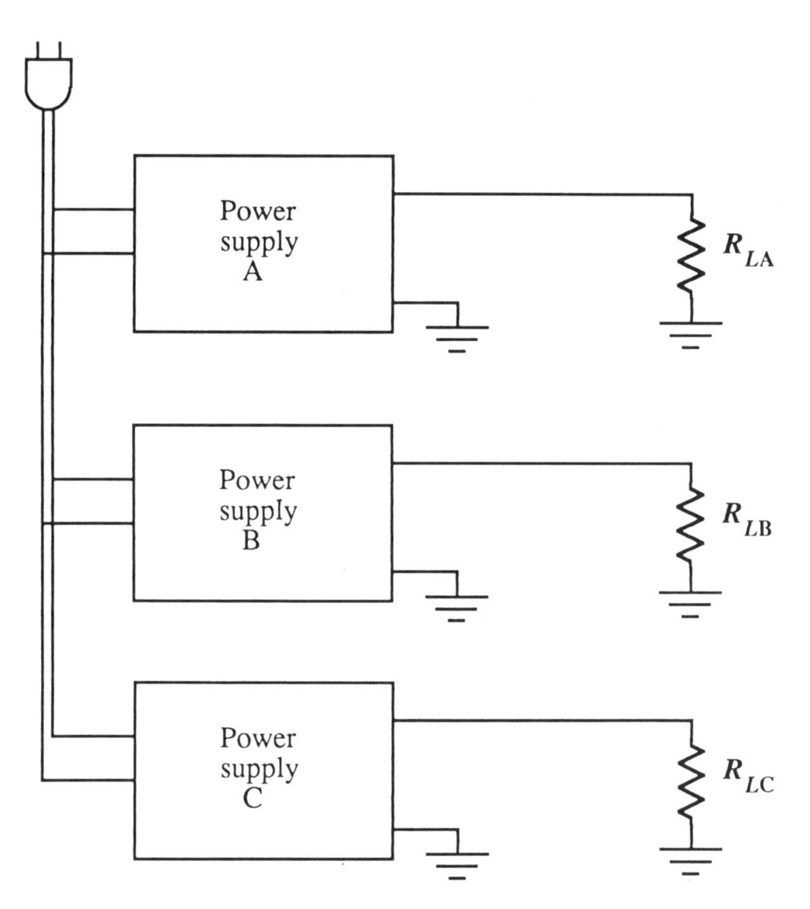

Figure 7.1 Schematic of three power supplies and loads.

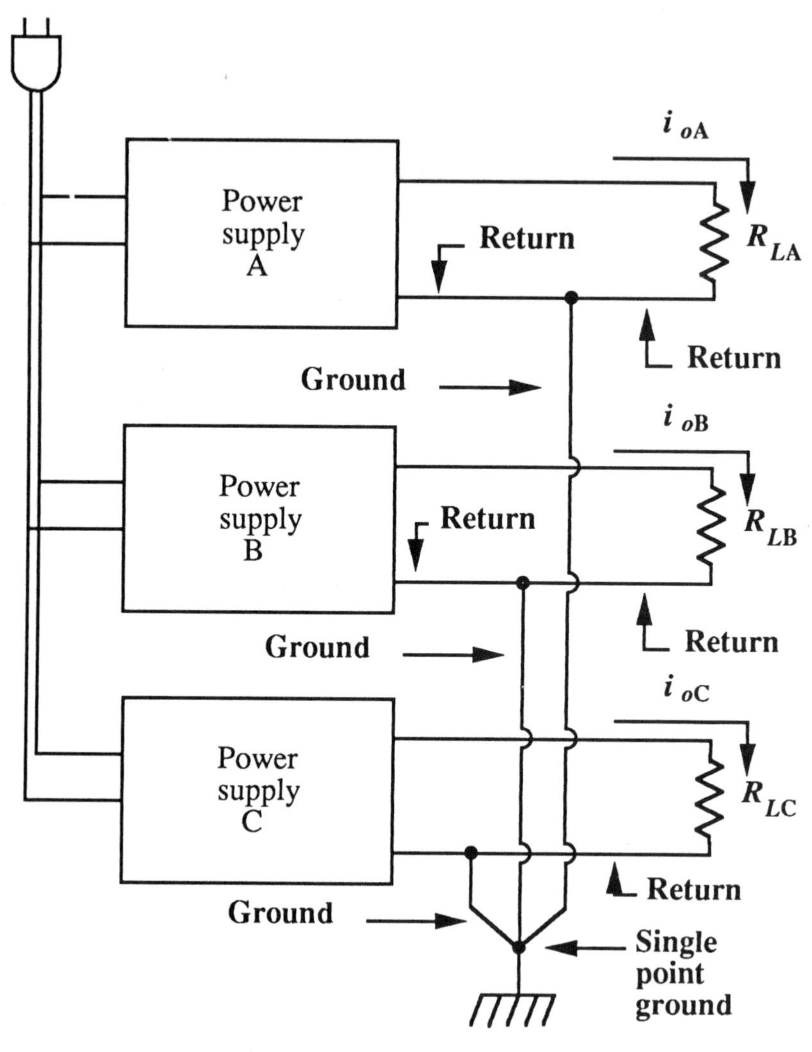

Figure 7.2 Grounds and returns.

and loads are connected together. The scheme of this assembly is clear. Three loads are connected to three power supplies that are all referenced to the same potential. The ground symbols in this schematic represent both grounds and current returns. There is no information about the layout. If layout information were shown in this schematic, the scheme of the functions of the circuit might be obscured.

Figure 7.2 is a quasiwiring diagram of the same circuit with the layout information included. The scheme of the function of this circuit in this diagram is somewhat obscured by the addition of the layout information. This diagram might be very good for communicating information to the layout person about the desired layout. The schematic in Fig. 7.1 would be better for communicating information to a lab technician about the function of the circuit.

We now return to the discussion about the differences in current returns and grounds. There are three returns and three grounds labeled in Fig. 7.2. The three grounds are connected to a single point ground. The three returns are current conduction paths for load currents i_{oA}, i_{oB}, and i_{oC} back to their respective power supplies. All current must return to the source. The grounds connect the three low-potential sides of the power supplies and loads to a common reference called the single point ground. For an ideal system, with no radiated fields present and no current from any other system flowing anywhere in this circuit, there will be no current in the grounds.

Key Idea 1

Grounds conduct no current in ideal circuits.

In real circuits there is always some amount of radiation and current (or differing potentials) from outside or internal influences. We want to have the ground paths of as low an impedance as possible so that the unavoidable interference currents cause as little difference in potential between the three circuit references as possible. Grounding and current returns are clearly two different functions.

The purposes for grounding are:

1. Increase safety of personnel and equipment
2. Reduce noise
3. Provide a signal voltage reference

All three purposes require minimizing the voltage difference between certain parts of a system.

7.2 SAFETY GROUNDS

The primary purpose of safety grounding is to protect personnel, equipment, and wiring from excessive voltage differences. The requirements for grounding are specified in article 250 of the *National Electrical Code.*

Safety grounding reduces the chance of electric shock to people using equipment in which the neutral wire integrity has been impaired.

For safety we want the circuit breaker to trip as soon as possible when a fault occurs (Fig. 7.3). The safety ground should be of low inductance and low impedance so that the fault current will rise quickly to trip the circuit breaker.

Ground fault interrupters are now commonly used to interrupt the electrical circuit when a fault current to ground exceeds a predetermined value that is less than that required to operate the overcurrent protection device of the supply circuit.

The maximum allowable chassis to chassis voltage (V_{hs}) for human safety is less than 10 V peak. The worst case fault currents (I_{fc}) should be determined. The impedance (Z_g) of the interconnecting ground between any two chassis should be designed to be:

$$Z_g \ll V_{hs}/I_{fc} \tag{7.1}$$

To get a feeling (no pun intended) for the 10 V peak limit for safety, we consider that a typical 9-V dc battery is very safe to handle and poses no safety hazard. Touching, with fingers, across the terminals of the 9 V dc battery will give virtually no electrical shock sensation whatsoever. A better electrical contact can be made by touching one's tongue across the 9-V dc battery terminals. A noticeable electrical shock will be sensed, but it will not be painful or hazardous to one's safety. This experiment is not recommended for those with pacemakers, but it probably would not be harmful.

7.3 GROUND GEOMETRIES

Ground Planes

Ground planes are low-impedance, low-inductance reference planes. A ground plane should be designed to be an equipotential

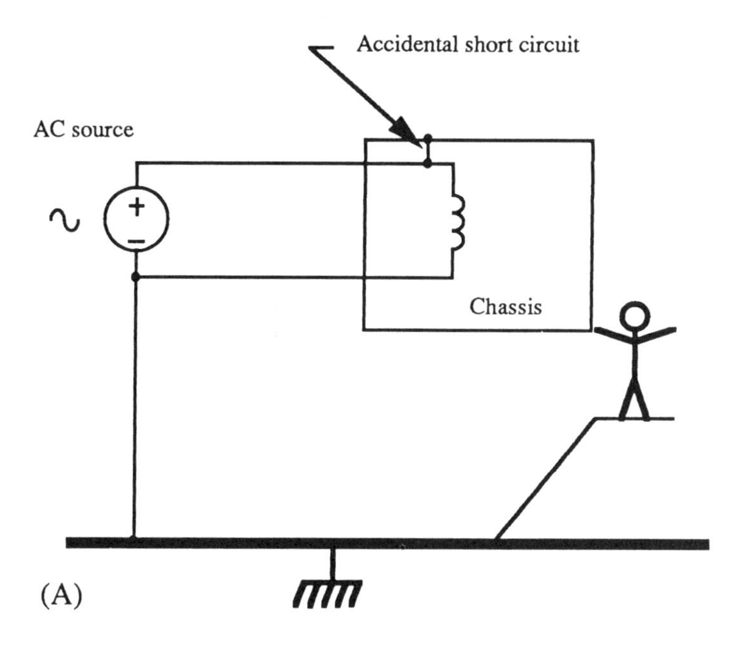

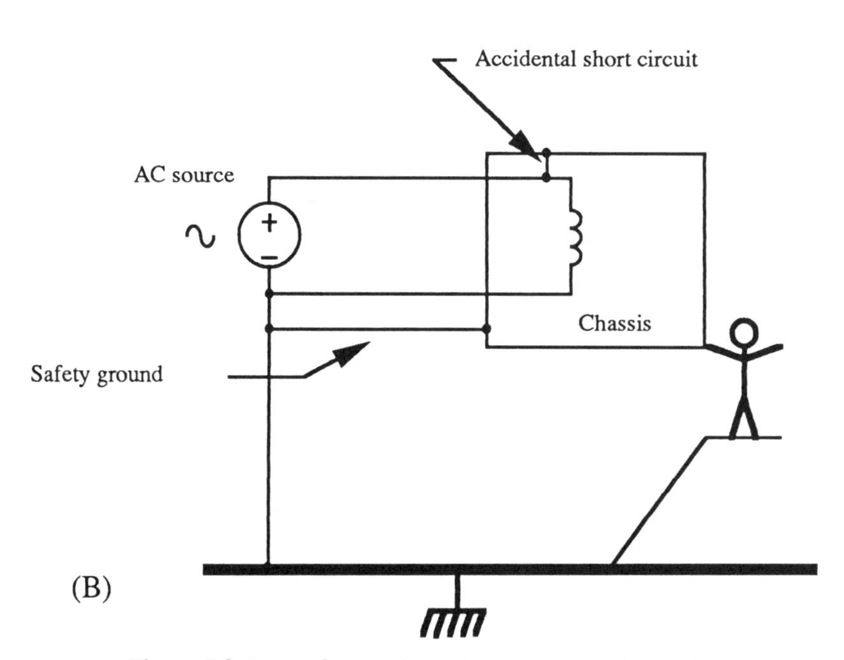

Figure 7.3 Loss of neutral, putting personnel across line.

surface. This means that there is to be virtually no current flowing in the ground plane. Oftentimes, current returns are mistakenly called ground planes because a current return is designed to be of as low impedance as possible. The lowest-impedance (Z_m) conductive paths are large planes separated by the smallest distance :

$$Z_m = (L/C)^{1/2} \tag{7.2}$$

In circuits where the currents are relatively small, a conductive plane can be used for both a reference plane and a return plane. This is actually the case in all ground planes because a conductor with no current in it cannot be part of an electrical circuit. The key words are "relatively small currents."

A good illustration of the use of a ground plane is the control circuit in a pulse width modulation (PWM) power supply. An example of a typical control circuit is shown in Fig. 7.4. There are six connections to signal ground. Four of the ground connections have very low current:

1. Current sense
2. Voltage sense
3. Over current sense
4. Timing capacitor (sets frequency of operation)

The current in these four grounds is typically in the milliamp range. The fifth ground is the PWM control IC current return and ground reference. The current in the PWM control IC ground is about 10 mA. The sixth ground is the current return for the drive transistor and primary of the drive transformer. The typical average current in a driver ground is approximately 100 mA. The peak currents are typically 600 mA.

The first five grounds can all be connected directly to the ground plane (Fig. 7.5) because they are of relatively low current levels. Because it has a relatively high current, the sixth ground (return) must be routed on a separate trace (see Fig. 7.6) to a single-point ground located on the ground plane where it connects to the main ground reference. This method of layout will isolate the high current circuits from the relatively lower-current circuits to reduce the effects of shared common ground impedances. Also, the reference to the ground plane potential in the sixth ground is not as important as in the first five grounds.

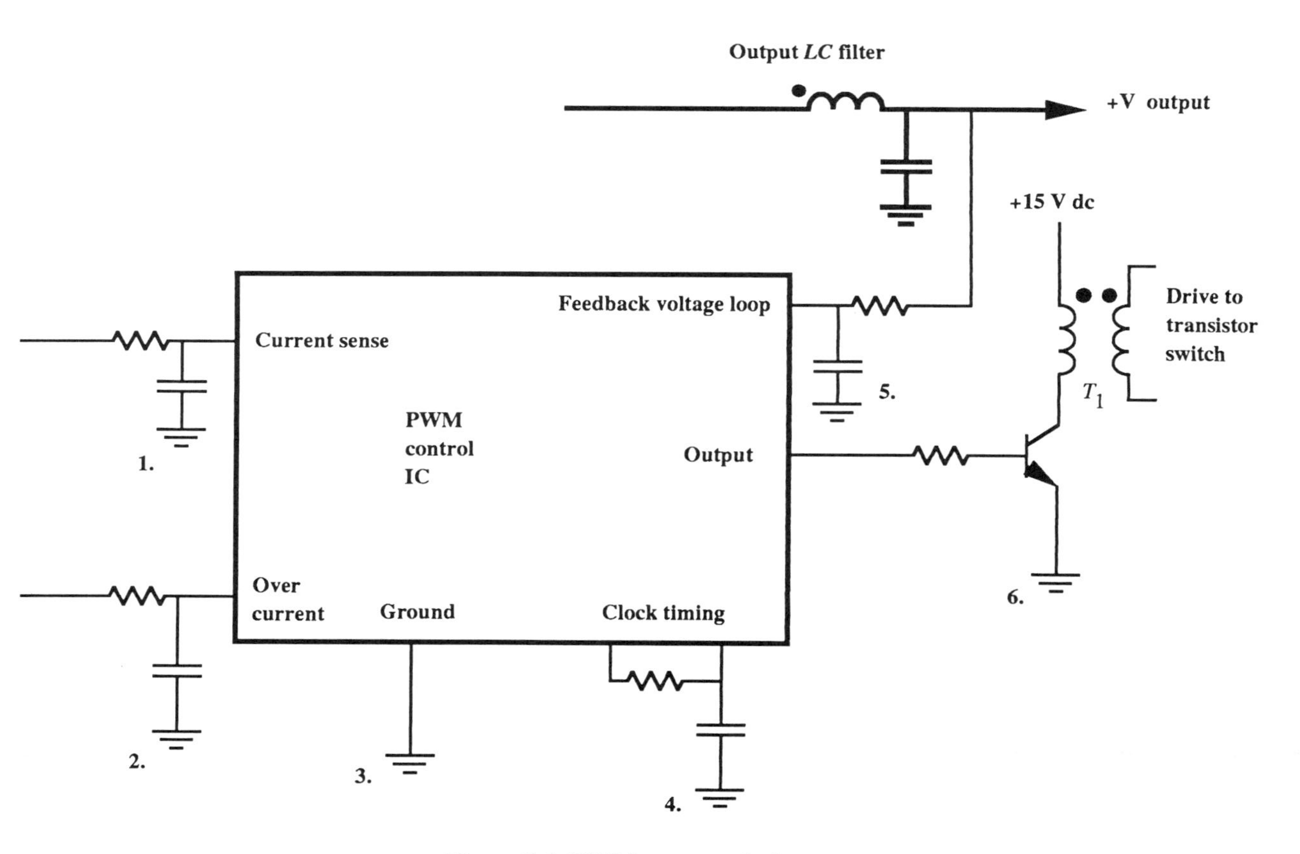

Figure 7.4 PWM on ground plane.

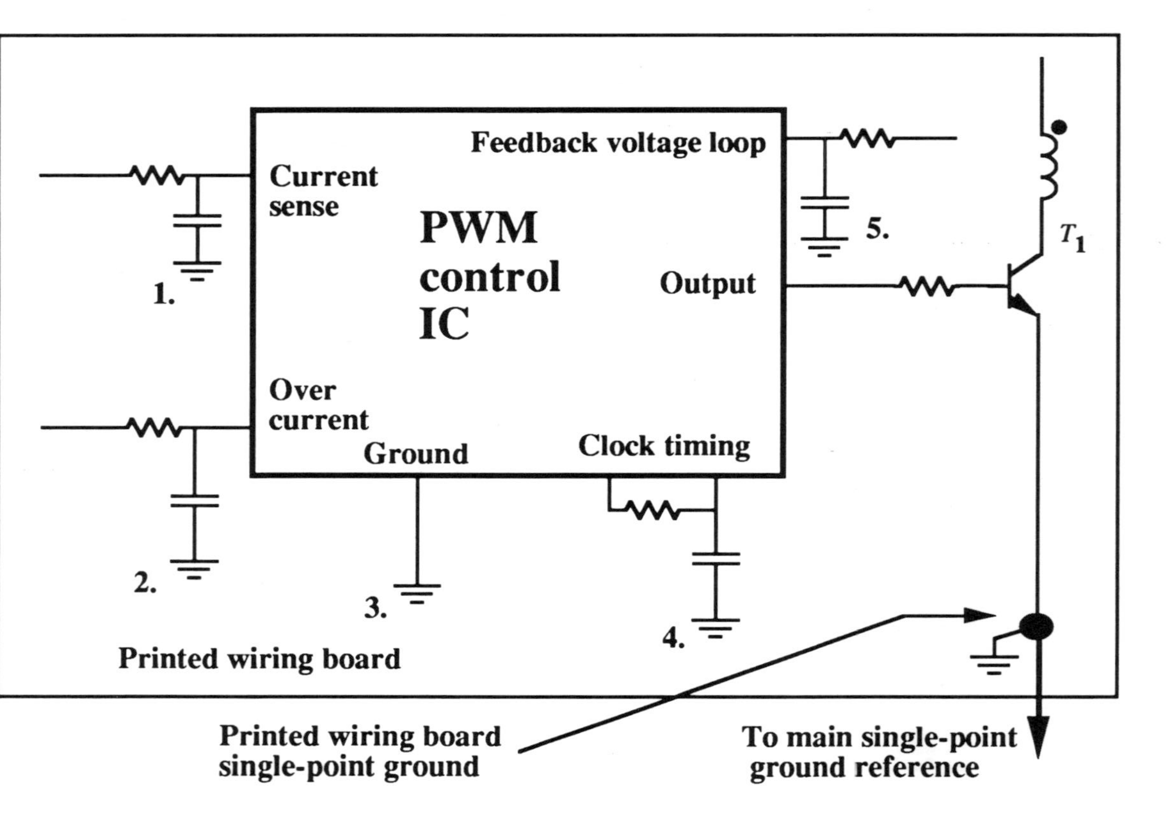

Figure 7.5 SPG on ground plane.

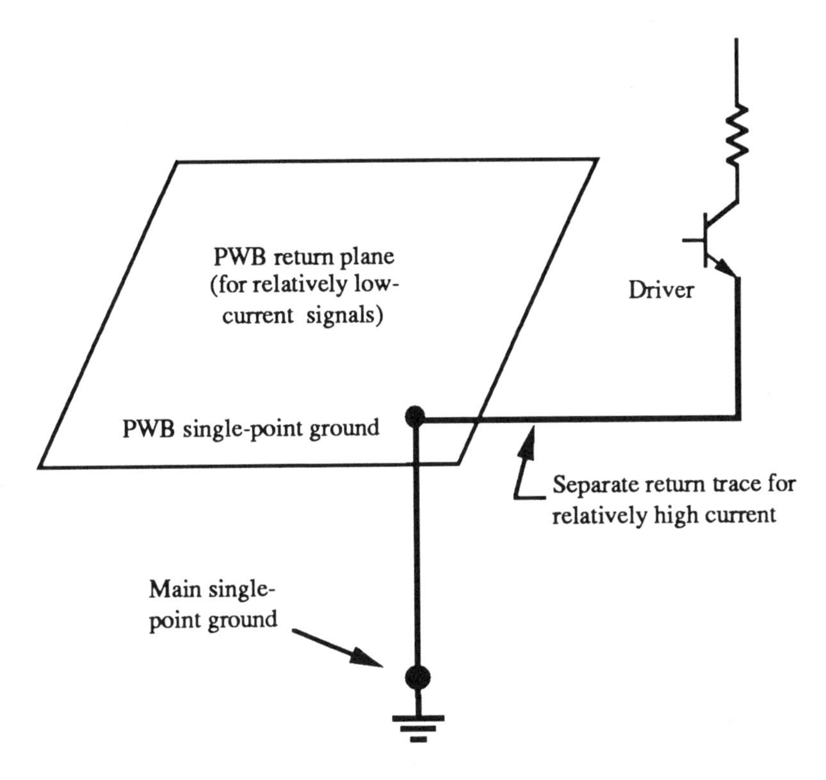

Figure 7.6 Diagram of ground circuit.

Shared Return Common Impedance / Single-point Grounds

Common mode currents are present in virtually all systems. Common impedance ground currents are one source of common mode currents. The proper use of single-point grounds can reduce the common current caused potentials created by shared ground paths. Each separate ground conductive path should "tree" into its relative ground reference as opposed to "daisy chaining." An example of the common impedance problem is shown in Fig. 7.7. The two grounds G1 and G2 should be at the same potential but they are separated by some distance. This example of electronic instrumentation could be for a car, ship, aircraft, or for communications between two buildings. The grounds G_1 and G_2 could be two different points on the car body or the hull of a ship or the fuselage of an airplane or could be the ground stakes in the earth for two buildings. The impedances Z_1 and Z_2 represent the characteristic impedance of wire or circuit traces or could be a filter designed to reject unwanted frequency components. The impedances Z_1 and Z_2 are shown as an inductor and capacitor, respectively, but could be swapped and the same academic point could be made from this example. The impedance between G_1 and G_2 is represented by Z_g. The current source I_{gc} represents a current flowing in a totally separate electrical circuit from the one under investigation. No matter how small Z_g is, there will be a potential V_g between G_1 and G_2 caused by the shared ground impedance and current I_{gc}. A differential EMI voltage, resulting from the common mode currents and the unbalances of the circuit impedances, will be impressed across the load Z_L.

In older cars all of the return current was conducted through the chassis so that there was no common mode current by definition. The current I_{gc} in Fig. 7.7 could be a headlight current of many amps causing volts of drop in the car chassis. A 1-V drop across a

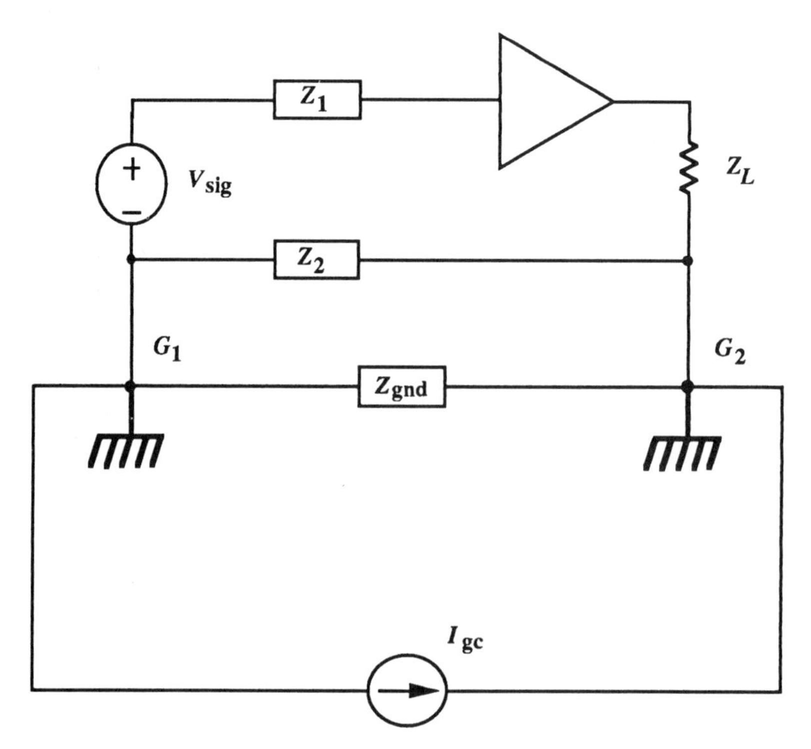

Figure 7.7 Common impedance problem.

computer ground circuit return will cause circuit logic upsets. For this reason, current returns have been added to automobiles where computers and/or electronic logic circuits are installed. The common mode current in the hull of a ship is often hundreds of amps. Currents in the earth can be relatively large and are not really very controllable. We must keep these potential problems in mind when designing with modern electronic circuitry.

7.4 GROUND DESIGN FOR PACKAGING ELECTRONIC CIRCUITRY

Grounds are connected to minimize the voltage differences in parts of a system. The voltage potential across a ground is the product of the current and ground impedance:

$$\Delta V_g = I_g Z_g \tag{7.3}$$

To minimize ΔV_g we must minimize I_g and Z_g. We minimize the ground current I_g by physically configuring the layout so that the least amount of current is created in the ground circuit. The ground impedance Z_g is minimized by designing current paths that have the lowest characteristic impedance that is practical. The characteristic ground impedance is:

$$Z_g = (L_g/C_g)^{1/2} \tag{7.4}$$

To minimize Z_g, it is necessary to minimize the ground inductance L_g and to maximize the ground capacitance C_g.

Differential Mode Layout

We have separated the concepts of current returns and grounds. This chapter is about grounding, but for completeness we will discuss the design of current returns here. The low-frequency considerations that are generally used as design criteria for current returns are the current carrying capabilities of the conductors. At frequencies of up to approximately 20 MHz, the current returns are designed to be of as low an impedance as possible. When the lengths of conductors is greater than approximately the wavelength divided by 20, transmission line and radiation considerations are necessary. The source impedance, line impedance, and load impedance should be matched to achieve maximum signal or power transfer. The wavelength of a 1-GHz signal is about 11 in. At 1 GHz, 1-in-long conductors should be designed with impedance matching techniques.

Common Mode Layout

The chassis or earth ground is the common mode current return path. We want thc smallest common mode currents possible. Fast-changing voltages referenced to the chassis or earth grounds capacitively inject current into the common mode path. All power and signal returns are generally grounded to the chassis reference at some point in a system.

For a single-ended circuit, the layout configuration that generally causes the least amount of current injection into the chassis or earth ground is shown in Fig. 7.8. The current return is layered between the signal or power line and chassis or earth ground. The current return potential changes very little relative to chassis or earth ground. The current flowing in the common mode loop of the current return and chassis or earth ground causes voltage potentials across the loop impedances. These current return and chassis or earth impedances are designed to be very small. The voltage potential generated is usually very small when compared to the signal or power line voltage changes, so very little current is injected. The signal and power voltage changes are shielded from the chassis or earth grounds. When signal or power line voltage changes cannot be shielded from the chassis or earth grounds by the current returns, it is desirable to maximize the distance between the signal or power lines and the chassis or earth ground. An increase in distance decreases the capacitance, which reduces the amount of current injected into the common mode ground loop.

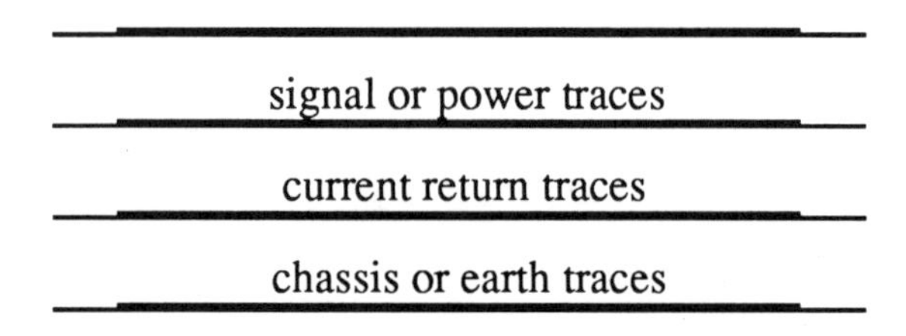

Figure 7.8 Recommended circuit trace arrangement.

Routing

The current returns should be routed first; the grounding scheme doesn't affect the route of return current. Routing current returns is the first priority; grounding is the second. Current returns and grounds should not be mixed. Their separation should be well defined.

Current Flows in Path of Least Impedance

Current will flow in the path of lowest impedance. The characteristic impedance of the ground path should be the lowest impedance path; otherwise the ground current will flow in unexpected parts of the system. Inductance usually determines the path of lowest impedance in a ground circuit.

Printed Wiring Board

It is very difficult to fabricate high-speed circuitry on single-layer printed circuit boards. A section of a single layer printed circuit board trace is shown in Fig. 7.9A. The signal and return traces are side by side. A section of a multilayer printed circuit board is shown in Fig. 7.9B. The signal trace and the return traces can be on adjacent layers. The traces in Fig. 7.9A and 7.9B are given to be the same length and area. The capacitance of the traces in Fig. 7.9B is greater than the capacitance of the traces in Fig. 7.9A because of the effective distance between the traces (plates of the capacitance). The permittivity of the epoxy circuit board is about 4.7 times that of free air. The electric field on one side of the traces in Fig. 7.9A is in air. The capacitance of the side by side traces would also be larger if they were on a multilayer printed circuit board. The inductance of the traces in Fig. 7.9B is less than the inductance of the traces in Fig. 7.9A because of the effective loop area of the current path. The characteristic impedance of the traces is:

$$Z = (L/C)^{1/2} \tag{7.5}$$

where

$$C = \sigma A/d \tag{7.6}$$

and

$$L = (0.4\pi N^2 \mu A_L)/l_m \tag{7.7}$$

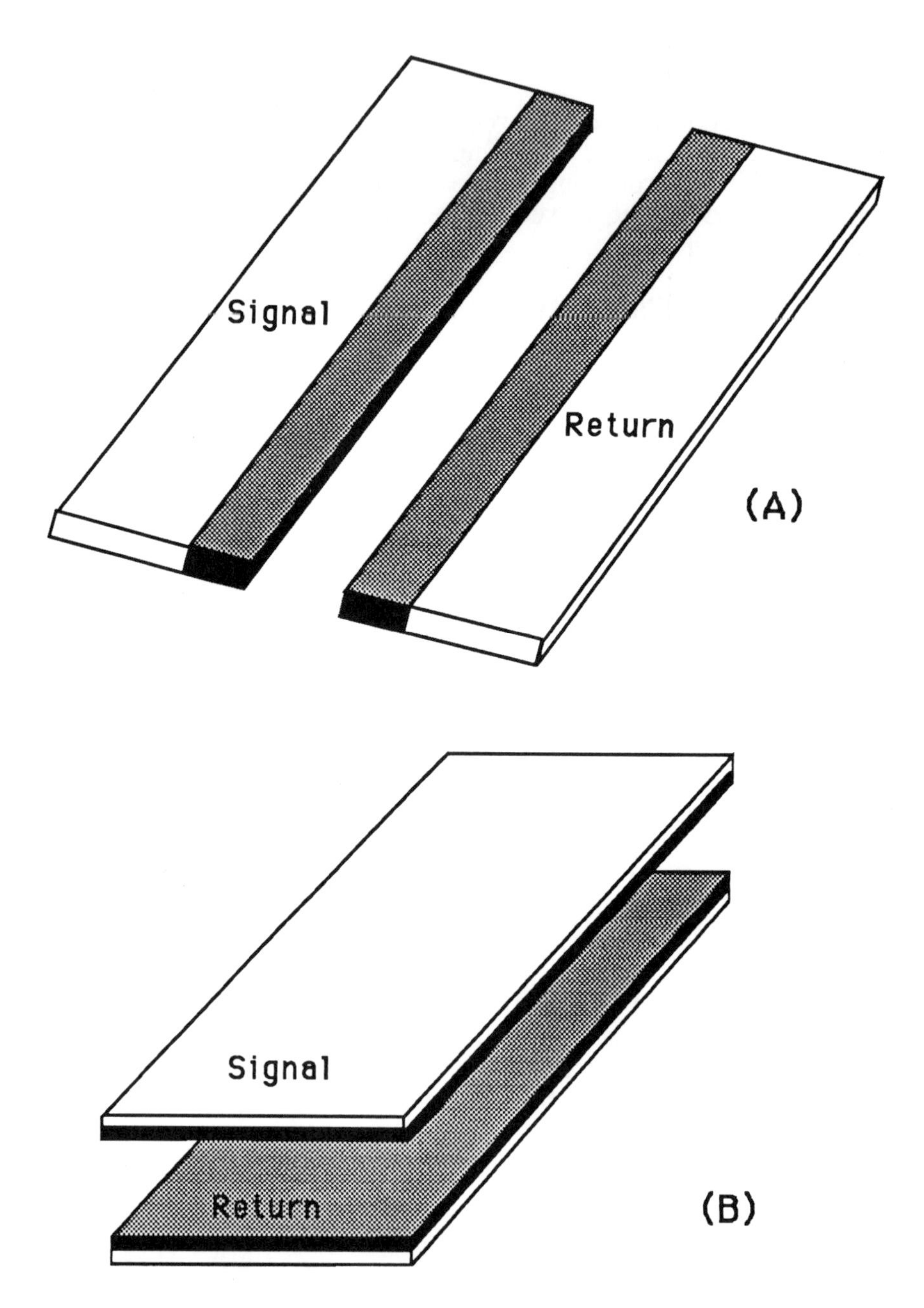

Figure 7.9 Multilayer PCBs facilitate lower characteristic impedance than single layer PCBs.

where:

σ = the permittivity in farads per meter
μ = the permeability in Henrys per meter
A = the area in square meters
d = the distance between plates in meters
N = the number of turns (one in this case)
l_m = the mean magnetic path length

The typical characteristic impedance for parallel traces on adjacent layers of a printed circuit board is 10 Ω, and the characteristic impedance of the side by side traces on a single layer board is approximately 100 Ω.

For very high-frequency layouts, strip line layout techniques are used where the trace widths and distance between traces is designed so that the characteristic impedance of the traces matches the load and source impedances.

On the two conductors in Fig. 7.9A and B, more current flows in the darkened adjacent areas because those areas are the paths of least impedance (L is smaller and C is bigger).

Cutouts on power and return planes can cause unexpected current loops. The self-inductance of circuit traces is determined by the current loop area. Current will flow in the path of lowest impedance. In Fig. 7.10, the ground plane has a cutout, over which the signal trace passes. The return current flows in the path shown in the figure. Without the cutout, the current loop area would be:

$$\text{Length} \bullet \text{width} = 7 \text{ in} \bullet 0.01 \text{ in} = 0.07 \text{ in}^2 \tag{7.8}$$

With the cutout, the current loop area would be:

$$7 \text{ in} \bullet 0.01 \text{ in} + 1 \text{ in} \bullet 1 \text{ in} = 1.07 \text{ in}^2 \tag{7.9}$$

Since the trace inductance is proportional to the current loop area, the inductance of the trace with the cutout is more than 10 times the inductance of the same trace without a cutout in the ground plane.

Analog and digital circuitry with a common reference potential (ground) should be connected at one point only. In addition, any two circuits that operate at different current levels or susceptibilities should have separate conductive return paths. A circuit with a low susceptibility will be logically upset or suffer signal to

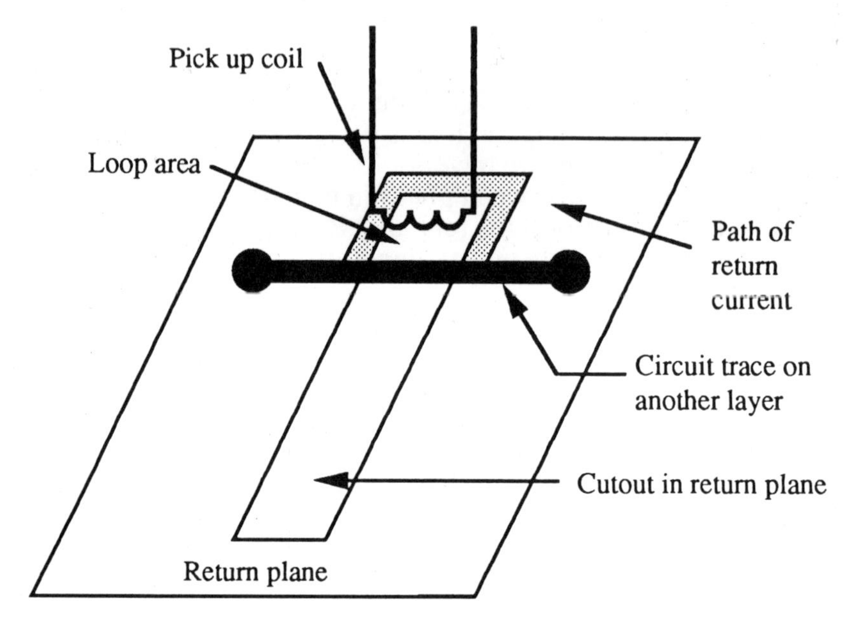

Figure 7.10 Unexpected current loops cause excessive wiring inductance.

noise problems when sharing a conductive return path with a circuit that has relatively large currents. The problem is caused by the difference in susceptibility levels. The current returns should be on separate conductive paths and then "treed" to a common ground reference.

Equipment

Bonding is the electrical interconnecting of conductive parts. The purpose of bonding is to maintain a common electrical potential in an assembly or system of electronic equipment. A prime example of bonding is the electrical connection made between a connector body shell and the chassis to which it is mounted. The bonding requirement is important for safety and EMI reasons. The typical military bonding requirement is less than 2.5 mΩ. This requirement often calls for machined or specially treated mating surfaces. Typical commercial bonding limits are 10 mΩ or less. All switch bodies, indicator bodies, bezels, etc., are bonded to the chassis in which they are mounted to prevent the occurrence of safety hazards and to reduce the apertures for radiation susceptibility or emissions.

Low-Impedance Path to Chassis

The EMI filters must have a low impedance conductive path to the chassis to be effective. Fig. 7.11 is a model of a ceramic capacitor, functioning as a common mode filter element, connected to a chassis using a 5-in wire. The reactance of the capacitor is 1.6 Ω capacitive and 1.26 Ω inductive. The inductive reactance of the 5-in wire is greater than 6.3 Ω. Since the reactance of the wire is much greater than the reactance of the capacitor, the wire imped-

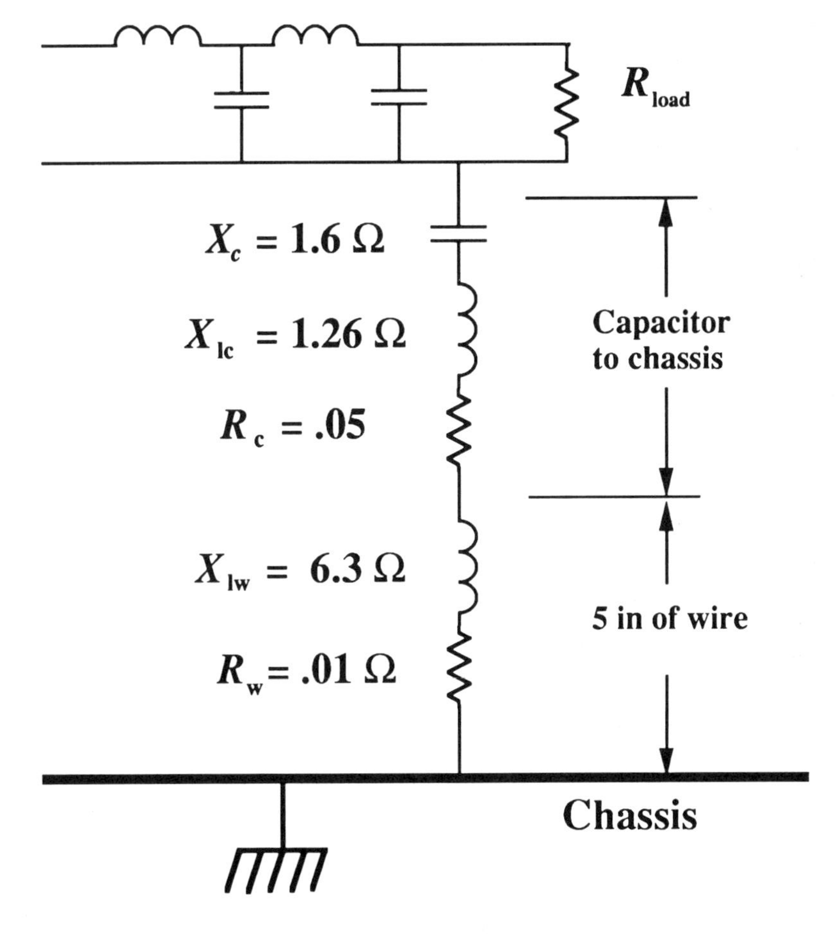

Fig 7.11 Five inches of wire is not a low-impedance path to the chassis.

ance dominates, and the capacitor is ineffective as a filter. To be effective this capacitor component should be mounted to the chassis. One lead should be connected directly to the chassis, the other should be short, and the circuit should wire to and away from the ungrounded capacitor lead.

System

An electrical system is an interconnected group of assemblies that perform a function or many functions as an integrated unit. The maximum allowable chassis to chassis voltage for human shock safety is less than 10 V peak. The maximum allowable chassis to chassis voltage for signal transmission is much less than the voltage of the interconnecting signals.

For analog signals, the signal to noise ratio (S/N) is the method of quantifying maximum allowable noise in a system. The signal to noise ratio is usually specified in decibels. Digital electronics noise limits are quantified by the use of noise margins. The noise margin, specified in volts, is the worst case minimum difference in input thresholds for the two input states, a 1 or a 0 that will cause a change in state of the output of a digital device. Noise amplitudes that exceed the noise margin will upset the logical state of the digital circuit.

Input to Output Isolation

Input to output isolation in circuits is often required so that proper grounding at the system level may be implemented. Input to output isolation is often required in power supplies. The power is magnetically coupled through the power supply transformer. Control signals between the primary side and secondary side are isolated magnetically, optically, or by very high-impedance circuits. The primary and secondary sides are grounded by two separate conductive paths. The difference in potentials of the two grounding systems is isolated by the transformer windings. Between the windings is the "give" point for the difference in the two system potentials. This isolation facilitates the connecting of the two ground systems together at only one point.

Earth Grounds

Buildings are required to be connected to the earth as a common ground. A large conductive metal stake is driven deep into the ground. There are six connections (see Fig. 7.12) that should be made at the earth ground stake:

1. Structural steel
   a. Metal frame
   b. Reinforcement steel in foundation
2. Water piping if conductive
3. Input power neutral
4. Neutral to the electrical equipment
5. The ground bus conductor
6. Electrical wiring conduit

Any additional staking should only be used to improve existing building ground. Additional stakes should be in the same place as the original building earthing stake. Two systems should tie together at only one point.

There can be large potentials between two ground stake positions on earth even though the stakes are in the same vicinity. Electronic communications between buildings generally requires signal isolation and separate ground systems on the input and output sides of the communications link. Differential signals are often used to reduce the net fields that affect transmission and reduce the importance of the ground reference potential.

7.5 SHIELDING

Is shielding part of grounding? No, shielding is not grounding, but it is a related subject and should be included here for completeness. For an electrostatic shield to be effective, it must be properly grounded. But, shielding is a subject related to radiated fields, which is not the subject of this book. The truth is that one must have an understanding of electromagnetics in general to understand what is conducted and what is not. The understanding of radiation and conduction is crucially integral to the knowledge of electromagnetics. Radiation can impinge upon an electronic assembly and be converted to conducted energy. Because shielding is only a closely related subject, we will include just a brief tutorial here.

Shielding is the placement of a material between a noise source and receiver to absorb, reflect, or bypass the interference energy. An electrostatic shield (Faraday shield) is a grounded conductive

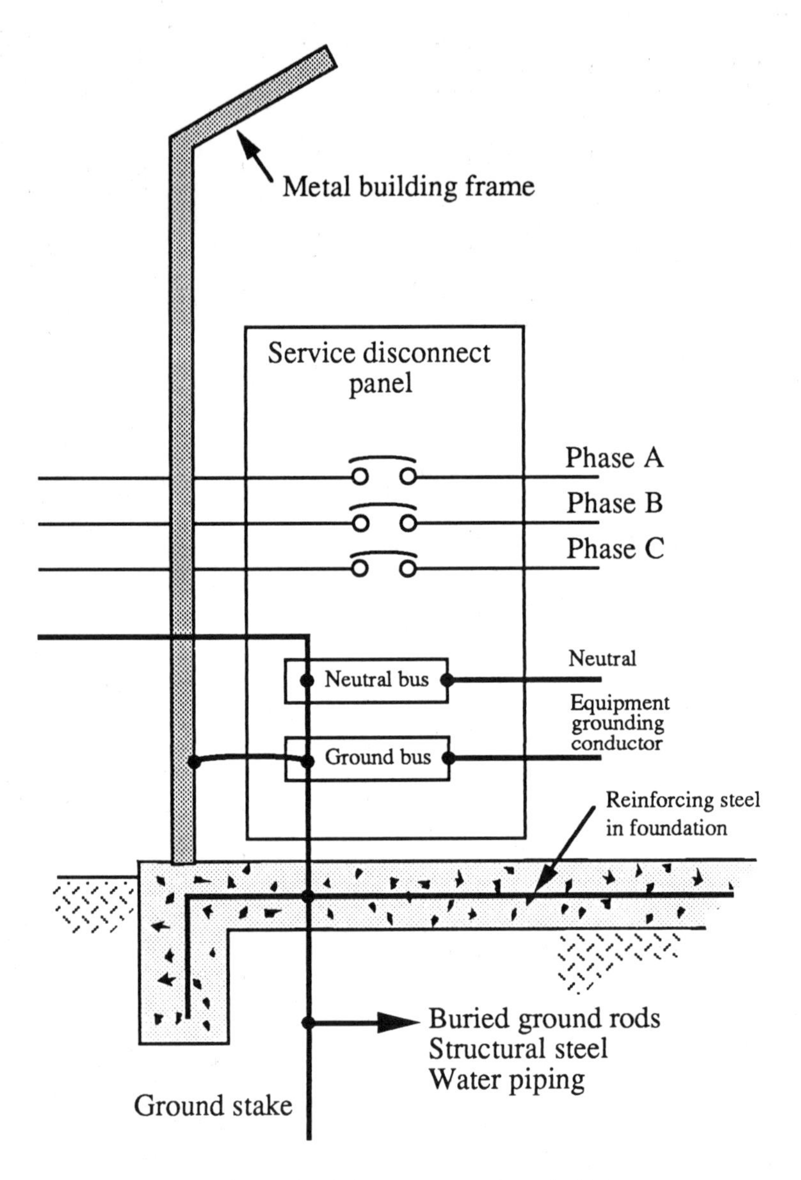

Figure 7.12 Earth ground for a building.

material placed between a noise source and receiver to absorb, reflect, or bypass electric near-field radiation. A magnetic shield is a highly permeable material placed between a noise source and receiver to absorb, reflect, or bypass magnetic near-field radiation.

An electromagnetic shield is a conductive and highly permeable material (iron or steel) placed between a noise source and receiver to absorb, reflect, or bypass electromagnetic far-field radiation.

Chassis

A conductive electronic chassis has two electrical functions in addition to the structural ones:

1. Electrostatic shield
2. Return path for common mode currents

Figure 7.13A is a schematic of a typical electronic circuit housed in a chassis. The differential current I_{dm} flows as shown and is a parameter with which we generally design. In high-frequency electronics we must be concerned with the common mode paths shown in Fig. 7.13B.

In equipment installations where the chassis is not bonded with a low-impedance path to the system ground reference, a conductive chassis becomes a capacitive coupling path to the whole circuit within the chassis. In these circumstances, the EMC performance would be better with a chassis constructed of plastic or some other nonconductive material.

For equipment with a conductive chassis bonded through a low-impedance path to the system ground reference, the chassis material and thickness of the chassis walls determine the penetration depth and hence the shielding effectiveness against electric field threats from outside of the equipment chassis (Fig. 7.14A).

The incident EMI wave is both reflected and absorbed by the chassis walls. The mismatch of impedances at both the outer and inner surfaces of the chassis walls causes reflection of a portion of the incident wave. The rest of the incident EMI energy must be absorbed in the chassis wall to totally shield the "clean area" within the chassis.

If the electric field strength has a sufficient amplitude to penetrate the chassis wall and induce a voltage on the inner side of the chassis wall (the clean area in Fig. 7.14B), the field will radiate into the clean area and become a threat to the electronic assembly inside the box.

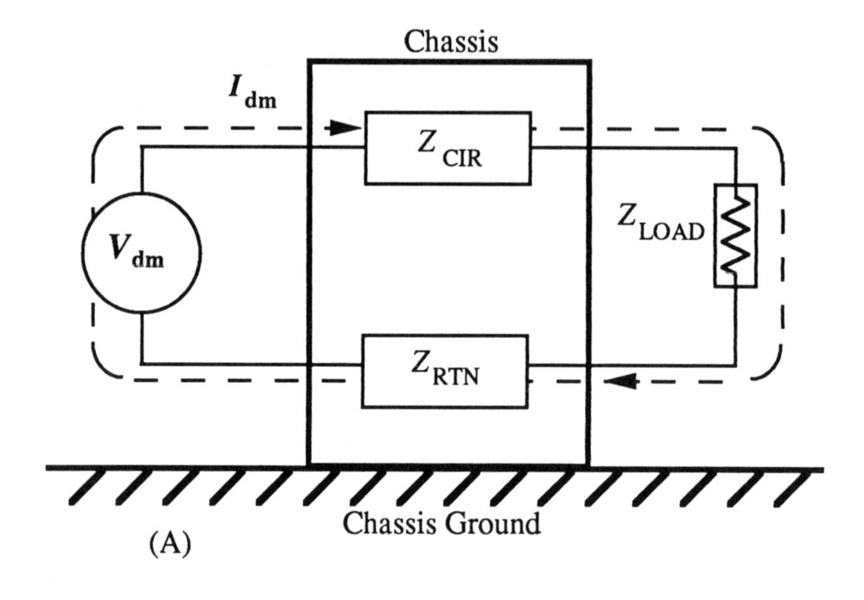

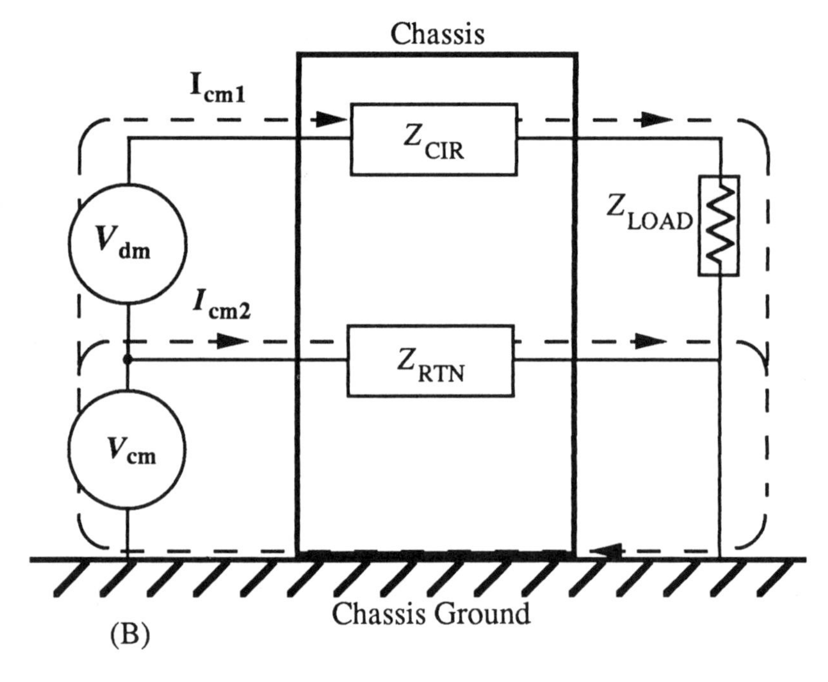

Figure 7.13 Differential and common mode paths.

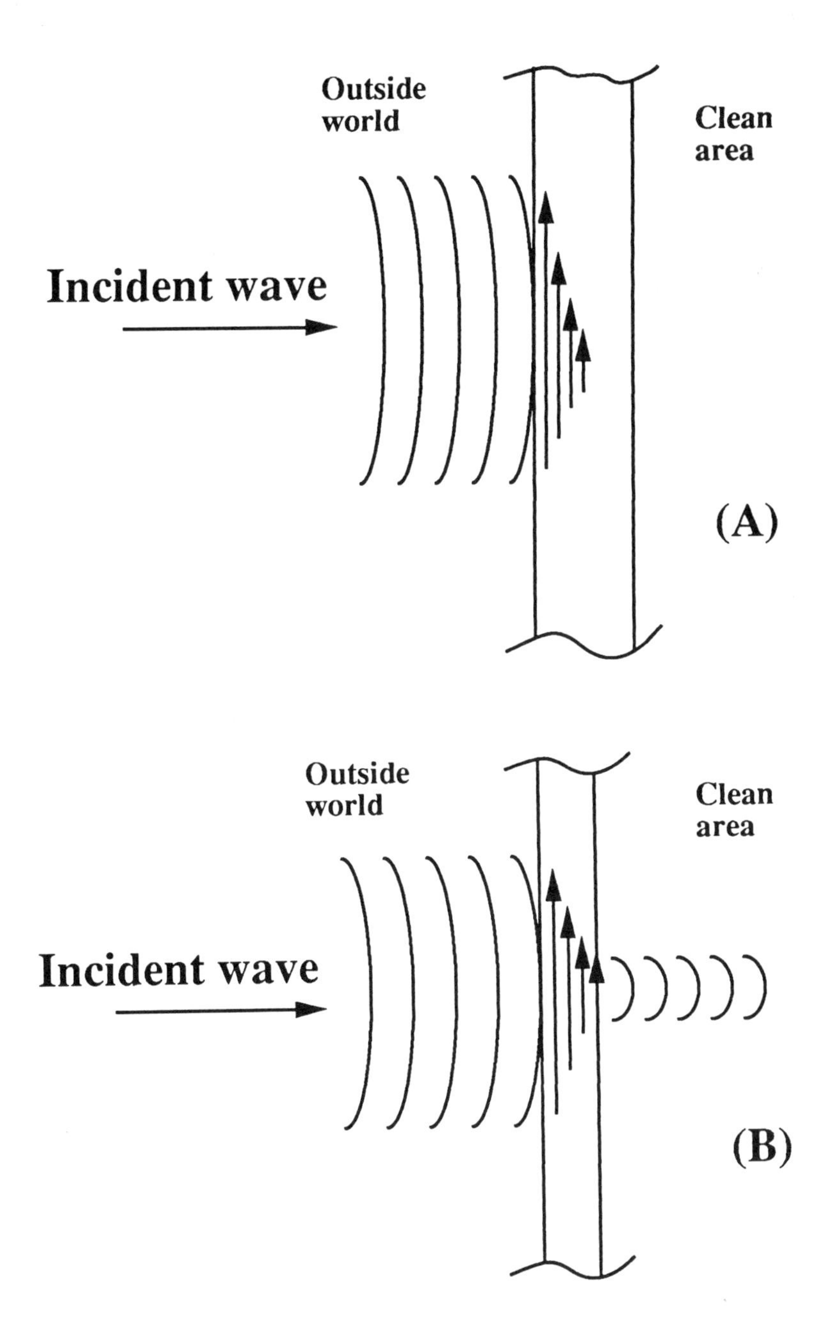

Figure 7.14 Penetration.

A Chassis Can be a Faraday Shield

The chassis can function as an electrostatic shield to shield an electronic assembly from electric fields. A effective shield can be fabricated by mechanically sealing the box with conductive gaskets and reducing all apertures large enough allow penetration into the chassis. The size of the maximum aperture determines to lowest frequency that can penetrate into the chassis clean area.

Figure 7.15 diagrams two electromagnetic apertures. The wave length of the lowest frequency that can penetrate an aperture A is:

$$\lambda_{max} = d_1/2\pi \tag{7.10}$$

The wave length of the lowest frequency that can penetrate aperture B is:

$$\lambda_{max} = d_2/2\pi \tag{7.11}$$

The area of the aperture is unimportant. The factor governing the lowest frequency of penetration is the maximum dimension in any direction. This of course assumes that the electrical field threat is not polarized.

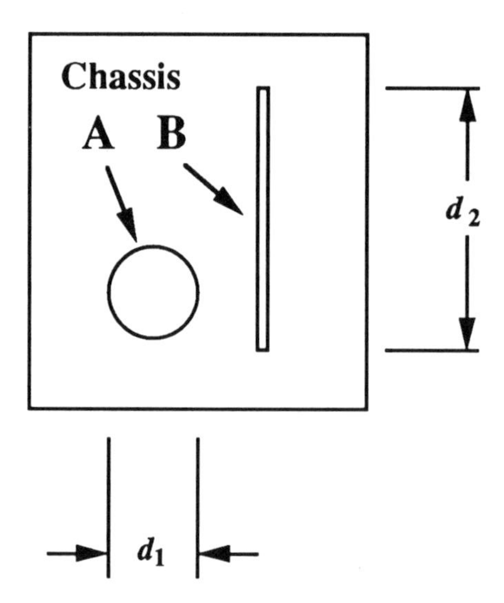

Figure 7.15 Round versus rectangle hole for penetration.

To be effective, electrostatic shields internal to the chassis must be grounded through a very low-impedance conductive path to the chassis. Faraday shields with a wire connection to chassis are ineffective and usually are detrimental because the inductance of the wire is a high impedance at the frequencies at which the shield is supposed to be effective. The shield actually forms a capacitive path around the filter.

Cable Shielding

Cable shielding is recommended for electronic signals with the following characteristics:

1. Small amplitude
2. Long run

As a quantified rule of thumb these are:

Signal level		Length of run
Micro volts	>	10 ft
Milliohm volts	>	100 ft
Volts	>	1,000 ft

Shielding effectiveness is the measure of the ability of the shield to reject electric field threats to signal lines.

Shielding

There are three basic types of shielding:

1. Capacitive
2. Radiative
3. Magnetic

Capacitive

In capacitive coupled problems, the shield should be placed between "plates" of the capacitive coupling path. A capacitive shield causes the noise current to bypass the circuit being protected. Capacitive shields should be electrically connected to the signal return at one point at the receiver.

Radiative

A high-frequency radiative shield such as a chassis should have a continuous connection to the major ground reference within the system. A continuous connection, such as welded seams or small bolt spacings, is required to stop high-frequency radiation with small wavelengths from penetrating the shield into the clean area.

Magnetic

To couple noise magnetically, two conductive loops are required, one to radiate and one to receive. By definition magnetic coupling can only couple differentially. Common mode energy cannot be magnetically coupled between separate circuits.

Magnetic shielding is usually not needed for most types of electronic circuitry. A large portion of the electronics industry uses 5-V dc digital logic such as TTL and CMOS. These types of circuitry operate at relatively high switching speeds. Magnetic fields are generally not a significant problem in electronic circuitry of this nature. For this circuitry to function properly the layout techniques require that the loop area of the signal and return circuit traces be minimized so that the inductance is reduced. A large layout inductance would slow the circuit operation so that logic errors would occur. Also, the +5-V dc power lines would be unable to supply a solid +5-V dc to the ICs, which would cause logic errors.

The inductance and field radiation from a circuit are directly proportional to the area of the circuit loop. The circuit layout for high speed digital circuitry is generally done on a multilayer PCB with the signal and signal return and the power and power returns parallel to each other on adjacent layers. Typical circuit loop areas must be small for proper operation of the logic. This type of layout minimizes magnetic radiation. For the same reason, the magnetic field susceptibility threshold is at a high level because small circuit loop areas do not couple well to magnetic fields. Some types of electronic circuitry in which magnetic shielding could be required are motor control circuits, relay circuits, automobile ignition circuits, and other circuits with highly inductive loads.

Low-frequency magnetic fields penetrate aluminum plate with very little attenuation. Magnetic shields must be constructed of a highly permeable material such as iron, steel, or some type of ferrous material. Eddy currents induced in permeable magnetic shields induce opposing fields in the shield to cancel the incident field. The attenuation constant is:

$$(\pi\mu\sigma f)^{1/2} \tag{7.12}$$

The skin depth is:

$$(\pi\mu\sigma f)^{-1/2} \tag{7.13}$$

The attenuation is 8.7 dB per skin depth. At 100 MHz, for example, the magnetic skin depth is 10 μm.

QUESTIONS

7.1 What is the purpose of electronic grounds?

7.2 What is the purpose of electronic returns?

7.3 Why should we make a distinction between a ground and a return?

7.4 Why is information about shielding included in a chapter about grounding?

7.5 Why should there be no ground symbols on a wiring diagram?

7.6 Is the artwork for a PWB a wiring diagram? What features are similar?

7.7 What is meant by the statement that (ideally) a ground conducts no current? What current properly flows in a ground?

7.8 What is the importance of bonding requirements in electronic circuits?

7.9 Input to output isolation is a very common requirement for power supplies and some electronic equipment. What is the importance of subassembly input to output isolation in regard to system level grounding?

8

EMI ANALYSIS

Analysis is the process of dividing an object into its constituent parts. Generally we examine the assembly of parts to learn more about how they function together. Simulation is the process of making models of objects and making the objects perform in the simulated environment to learn more about the objects' behavior. In this book we use the term "EMI analysis" as a general term for the processes of analysis and simulation. The term "component" is defined here as electrical or electronic parts such as:

Resistors
Capacitors
Inductors
Baluns
Transformers
Diodes
Transistors

The term "element" in this book is defined as the resistance or inductance of a capacitor. For example, a capacitor (a component) is modeled as a series connection of three elements:

Capacitance
Inductance
Resistance

In a computer simulation a distributed capacitance is modeled by a lumped equivalent capacitor, a distributed inductance is modeled by a lumped equivalent inductor, and a distributed resistance is modeled by a lumped equivalent resistor.

Usually when doing EMI analysis, the time domain waveforms are converted into the frequency domain. In the computer, the threat waveform is injected into a model of the emissions path, the filter model, and the test setup model. The resulting current in the power lead and the return lead are measured and compared to the limits of the specified standard.

This chapter shows how spectrum generators can be created in SPICE with simple circuits and ac analysis. The spectrum can be either known by previous experience, found in publications, or computed by the use of the FFT. Some of the most common waveforms are included. The data from the presented SPICE EMI analysis represents the envelope of the EMI waveform. The differences and interpretations of the computer data versus the data from the spectrum analyzer data taken in the laboratory are covered in depth. The importance of correlation of hand calculations, computer data, and laboratory measurements are also presented. Approximate hand calculations, using a calculator, can be made at any frequency. Hand calculations are a good way to double-check computer analysis or laboratory measurements. Some very simple equations are included with some examples of their applications.

8.1 EMI MODELING

Probably the simplest example of EMI analysis is the analysis of the differential conducted emissions of a switching power supply output. All of the essential aspects of the process are involved. After the process is well understood, we will apply this skill to some useful examples. The buck convertor topology (Fig. 8.1A) is chosen for this first example because it is very common and also a very simple topology; therefore we may focus on learning the simulation analysis process.

The waveform of Fig. 8.1B is the plot of the voltage at node V_D after the circuit has reached steady state conditions. When the switch SW_1 closes, the voltage at V_D is 15 V. When the switch is opened, the dotted side of L_1 goes negative and is clamped to 0.7 V below ground. Typical FET switches used for SW_1 have rise and fall times of approximately 100 nS. The voltage at V_D alternates between 15 and -0.7 V. This is a rectangular wave with fast rise and fall times that could cause a great deal of EMI if not properly filtered. This is an inherent internal EMI source common to all switching power supplies. The filter, L_1 and C_1, is placed in the circuit path to average the output waveform and to attenuate the EMI noise. The load is modeled as a pure resistance.

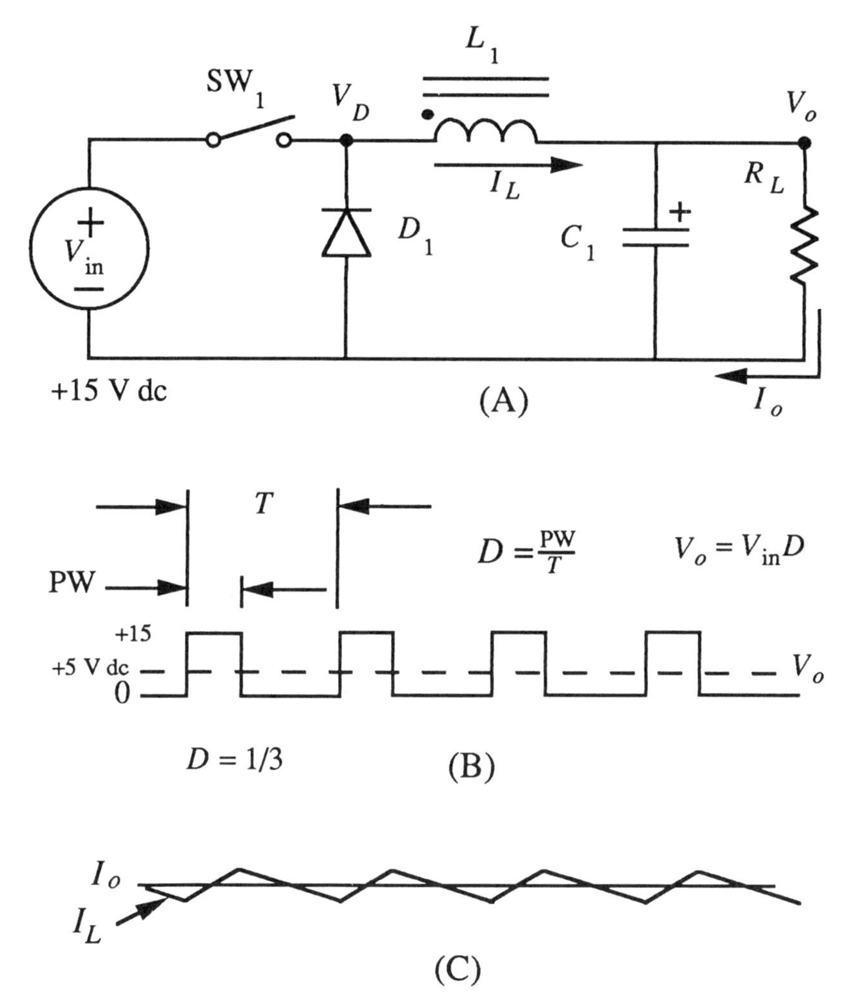

Figure 8.1 Simple buck convertor and its waveforms.

The voltage at V_o is equal to the average value of waveform in Fig. 8.1B. If D, the duty cycle, is 1/3 and V_{in} is 15 V dc, V_o is +5 V dc. When the switch is ON, the voltage across L_1 is 15 - 5 = 10 V. The current I_L ramps up at delta $I = V\delta t/L$. When the switch shuts off, the dot side of L_1 "flies" down and is clamped to 0.7 V dc by D_1, the catch diode. The other side of L_1 is held at V_a (+5 V dc). The inductor L_1 in this state has V_o + 0.7 V dc = 5.7 V dc impressed across it. The current I_L ramps down at $\delta I = V\delta t/L$ (Fig. 8.1C).

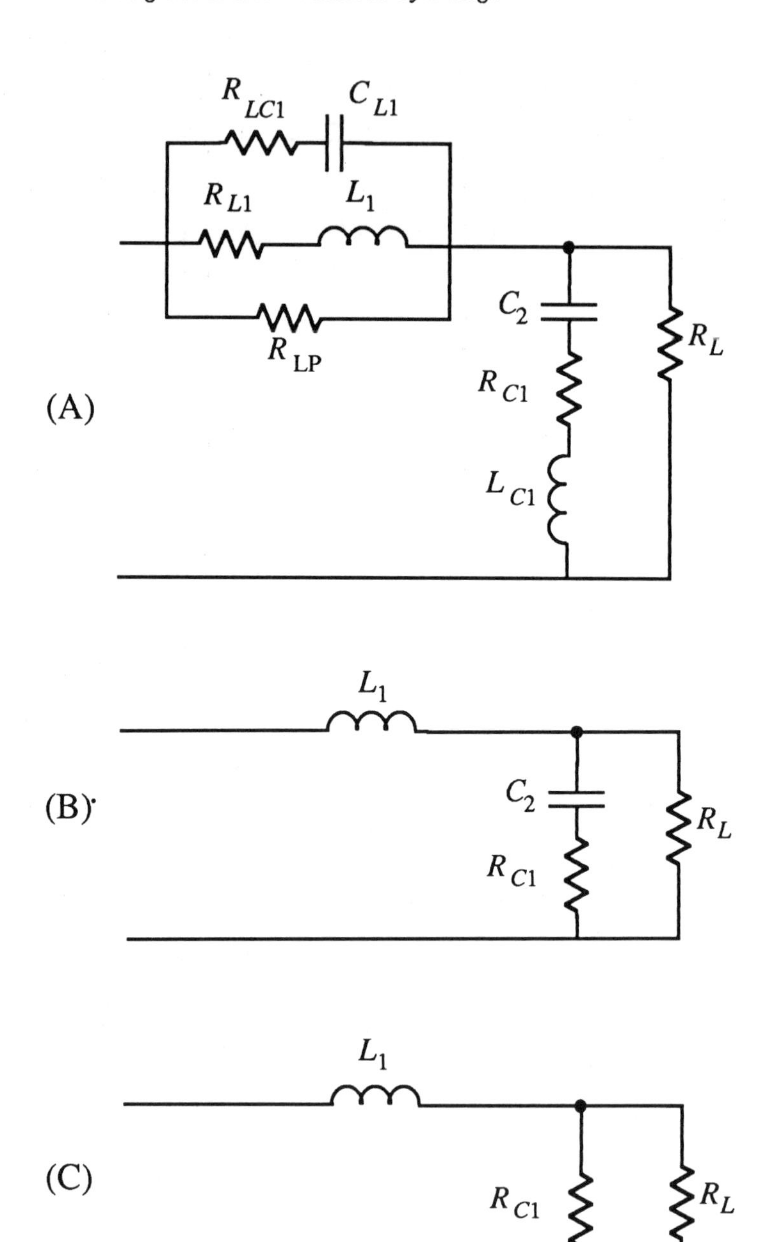

Figure 8.2 The reduction of *LC* filter models to "key players" at the fundamental switching frequency.

A first-order model of L_1 and C_1 is shown in Fig. 8.2B. The equivalent resistance R_{C1} is the ESR of capacitor C_1. For low-voltage power supplies of less than a few hundred volts and typical switching frequencies between 100 and 300 kHz. the reactance of C_1 is designed to be small when compared to the ESR R_{C1}. The simple model can be reduced to Fig. 8.2C.

The attenuation (Attn) is approximately R_{C1} in parallel with R_L divided by ωL_1 (see chapter 6). The largest Fourier component of waveform A is the fundamental. In the frequency domain, the fundamental voltage V_{pk} is 2AD. At the fundamental frequency, the EMI current (referenced to db above a microamp) in R_L is:

$$i = 20 \log\left(\frac{\left(\frac{V_{pk}\ \mathrm{Attn}}{R_1}\right)}{1e^{-6}}\right) \tag{8.1}$$

$$\mathrm{Attn} \approx (R_{C1}\ \mathrm{parallel}\ R_L)/2\pi L \tag{8.2}$$

$$V_{pk} = 2AD \tag{8.3}$$

$$(R_{C1}\ \mathrm{parallel}\ R_L) = 1/((1/R_{C1}) + (1/R_L)) \tag{8.4}$$

where R_L is the load resistance, R_{C1} is the capacitor equivalent series resistance, and D is the duty cycle.

A very good model of the *LC* filter is shown in Fig. 8.2A. This model will generally be used for computer modeling. At the fundamental switching frequency, the key players of the model are shown in Fig. 8.2B. A typical value for a 100-μF tantalum capacitor ESR (R_{C1}) is approximately 0.1 ω. For a switching frequency of 100 kHz the capacitive reactance is:

$$X_C = 1/(2\pi fC) = .016\ \Omega \tag{8.5}$$

$$R_{C1} >> X_C \tag{8.6}$$

therefore the simple model reduces to the model in Fig. 8.2C.

Frequency Domain Characterization of the Time Domain Waveform

The waveform in Fig. 8.1B is shown with more detail in Fig. 8.3A. The frequency domain version of the waveform in Fig. 8.1B is included as Fig. 8.3B.

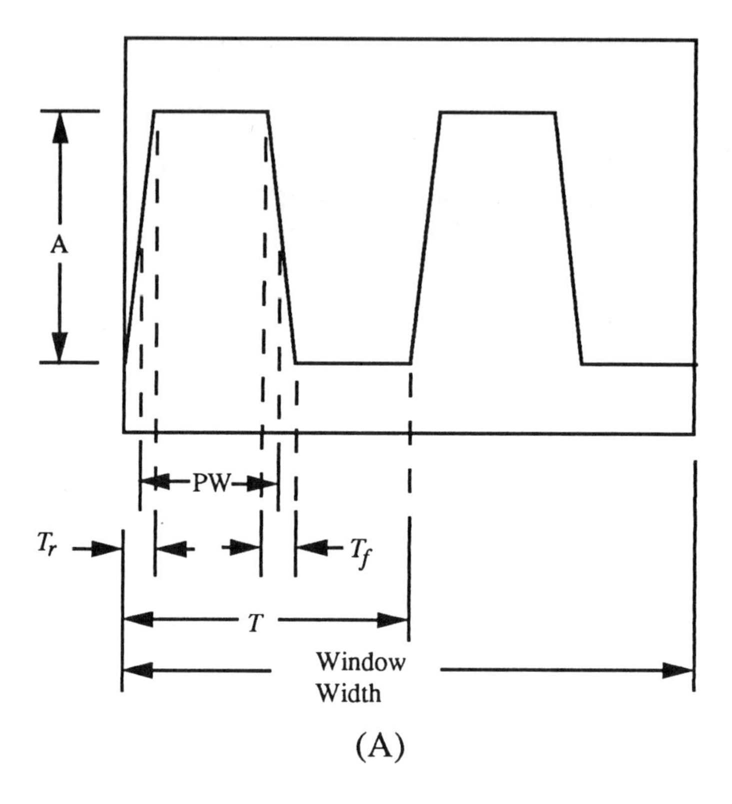

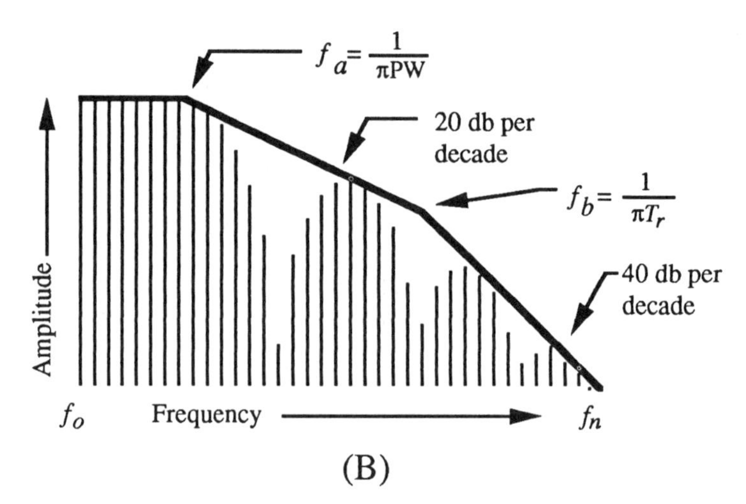

Figure 8.3 Time domain to frequency domain conversion.

The bold line in Fig 8.3B is the envelope of the harmonics of the waveform of Fig. 8.1B. We are looking at only the dc and positive frequency components of the frequency domain representation of waveform A. The harmonics above the switching frequency are considered to be EMI. The envelope is the worst case amplitude at any frequency. The envelope of the amplitude is easy to characterize and is very useful in quantifying and characterizing EMI spectra.

The period T of the repeated portion of the waveform causes spectra at $1/T$, $2/T$, ..., n/T. The amplitudes of the spectra are determined by the average value of the waveform or, in this case, twice the time domain waveform amplitude times the duty cycle (D). At the frequency of $1/(\pi PW)$ the envelope reduces in amplitude at the rate of 20 dB per decade with increasing frequency. At the frequency of $1/(\pi T_r)$ or $1/(\pi T_f)$, the envelope reduces in amplitude at the rate of 40 dB per decade with increasing frequency. The faster of the rise or fall time in the time domain causes and dominates the 40-dB rolloff in the frequency domain.

The frequency domain characteristics in all cases are similarly affected by time domain features of all wave shapes. The key time domain parameters are the fundamental waveform period, the average value, and the rate of the fastest of the rise or fall times. Fast rise or fall times cause the spectrum to contain higher order harmonics. The more sinusoidal in appearance the time domain waveform is, the less high order harmonics are involved. For instance, a triangle waveform has very few high order harmonics.

8.2 EMI ANALYSIS USING SPICE

We will now present a method of EMI analysis for conducted emissions. EMI analysis can be performed by using any standard SPICE ac analysis with the ac spectrum input contoured to be equivalent to the spectrum of the noise source. The ac spectrum input is injected into the circuit at the node where the EMI generator creates noise in the circuit. The resulting current in the load is the conducted emissions. This process is very easy to do on the computer and is a very powerful tool, as will be shown.

In SPICE ac analysis, a voltage source insertion point is chosen and a spectrum as in Fig. 8.4 is swept through the specified frequency range. The ac source amplitude is the same throughout

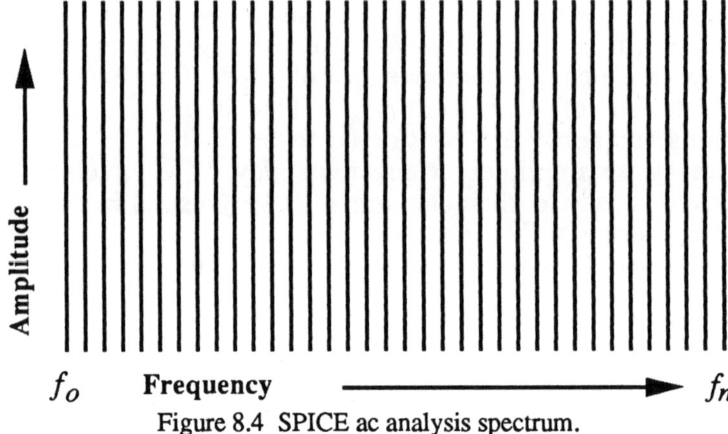

Figure 8.4 SPICE ac analysis spectrum.

the chosen spectrum. At each frequency, the SPICE program calculates the resulting currents or voltages and the chosen measurement points in the circuit. Let's now consider the circuit in Fig. 8.5 which shows two ideal cascaded operational amplifier (op amps) invertors, each configured for a gain of 1. The signal at V_o is the same as the signal at V_{in}. The ac analysis input spectrum can be injected at V_{in}. The output at V_o is the SPICE ac analysis spectrum. Now let's add two capacitors to the circuit as shown in Fig. 8.6.

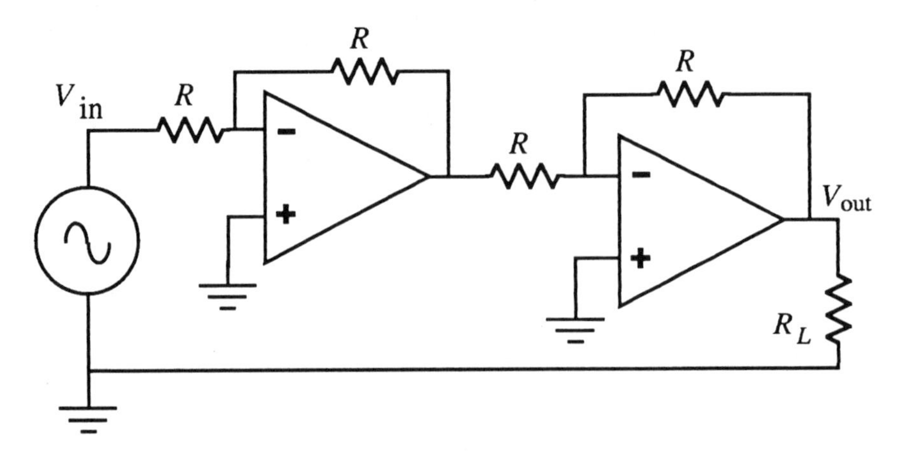

Figure 8.5 The cascaded gain of two op amp invertors is one.

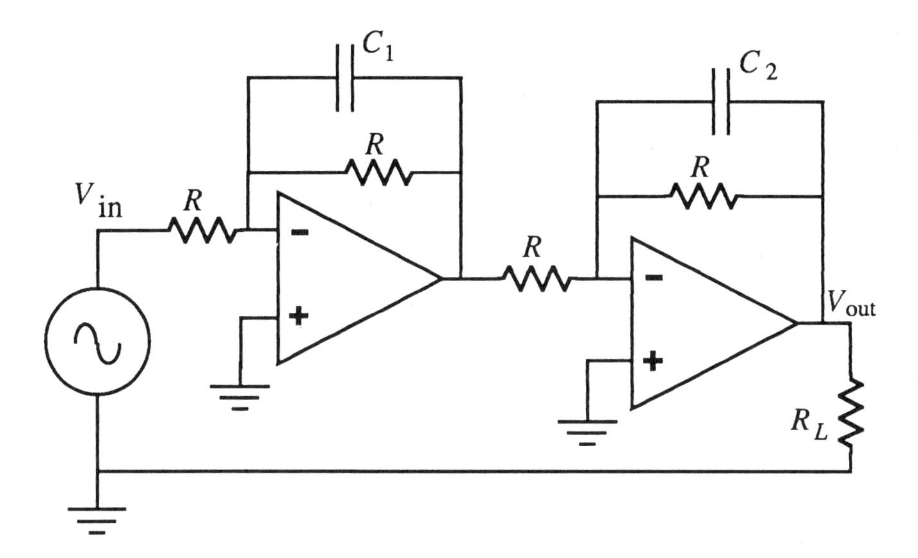

Figure 8.6 SPICE EMI spectrum generator.

The spectrum of the output at V_o with the added capacitors is shown in Fig. 8.7. The frequency spectra is linearly spaced. The frequency spacing is determined by the SPICE command parameter TNUM (number calculation points in the frequency domain). The low-frequency amplitude will be equal to V_{in}. At the fre-

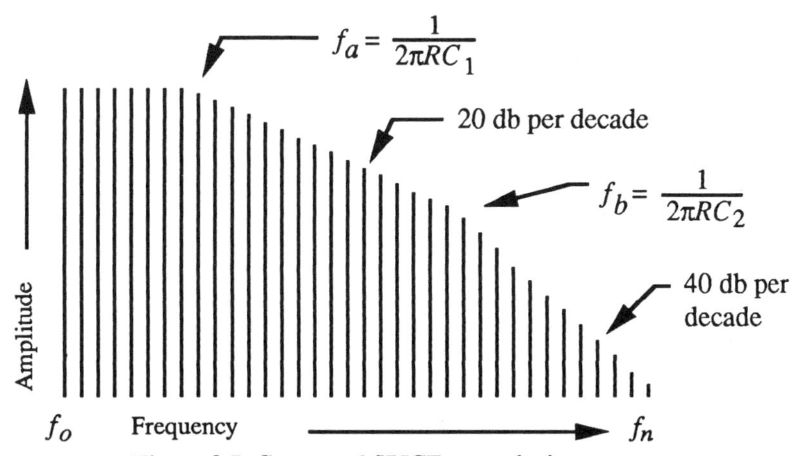

Figure 8.7 Contoured SPICE ac analysis spectrum.

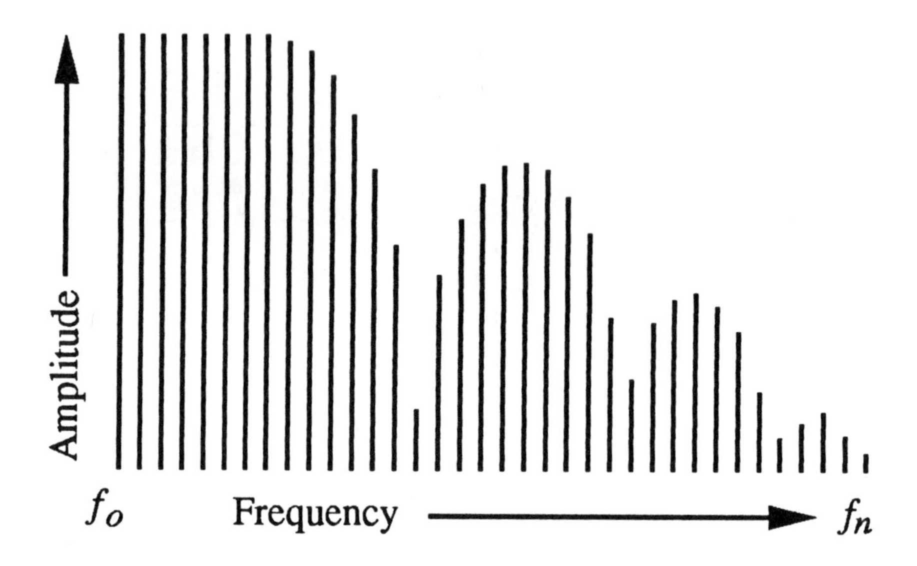

Figure 8.8 SPICE frequency domain representation of the waveform of Fig. 8.3A.

quency of $1/(2\pi RC_1)$ the amplitude will reduce at a rate of 20 dB per decade. At the frequency of $1/(2\pi RC_2)$ the amplitude will reduce at a rate of 40 dB per decade. Let's choose $1/(\pi \text{PW})$ equal to $1/(2\pi RC_1)$ and $1/(\pi T_r)$ equal to $1/(2\pi RC_2)$. We now have a simple EMI spectrum generator created in SPICE. The envelope of the spectrum at V_o is the same as the envelope of the frequency domain representation of the time domain waveform of Fig. 8.3.

The SPICE frequency domain representation of the waveform of Fig. 8.3A is shown in Fig. 8.8. This is of course only the positive frequencies. The negative frequency domain spectrum is the horizontal mirror image.

Envelopes of the actual frequency domain representation of the waveform of Fig. 8.3A are shown in Fig. 8.9. The straight line envelope can be considered as a worst case representation of the spectrum. The straight line envelope for the actual spectrum is identical to the SPICE contoured ac analysis spectrum. The difference between the actual and the SPICE frequency domain representations are shown in Fig. 8.10. The shaded area is the difference. The worst case straight line envelopes are the same.

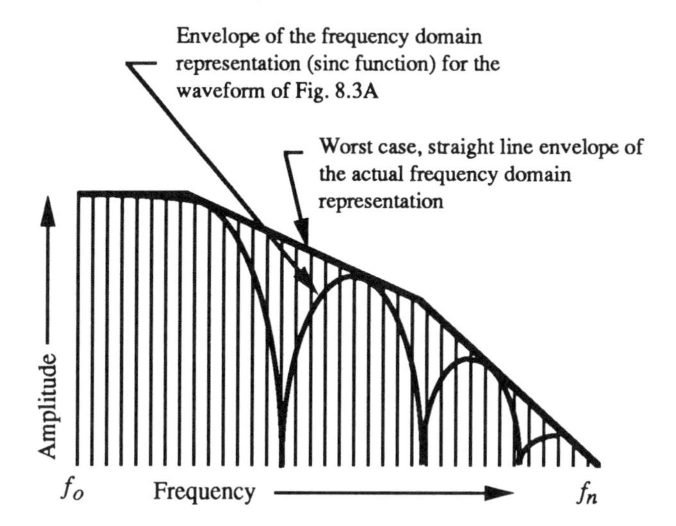

Figure 8.9 Contoured SPICE ac analysis spectrum envelope.

Consider the model in Fig. 8.11 using SPICE ac analysis. At node V_{EMI} we are injecting the EMI noise spectrum into a single section *LC* filter with a load R_L. A current probe is shown for measuring the resulting EMI emission in the load. The current probe is referenced to decibels above 1 μA.

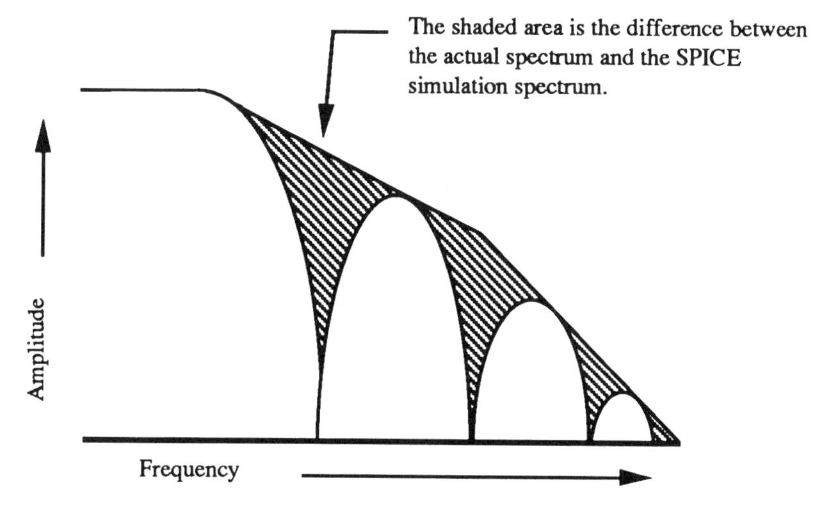

Figure 8.10 Difference between the actual spectrum and the SPICE simulation spectrum.

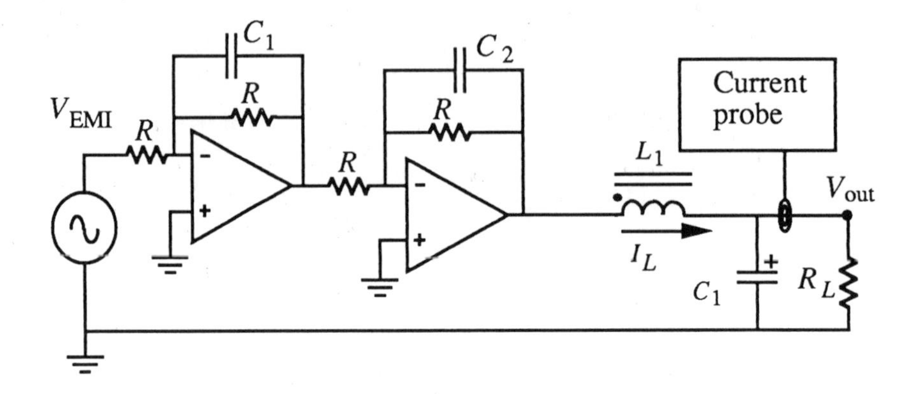

Figure 8.11 Simple SPICE EMI analysis model.

A typical military standard plot of limits for conducted emissions is shown in Fig. 8.12. The measured frequency domain current is plotted in decibels referenced to 1 μA. This requirement covers the frequency range from 15 kHz to 50 MHz. A typical variation to the limits allows the 15-kHz amplitude to be relaxed (maximum allowable is increased) by a factor of 20(log (load current)). A straight line is then drawn from the new 15-kHz ampli-

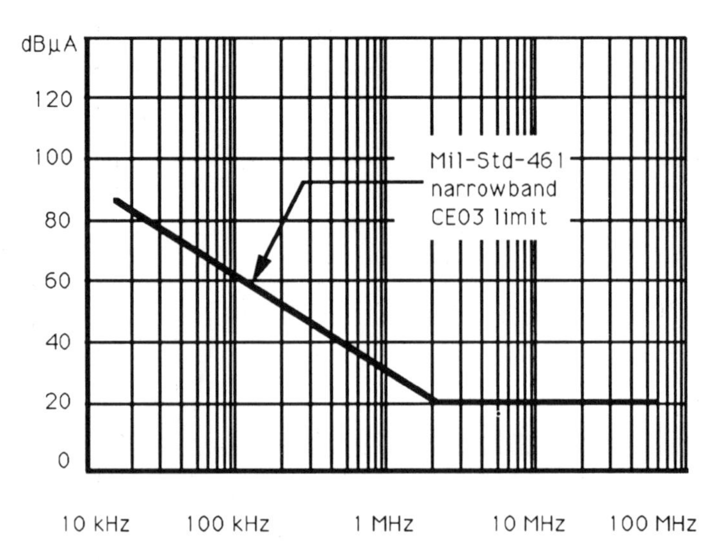

Figure 8.12 Mil-Std-461 CE03 conducted emission limits.

tude to the 20-dB amplitude at 2 MHz. The allowable maximum amplitude from 2 to 50 MHz is generally not relaxed. The allowable amplitude in this frequency range is 20 dB above 1 μA (or 10 μA). This is a rigorous requirement that forces the designer to pay attention to details and use good layout and circuit design practices. When done properly, this requirement is not difficult to meet. Mil-Std-461 CE01 (shown in Fig. 8.13) is another typical military requirement for low-frequency conducted emissions. A comparison of some of the common military and commercial conducted emissions requirements is included in chapter 9.

EMI Noise Generators at Slopes Other than 20 or 40 dB per Decade

Most simple and common waveforms can be characterized with 20 dB per decade and 40 dB per decade roll offs as was done in the previous example. But not all waveforms are simple and some roll off at combinations other than 20 dB per decade. For a 80-dB-per-decade roll one simply cascades four of the 20-dB-per-decade op amps, but for 5, 10, or 15 dB per decade the SPICE EMI spectrum generator is a little more complicated.

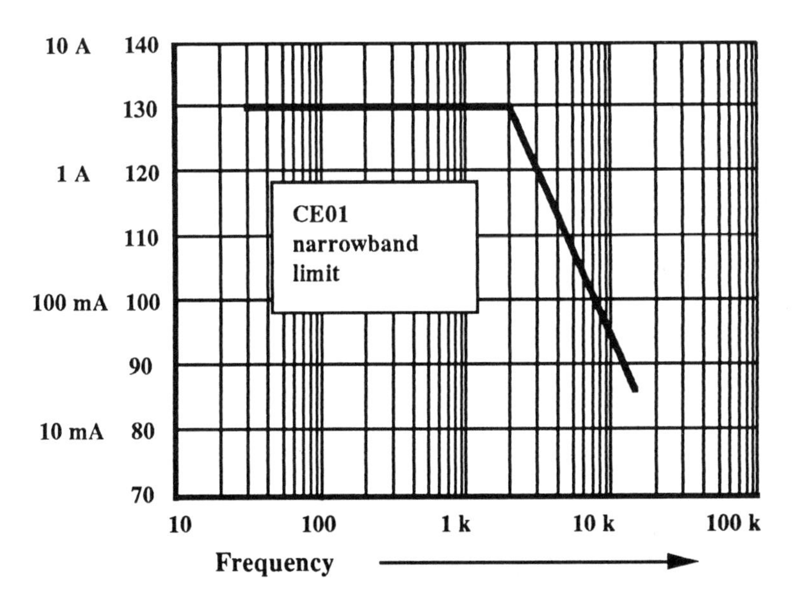

Figure 8.13 Mil-Std-461 CE01 conducted emission limits.

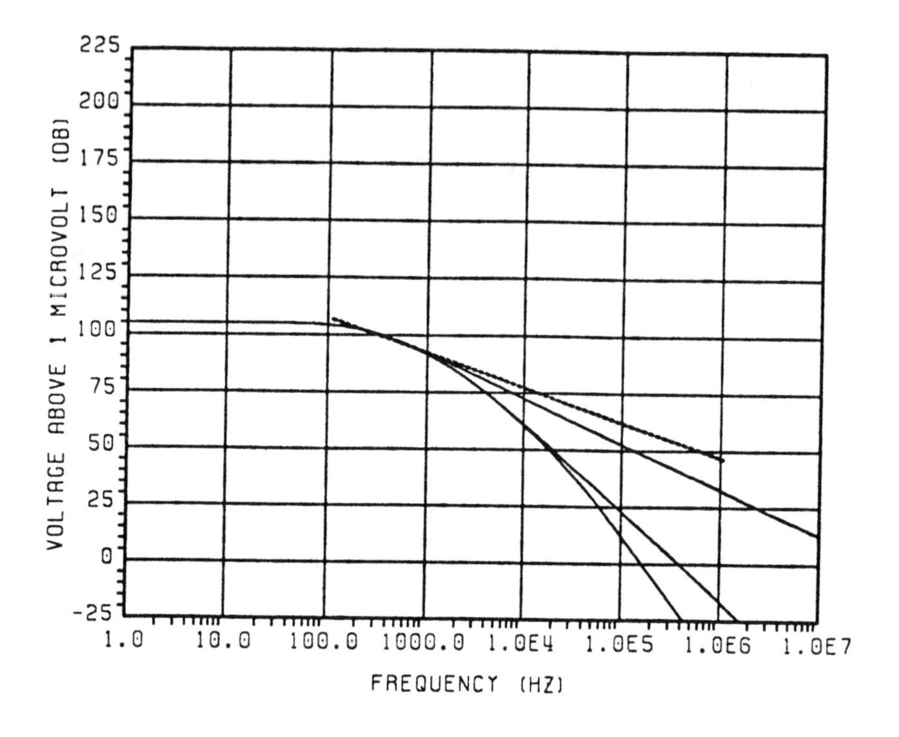

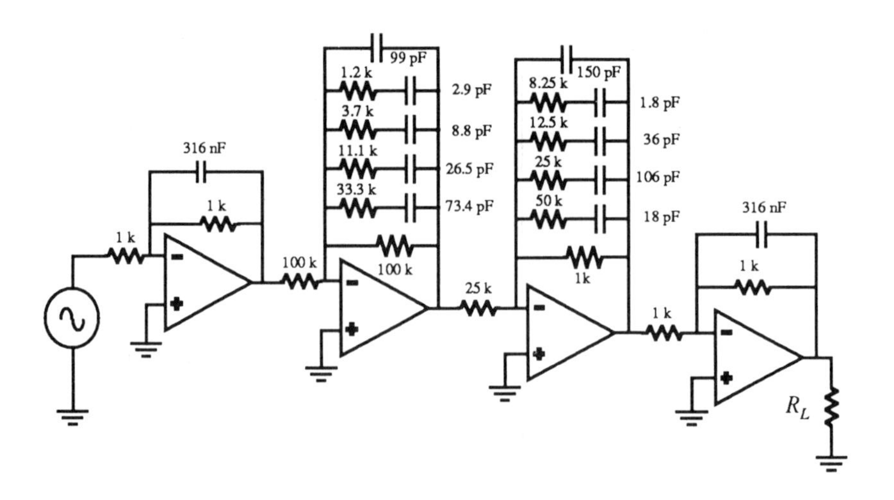

Figure 8.14 Fifteen-dB-per-decade EMI generator.

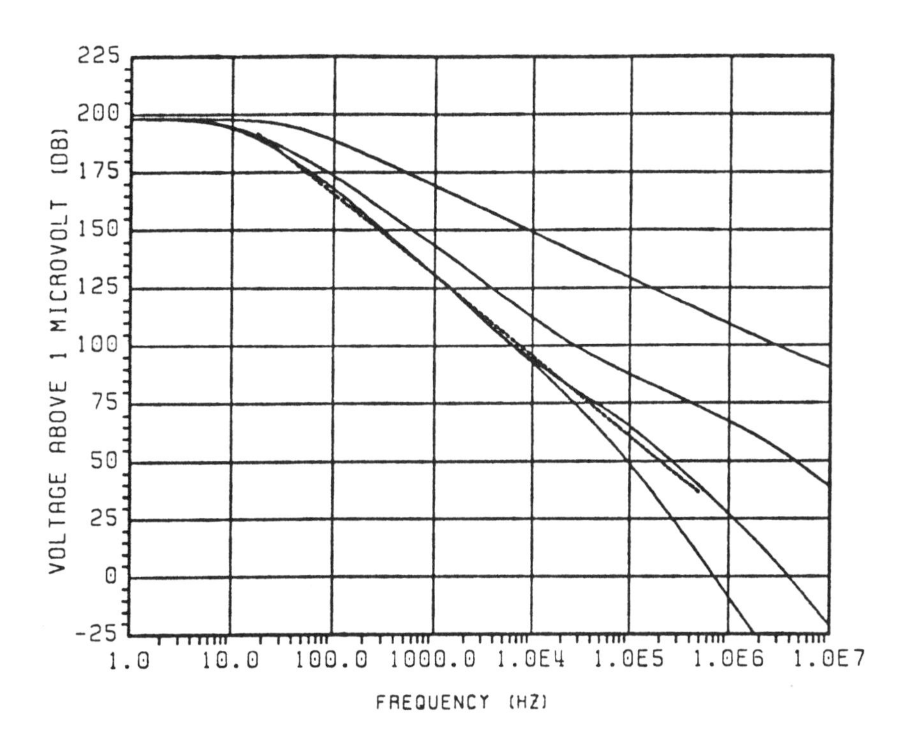

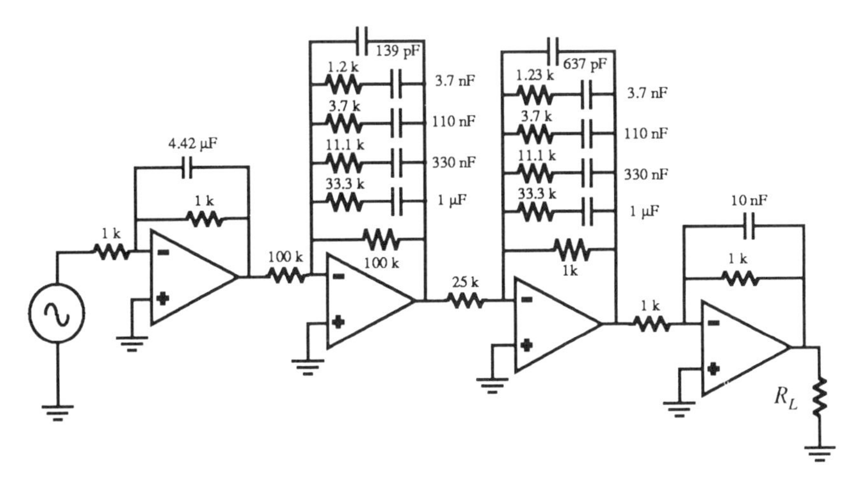

Figure 8.15 Thirty-five-dB-per-decade EMI generator.

A 15-dB-per-decade generator is presented in Fig. 8.14. Many pole-zero combinations are included in the feedback loop to contour the rate of attenuation with increasing frequency. The rate of attenuation depends upon the ratio of the capacitor and resistor pole-zero combinations. The pole-zero combinations must also be separated from each other so as to create smooth curves. This can be a process of trial and error, which is time consuming but can be rewarding when very accurate simulation of an uncommon waveform is necessary. Most common waveforms can be simulated in the frequency domain with very simple spectrum generators.

Combinations of 5-, 10-, 15-, and 20-dB-per-decade op amp EMI spectrum generators can be used to achieve any desired spectrum. This process involves painstaking hard work, but when possible, close approximations may be chosen for economic reasons.

Figure 8.15 shows a 35-dB-per-decade EMI generator and its corresponding frequency domain response. The four solid line plots are plots of each op amp EMI generator section. The dotted line is the composite of the four cascaded generators. Using parallel circuits, amplitude adjustments and summing any desired frequency domain spectrum can be achieved.

QUESTIONS

8.1 What is the difference between the terms "elements" and "components" as defined in this text?

8.2 What are the differences between the frequency domain spectrum of an actual time domain rectangular wave and a SPICE simulated frequency domain spectrum as generated in this chapter?

8.3 What are the major differences between ac analysis and EMI analysis?

8.4 The *LC* filter example in this chapter performs two major circuit functions. What are they?

8.5 What are the frequency ranges of interest for the two functions in question 8.4? For this circuit, would the significant elements (players) be different when simulating time domain power conversion or frequency domain EMI?

PART II

ADVANCED CONDUCTED EMISSION DESIGN

9

EMC REGULATIONS

Many rules and regulations have been developed to limit emissions from devices and control susceptibility levels. Most commercial regulations are based on the recommendations of the International Special Committee on Radio Interference (CISPR), a subcommittee of the International Electrotechnical Commission (IEC). Each country develops its own rules and assigns enforcement to a regulatory body. Examples of regulatory bodies include the FCC in the United States, the FTZ/VDE in Germany, the VCCI in Japan, and the DOC in Canada. Many of the requirements are new and they are evolving constantly as more is learned about the EMC field.

Military EMC requirements are based on Mil-Std 461A,B,C, Mil-Std 462, and Mil-Std 463, which are U.S. Department of Defense issued standards. These requirements differ substantially from commercial regulations because of the different objectives for each: compatible operation of varied systems under complex battlefield conditions for the military and protection of TV, radio, and other communications services for commercial endeavors. There are a number of measurement units that are widely used in the EMC industry to designate the amplitude of emissions. Commercial conducted emissions are usually measured in decibels above 1 μV (dB/μV) while military conducted emissions are measured in decibels above 1 μA (dB/μA). Radiated emissions are measured in electric field strength, decibels above 1 μV per meter (dBμV/m) and magnetic flux density, decibels above 1 picotesla (dBpt). Broadband emissions are measured against a reference bandwidth, usually 1 MHz.

Portions of chapter 9 are excerpted from the Hewlett-Packard *EMC Testing: The Spectrum Analyzer Solution* (A "One-Day Seminar" textbook copyrighted by Hewlett-Packard Co. in 1983).

9.1 FCC

The FCC administers the use of the frequency spectrum in the United States and its *Rules and Regulations* cover many products and services. Volume II of the Regulations contains four parts dealing with intentional and incidental use of the spectrum. Part 15 governs emissions from radio and television receivers, low-power communication devices such as radio-controlled garage door openers and models, field disturbance sensors, auditory training devices, Class I TV devices, and computing devices.

Computers and other electronic devices that generate and use tuning signals or pulses at a rate greater than 10 kHz and use digital techniques are considered computing devices (Fig. 9.1). These devices are further classified according to use. Class A devices are those marketed for use in a commercial, industrial, or business environment while class B devices are those marketed for use in a residential environment.

To prevent interference to communications from such devices the FCC has adopted technical and administrative specifications which are contained in Vol. II, Part 15, Subpart J of the FCC *Rules and Regulations*. These rules set forth the limits, measurement procedures, and compliance requirements that apply to computing devices. Computing devices excluded from Subpart J include those

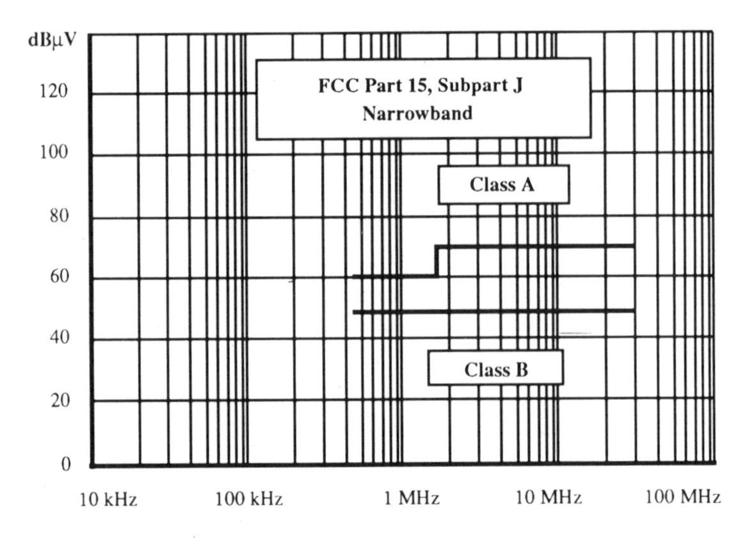

Figure 9.1 FCC RF narrowband conducted emission limits.

used in transportation vehicles, industrial, scientific, medical equipment, and appliances. RF devices such as radio transmitters and receivers subject to other FCC rules are also excluded.

Since the overall objective of the rules is to prevent harmful interference to radio navigation, safety, and radio communication services, the FCC has a non-interference requirement. This states that even if all technical specifications are met, a device may be required to cease operation if it is causing harmful interference and it is in the public interest to do so.

The FCC has established separate limits for conducted and radiated emissions and for Class A and Class B devices. The limits for Class B devices are more stringent because residential users are less able to cope with harmful interference. The distances referred to on the radiated limits mean the distance between the measuring antennas and the closest point of any part of the device under test.

The FCC *Rules and Regulations* specify conditions under which the emission tests must be made and procedures for making the tests. The most important objectives of the procedures is to create an environment that assures valid, repeatable measurement results. The FCC may call in equipment for testing at its site, so it is desirable for manufacturers to replicate the FCC conditions and procedures to ensure repeatable results.

Conducted tests are made using a Line Impedance Stabilization Network (LISN) as the transducer. The test may be made in a shielded enclosure. Power lines to all peripherals must be tested and the equipment must basically be configured and exercised as it would be under normal operating conditions.

Radiated tests are made using linearly polarized antennas as transducers. Tests are normally made on an open field site although semianechoic enclosures may be used if results are correlated to an open site. Test distances range from 3 to 30 m. Antenna orientation and height and equipment configuration must be varied to obtain the maximum emission level.

Measurements of both radiated and conducted emissions may be made with either radio noise meters or spectrum analyzers using peak and/or quasipeak detection. The bandwidths to be used are specified according to the frequency range being tested.

Compliance procedures vary depending on the type of device. The manufacturer must verify that Class A devices comply with the regulations and test results must be kept on file. The FCC may request submittal of a sample unit or representative data. Some Class B devices (electronic games, personal computers, personal computer peripherals) must be certified to be in compliance by the Commission and this is done through an examination of manufacturers' test results and in some cases by testing of a sample at the FCC test site. Other Class B devices may be verified by the manufacturer. Class A devices must have a label warning users that operation in a residential area may cause interference requiring corrective action. Class B labeling simply states that the device is certified to comply with Class B limits. All Class A and B devices manufactured after October 1, 1983, must meet these requirements.

9.2 VDE

The West German EMC requirements are the strictest and most well-defined in Europe and products designed in compliance with West German criteria are likely to meet requirements in most European countries. FTZ is a technical agency of the German Postal Services which has regulatory power regarding EMI. The VDE is a non-governmental body that prepares standards and performs certification testing. VDE standards are the basis for most FTZ regulations.

The VDE regulations are based on CISPR recommendations and are similar in many ways to FCC rules (Fig. 9.2). This enables manufacturers to develop design and test criteria which meet the requirements of both. VDE0871 is concerned with emissions from industrial, scientific, and medical equipment (ISM) that generates or utilizes discrete frequencies or repetition frequencies above 10 kHz. The requirements are similar to FCC Part 15, Subpart J except that the VDE rules include computing devices and ISM equipment while the FCC rules exclude ISM equipment.

There are three types of permits available to operate equipment generating or using electromagnetic energy at frequencies greater than 10 kHz. A General Permit may be granted for a specific type of equipment if the product meets VDE Limit B and a few additional requirements based on a type test. Products that only meet Limit A must operate under an individual permit for each user. An individual permit may also be granted for specific installations that are tested onsite.

Type testing is done at the VDE Test Station or at VDE approved test sites. The Radio Protection Mark of the VDE is attached to products that meet the General Permit requirements.

9.3 MIL-STD-461

Since 1943 there have been over 100 EMC related regulations and specifications issued by military agencies. Many of these are now obsolete or have been superseded by later regulations. The most important military standards for manufacturers of electrical and electronic equipment are the Mil-Std 461/462/463 series (Fig. 9.3). This standard is designed to force consideration and implementation of interference controls on equipment and provides a consistent procedure for evaluating EMC. The basic philosophy behind Mil-Std 461/462/463 is to introduce interference-reducing design techniques early in product development to reduce the need for after-the-fact EMC repairs. This is accomplished through the imposition of both design and test requirements.

Mil-Std 461C is the latest revision of the standard now being used. However, Mil-Std 461 A and B are still being applied on a case-by-case basis for some products. Mil-Std 461A/B/C contain limits for the various types of tests and define specific requirements for different classifications of equipment. Mil-Std 462 defines the procedures to be used for the tests specified by Mil-Std 461A/B/C. Mil-Std 463 provides definitions of common EMC and EMI terms.

There are 21 separate test methods defined by Mil-Std 461. Along with the test methods, the standard designates nine equipment classifications that have different EMC requirements. The nine classes are divided among three major categories: class A (equipment and subsystems for use in critical areas such as aircraft, spacecraft, ground facilities, surface ships, and submarines), class B (equipment and subsystems supporting class A but not located in critical areas such as electronics shop maintenance and test equipment), and class C (miscellaneous, general purpose equipment not usually associated with a specific platform or installation).

To determine which tests and limits are applicable to a particular product, first determine the classification of the equipment and then refer to the appropriate section of the standard. All test methods do not apply to all categories and some requirements are determined on a case-by-case basis.

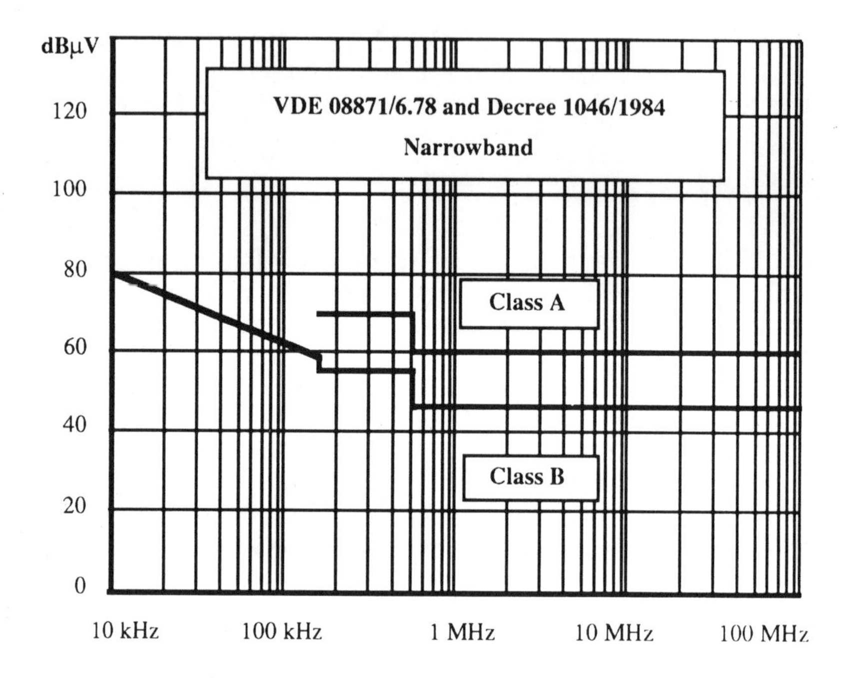

Figure 9.2 VDE RF narrowband conducted emission limits.

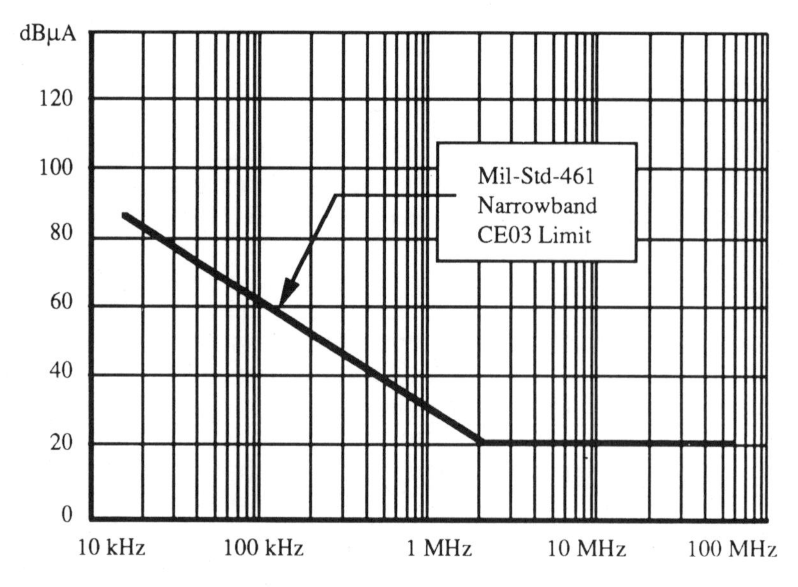

Figure 9.3 Mil-Std 461 CE03 RF narrowband conducted emission limits.

The applicability of Mil-Std 461/462 requirements is specified in each individual equipment or subsystem specification, contract, or order. The determination of the emission and susceptibility requirements to be used in each case is made by examining the type of equipment, its mission, and its intended installation. In situations where the requirements in the standard are not adequate for a particular procurement, they may be tailored to better achieve the necessary EMC performance.

The documentation required for Mil-Std 461/462 compliance consists of an Electromagnetic Interference Control Plan, Qualification Plan, Test Procedures, and Test Report. The Control Plan specifies the design procedures and techniques to be used to achieve the required EMC performance. The Qualification Plan describes how the contractor intends to prove that the equipment meets contract specifications. The Test Procedures specifically detail the test setups using Mil-Std 462 as a guide line and provide a step-by-step procedure and data sheets to record the results. The test report includes a log of the performance of the tests, a data package of all the test results, and a summary report of the resolution of all of the test problems or unplanned events.

Susceptibility, a long neglected aspect of EMC, has begun to assume major importance. From 1980 to 1982 the FCC was presented with about 300,000 complaints of interference to consumer-owned electronic equipment. As a result of the increasing incidence of interference events, the U.S. congress has given the FCC authority to regulate susceptibility as well as emissions. Although there are no all-encompassing commercial susceptibility regulations, many equipment manufacturers have implemented susceptibility testing to internal standards. Military standards have included susceptibility requirements since 1950 and susceptibility testing is an integral part of Mil-Std 461.

Although some special equipment is required for susceptibility testing, many of the tests are essentially the reciprocal of the emission tests. The criteria for passing a susceptibility test is the ability of the equipment under test to function in a specified electromagnetic environment without circuit upsets.

The fact that Mil-Std 461 susceptibility testing levels are as much as 100 dB higher than corresponding emission limits may seem inconsistent to some. However, the susceptibility levels are representative of the electromagnetic ambient encountered in military environments. The high-level ambients are not the result of unintentional emissions from equipment governed by the emission specifications but are a product of intentional emissions by communication and radar equipment. Mil-Hdbk 235 specifies radiated electromagnetic environment considerations for military applications and is used to determine appropriate susceptibility limits. Levels as high as 200 V/m are required for some environments. Many commercial manufacturers use in-house levels and test methods. The test levels range from 1 V/m to 200 V/m.

Another reason for the difference in emission and susceptibility levels (worst case emissions are less than 0.1% of the worst case susceptibility level) is that most EMI testing is done at nominal input voltages, room temperature, standard humidity, etc. The conducted EMI emissions can vary greatly with input power variations and the external environmental conditions. To test for the worst possible combination of operating parameters, an extraordinary number of tests would have to be performed. In the interest of economy, the tests are usually done under nominal operating conditions and rely on the worst case factor of a 1000 as the difference between emissions and susceptibility levels to assure a safe engineering margin.

The Japanese VCCI Class 1 and Class 2 EMI standards for conducted emissions are shown in Figs. 9.4 and 9.5.

The American organization Radio Technical Commission for Aeronautics (RTCA) limits for airborne equipment are shown in Fig. 9.6. The EMI standards presented here are some of the more

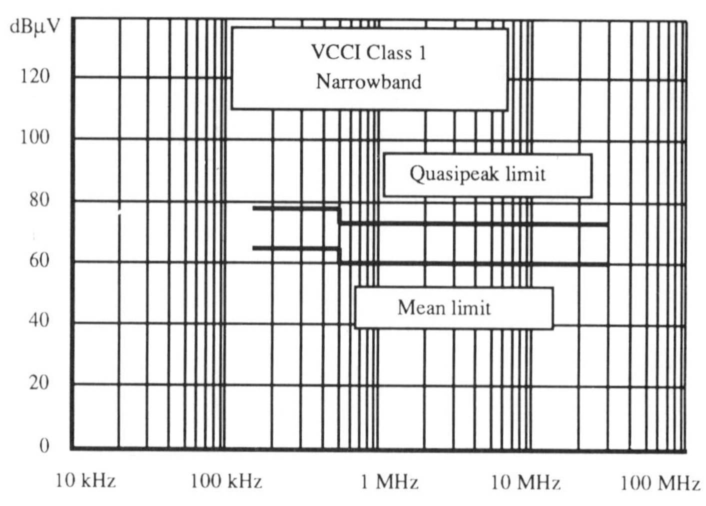

Figure 9.4 VCCI RF narrowband conducted emission limits Class 1.

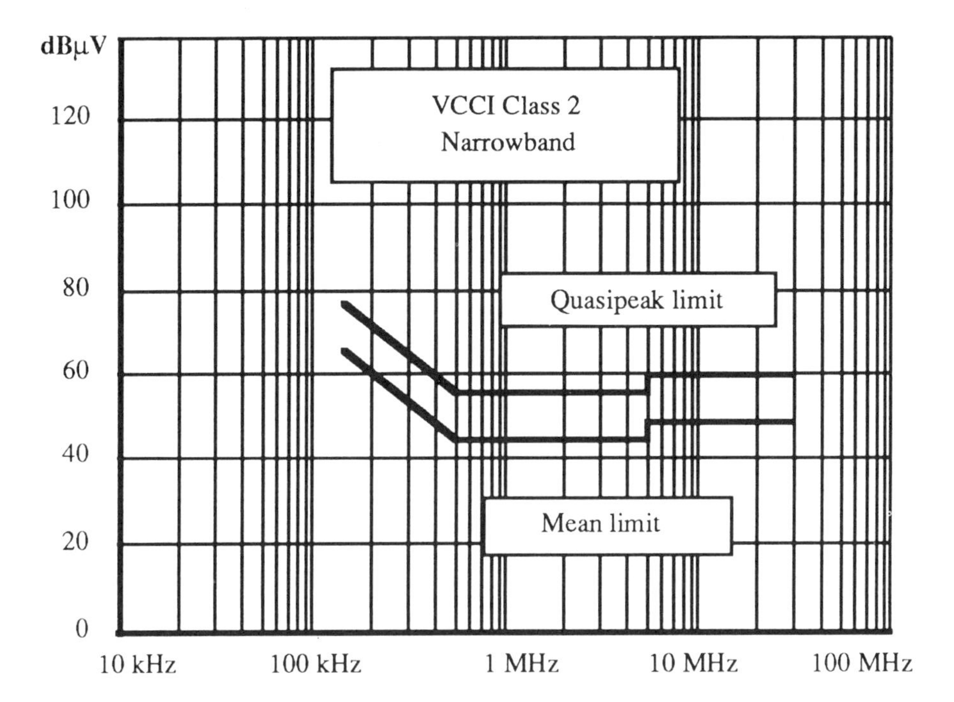

Figure 9.5 VCCI RF narrowband conducted emission limits Class 2.

common or more important requirements, but there are many other EMI standards for controlling conducted emissions.

9.4 VOLTAGE/LISN MEASUREMENT METHOD

The basic test setup for the voltage referenced measurements (voltage/LISN) is shown in Fig. 9.7. The power source impedance is modeled as 0.1 Ω. The power and return lines from the power source are connected to the unit under test (UUT) through line stabilization networks (LISNs). The LISNs are grounded through their cases with faying surfaces to the chassis ground. The chassis ground reference is generally a copper plate that covers the entire test bench. The LISN provides a constant impedance from 10 kHz to 1 GHz. A detailed description of the LISN is presented in chapter 6 and can be referred to if a refresher is needed. A capacitively coupled output (shown in Fig. 9.8) is provided on each LISN for measurement purposes. An EMI receiver or spectrum analyzer is

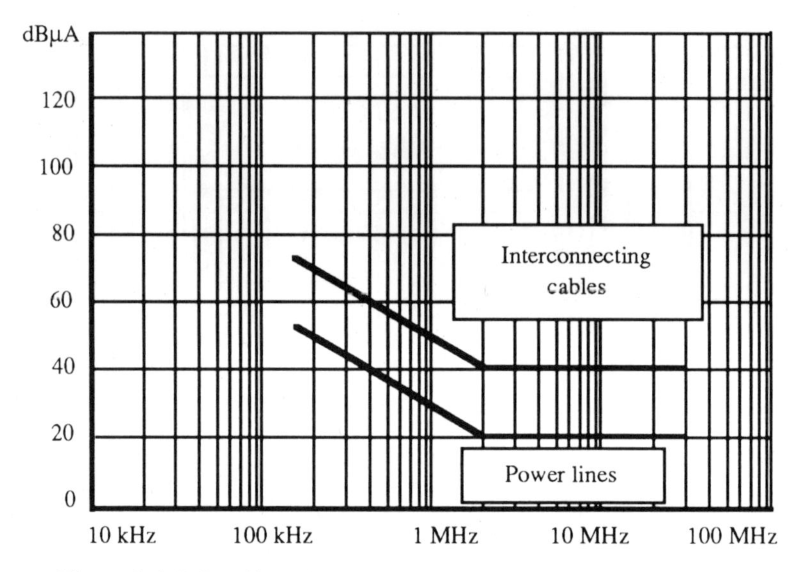

Figure 9.6 DO-160B RF narrowband conducted emission limits.

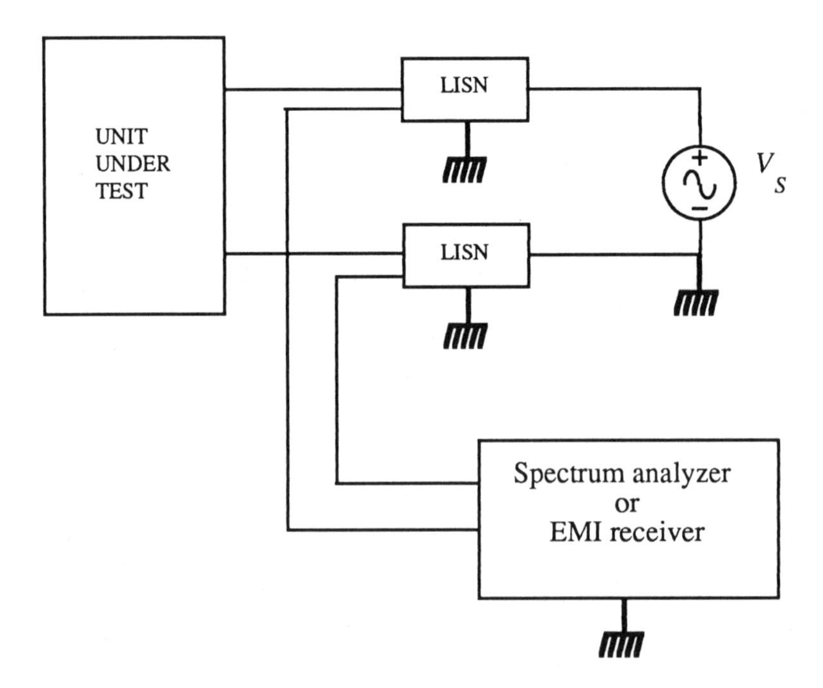

Figure 9.7 Test setup for voltage/LISN method for EMI measurement.

connected to the LISN outputs. A simplified equivalent model of the test setup, for frequencies greater than 10 kHz, is shown in Fig. 9.9. The LISN on the return line is neglected. For EMI sources (Z_{EMI}) with impedances much greater than the LISN impedance (Z_L), the EMI source operates as a current source that produces the measured EMI voltage across the 50-Ω impedance of the LISN (the power source impedance Z_S is neglected). If the EMI source impedance is small compared to the LISN impedance, the EMI source operates as a voltage source so that the output EMI current is determined by the 50-Ω source impedance of the LISN. The measured EMI voltage is determined by the LISN (part of the test setup). The voltage/LISN-type EMI measurement does not test low-impedance EMI sources on the same basis as high-impedance EMI sources. Low-impedance EMI sources are generally more difficult EMI problems than high-impedance sources. The voltage reference (voltage/LISN) measurement method affects low-impedance circuit measurements.

9.5 CURRENT/CAPACITOR MEASUREMENT METHOD

The military EMI conducted RF standards are based on current/capacitor measurements. A simplified model of a Mil-Std 461/462 test setup is shown in Fig. 9.10. The test setup is much like the voltage/LISN-type setups, but 10 μF feedthrough type

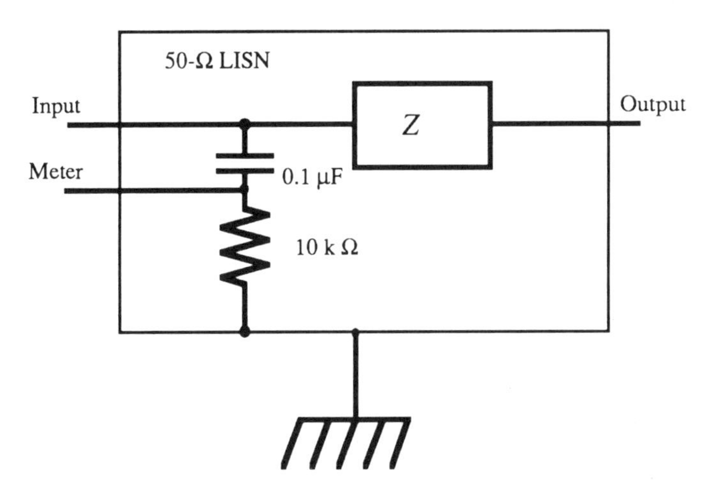

Figure 9.8 LISN meter output for measurement.

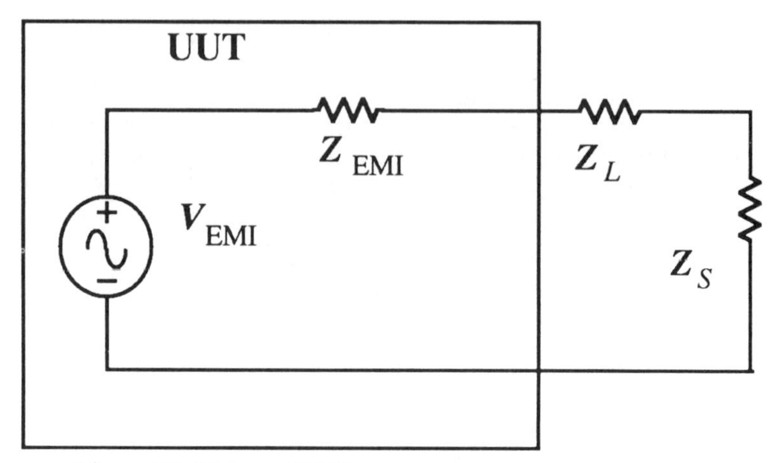

Figure 9.9 Voltage/LISN measurement equivalent model.

capacitors are connected between the power and return lines and ground instead of the LISNs. The feedthrough capacitors are virtually short circuits (less than 1 Ω) to the chassis for frequencies above 10 kHz. A simplified model of the current/capacitor method of EMI measurement is shown in Fig. 9.11. The UUT EMI source impedance determines the amount of measured EMI current flow. The test setup has much less influence on the EMI measurement with the current/capacitor method than with the voltage/LISN method.

9.6 A COMPARISON OF SOME OF THE RF CONDUCTED EMISSIONS STANDARDS

The differences in the voltage/LISN (FCC, VDE, VCCI, etc.) measurement method and the current/capacitor (Military Mil-Std 461/462) measurement method make a comparison very difficult. The FCC, VDE, and VCCI standards all use the voltage/LISN method but have different limit levels and slightly different test setup requirements for receiver bandwidths and relaxations under certain circumstances. One leading EMI consulting company compared Mil-Std 461 limits to FCC and VDE limits with a conversion factor to convert decibels above 1 μA to decibels above 1 μV. The rationale for the conversion was that 1 μA through the 50-Ω impedance of the LISN will create 50 μV across the LISN. Since $20(\log(50)) = 34$ dB, 34 dB was added to the Mil-Std 461

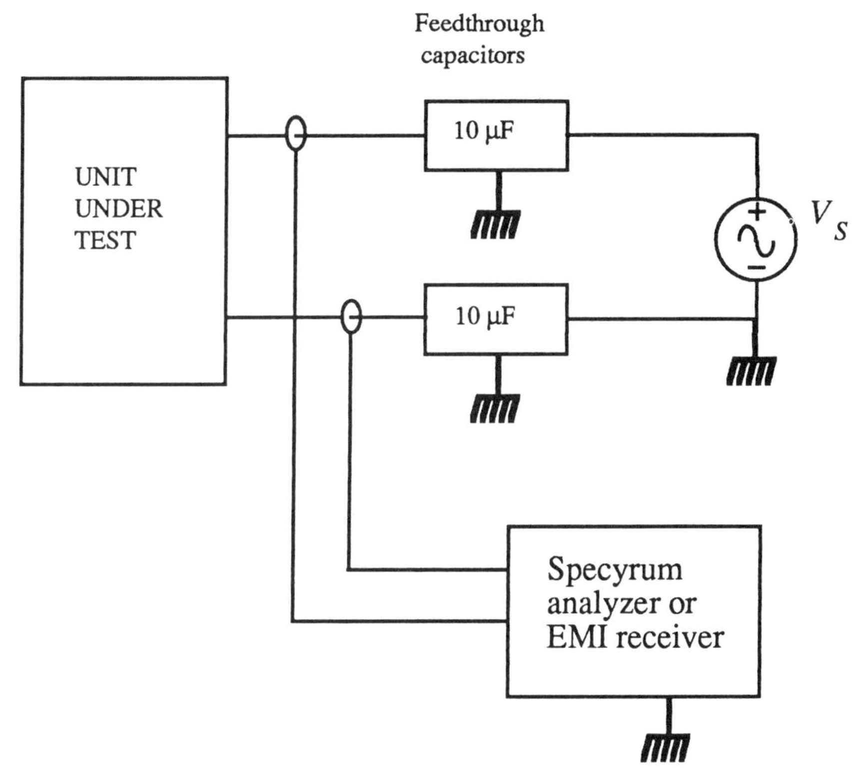

Figure 9.10 Test setup for current/capacitor method of EMI measurement.

limit across the whole CE03 frequency range as shown in Fig. 9.12. For high-impedance (much greater than 50 Ω) EMI sources, this comparison is fairly accurate. For low-impedance EMI sources, the comparison is not valid.

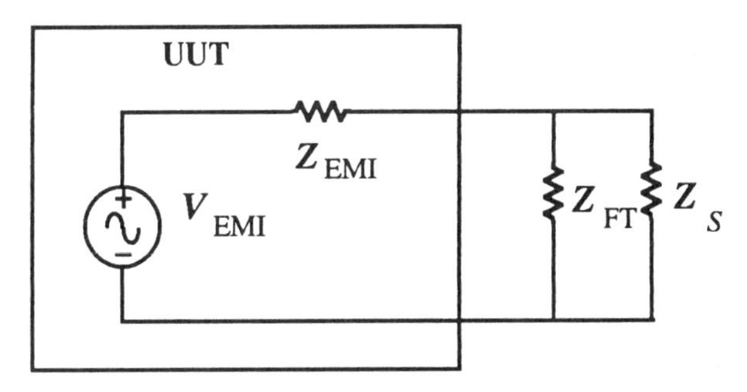

Figure 9.11 Current/capacitor measurement equivalent model.

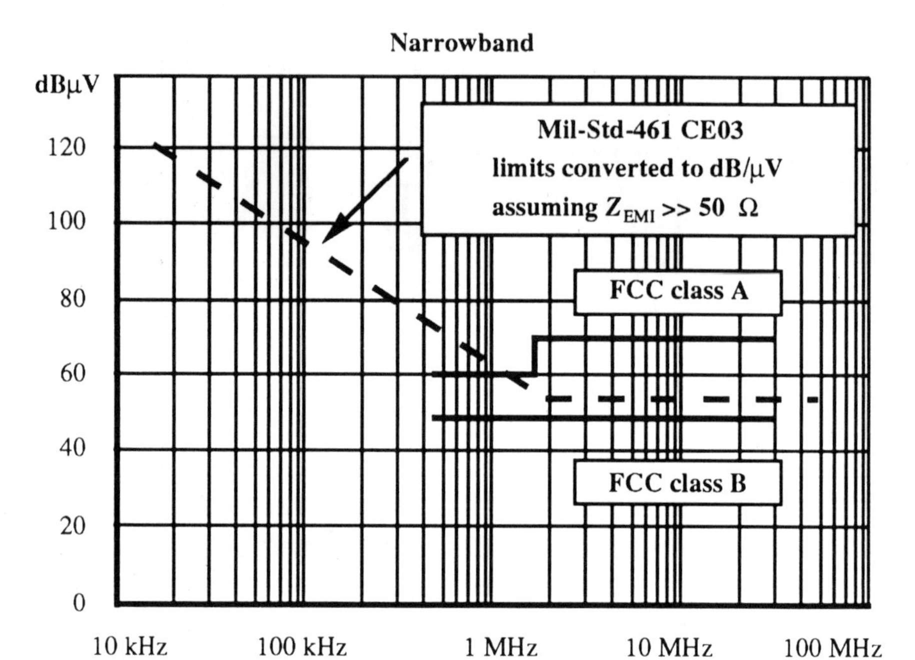

Figure 9.12 Comparison of FCC and Mil-Std-461 conducted emissions limits.

The Radio Technical Commission for Aeronautics EMI standard DO-160B allows for the use of either the voltage/LISN or the current/capacitor method with separate limit levels for each method. The standard states that the current/capacitor measurement method is preferred for category Z equipment. Equipment primarily for operation in systems where interference-free operation is required is identified as category Z. Category Z is the most stringent limit in DO-160B.

QUESTIONS

9.1 Why is there a need for controlling EMI? Why is there a need for controlling conducted emissions?

9.2 What are the two purposes for LISNs in EMI tests?

9.3 What are the ramifications of choosing to measure voltage or current for EMI emissions limits?

9.4 Why is it difficult to compare conducted emissions standard limits when one uses the current method and the other uses the voltage method of measurement?

9.5 What are the basic differences in the objectives of the commercial and military EMI standards?

10

SWITCH MODE POWER SUPPLIES

Power supplies are probably the major source of conducted emissions in modern state of the art electronics today. Switching power supplies are the most efficient dc power convertors. Switchers, by their very nature of switching, create a large amplitude of noise spread over a wide range of frequencies. The advantages of switching power supplies in most cases far outweigh the disadvantages of the associated EMI filtering that goes along with them.

We will present an overall view of a generic power supply and comments about each block within the power supply. Within a power supply there are four identifiable levels of EMI considerations: system, circuit, subcircuit, and component level. The fundamentals are the same at all levels, but the details at each level must be understood for the appropriate application of the electronic fundamentals. A block diagram of a typical power supply is shown in Fig. 10.1.

10.1 TYPICAL POWER SUPPLY BLOCK DIAGRAM

Input Common Mode EMI Filter

The common mode EMI interface filter is the classic EMI filter. Common mode chokes (baluns) and feed-through capacitors are used to reject and bypass common mode noise. The interface EMI filter is generally mounted in a location that permits the chassis wall to be used as a barrier to reduce the capacitive coupling around the filter so as to perform well at high frequencies in the range of a few megahertz to beyond 50 MHz, the upper limit for CE03.

Sometimes in the interest of simplicity and economics, attempts are made to design EMI filters, without Faraday shields, to be

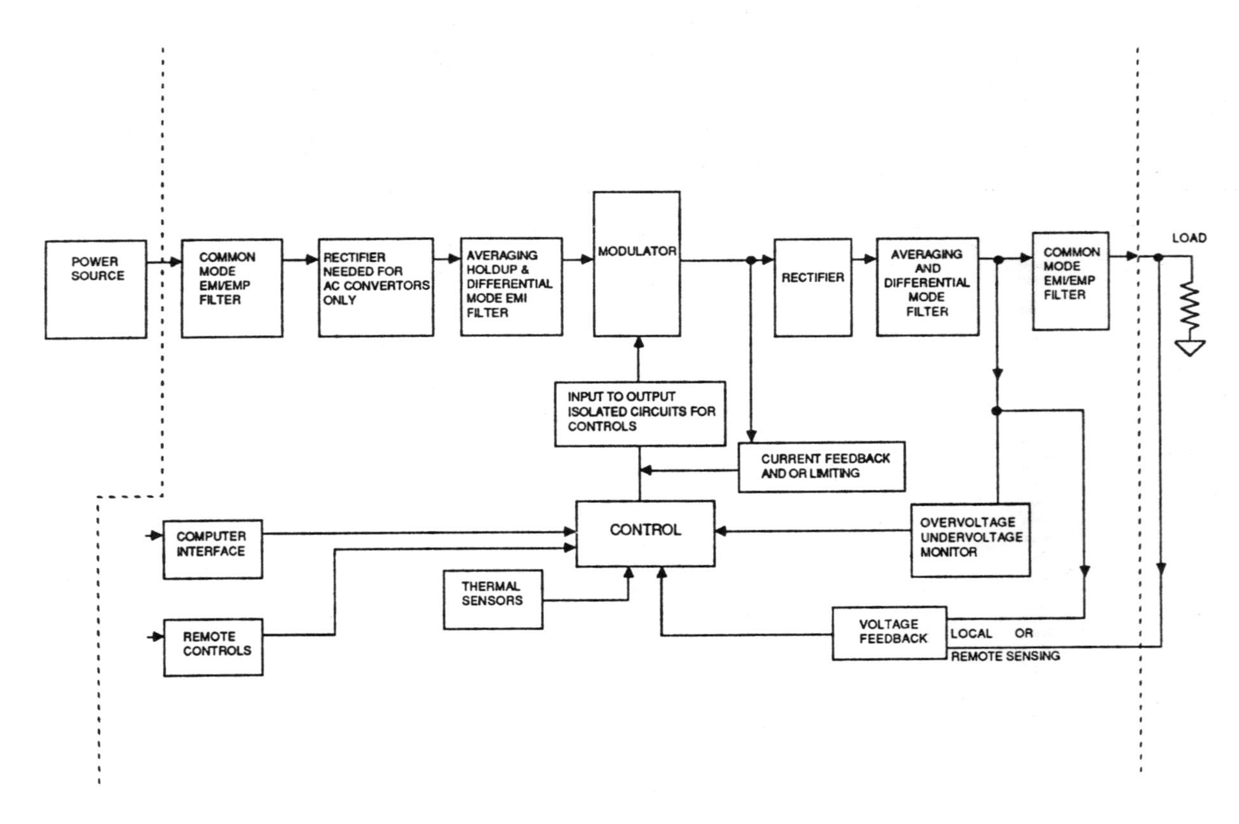

Figure 10.1 Generic switch mode power supply block diagram.

mounted on circuit boards along with the circuit to be filtered. This is not recommended in general unless the filter and circuit to be filtered are physically and electrically isolated, and there are low-impedance accesses to the chassis ground.

Generally the filters on printed circuit boards using standard inductors and capacitors resonate at frequencies well below the high-frequency range of military or commercial conducted emission limits. Near-field radiation is coupled around this method of filter layout, making the attenuation of the filter ineffective. The attenuation of frequencies above 10 MHz is done best at the input and output interfaces. Small components with special construction geometries for high-frequency applications are generally used. They are totally enclosed in a Faraday shield that is bonded to the chassis. The bonding resistance should be less than 2.5 mΩ to be effective as a low-impedance path to chassis. These classic EMI filters attenuate both outgoing and incoming noise. Some of the special types of inductors and capacitors used in interface EMI filters are baluns, beads, and feed-through capacitors.

Filter connectors are another interface EMI filter method that is becoming very popular. Lossy pi filters are fabricated on each pin of the filter connector to provide insertion loss at high frequencies. The filter connector is mounted in a physical position that takes advantage of the chassis wall in which it is mounted. The filter connector provides attenuation of unwanted high-frequency differential and common mode currents.

Rectifier

The input rectifiers in ac to dc convertors convert ac power to pulsating dc. The rectifiers' current switching causes damped sinusoids to "ring" at the natural frequencies of the circuit and layout. The diode switching occurs at twice the frequency of the power source. This low-frequency switching noise appears as broadband noise in the RF conducted emissions tests because of the close spacing of the upper harmonics relative to the wide bandwidth of the EMI receiver filter in broadband mode. Low frequency EMI problems in the AF conducted emissions tests will occur if the input inductor has too small of an inductance to spread the conduction angle so that harmonic distortion will not be severe.

Averaging, Hold-up, Differential EMI Filter

In single-phase applications the averaging, hold-up, differential EMI filter is necessary to smooth the pulsating dc into a wave shape that is mostly dc with some small ripple riding on top of the waveform. In three-phase applications, the phases overlap so that averaging is generally not necessary. When a particular hold-up is specified in the requirements, the capacitance of this filter must be increased to a value that will allow it to store enough energy to power the convertor through a hold-up period when the power input is interrupted. This low-pass *LC* filter must be designed to reject differential noise going out of the power supply as conducted emissions and coming into the power supply as conducted susceptibility.

Input to Output Isolation

Magnetics and opto-couplers are usually used to isolate the input from the output. This is generally needed for system grounding requirements. The control circuit is typically grounded on the secondary side but can be grounded on the primary side, especially in low voltage dc input systems. The power transformer in the modulator is often chosen to be the "give point" that isolates two different ground points in an electrical system grounding scheme. Power supply signals that transverse this isolation must also be isolated magnetically or optically or by some other isolation method.

Modulator

The modulator is the heart of the convertor. The three fundamental topologies are buck, flyback, and boost. Many versions and combinations of these are possible and are in use. This is the switch of the switch mode convertor. The modulator is the major EMI source in a switching power supply.

Rectifier

Rectification is needed here to remove the unwanted parts of the chopped waveform and decouple the power transformer so that it can reset. In the buck convertor, a "catch diode" is used to maintain continuous current in the output inductor during the OFF time of the power switch.

Averaging and Differential Mode Filter

The chopped output waveform from the modulator is averaged by the output *LC* filter to provide a smooth dc voltage for the output. The output impedance of this filter must be small compared to the load impedance. This filter should also be designed to provide differential EMI attenuation.

Overvoltage-Undervoltage Monitor

This circuit is a supervisor; it is not necessary for convertor operation but is usually needed to protect sensitive loads and to protect output components within the convertor from overvoltage stress and possible catastrophic failure. Current limiting is also sometimes included in this circuit, especially in multiple output situations.

Output Common Mode EMI Filter

The function of this filter is the same as the function previously mentioned for the input common mode EMI filter.

Control

There are many types of PWM controls. Voltage mode, voltage feed-forward mode, and current mode are three types of control for constant frequency pulse width modulated systems. The control is the "brain" of the power supply. Current mode is the state of the art control method using a two-loop feedback system. This system regulates the output depending upon both voltage and current. There are many other types of controls, such as frequency modulation or pulse stealing for resonant-type convertors.

Computer Interface

These signals are supervisory controls and monitors such as shut downs, overrides, and/or bit monitors for status.

Remote Controls

These signals are similar to the computer interface but may come from simple external electronic circuits.

Thermal Sensors

The thermal sensors usually monitor the convertor heat sinks and/or areas sensitive to high heat and shut the convertor down or limit the output in some way when an overtemperature condition exists.

Voltage Feedback

The output voltage is monitored and compared to a reference and the error is amplified and causes the control to change the duty cycle to correct the error. Voltage sensing can be either local, internal to the convertor, or remote (sensing at the load).

Current Sensing

In power convertors using current mode control, the current is sensed and applied as a feedback signal to the control. Currents are also sensed for monitoring overcurrent conditions that could lead to a failure. The convertor can be entirely shut down or the current can be limited to a safe value. "Hiccup mode" current limiting shuts the convertor down for a set time with the occurrence of an overcurrent condition. After the preset OFF time, the convertor restarts. At the restart, the convertor shuts down again if the overcurrent condition has not been removed. During Hiccup mode operation, the output current duty cycle is so small that power supply and interconnected circuitry are protected from any damage.

10.2 TYPICAL SWITCH MODE POWER SUPPLY EMI PROBLEM AREAS

The typical problem areas for passing tests for conducted emissions requirements are graphically shown in Fig. 10.2. Typically conducted emissions problems below 2 MHz are differential mode problems and above 2 MHz they are typically common mode problems. In differential mode at high frequencies, the circuit layout capacitances (signal to return lines) and inductances tend to filter the conducted emissions. The high-frequency common mode conduction paths typically bypass around inductors

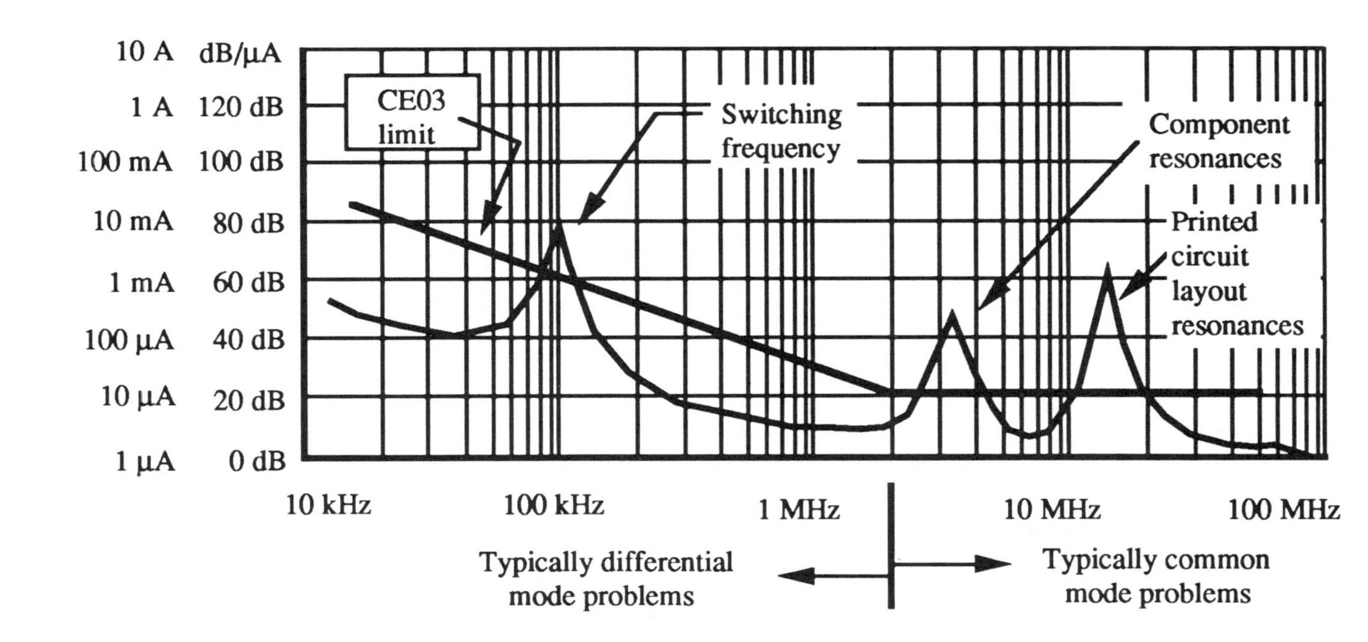

Figure 10.2 Typical switch mode power supply EMI problem areas.

and filters through the parasitic capacitances. The low-frequency common mode currents are generally small for two reasons:

1. The common mode conducted emissions are often capacitively coupled, which reduces the low-frequency components.
2. The circuit inductances filter the low-frequency common mode components.

The repetition rate of signals or waveforms with fast rise and fall times is generally the frequency at which the most significant differential EMI threat occurs. Properly designed differential filters will attenuate the fundamental and harmonic frequencies well up to the frequency range of about 2 MHz. In the frequency range of 2 to 20 MHz, the component resonances reduce the differential filters' ability to attenuate EMI. The inductors become capacitors and the capacitors become inductors. In the 10-to-50 MHz frequency range, the layout inductances and capacitances from the interconnecting printed circuit board traces and wires resonate.

Differential filters designed to attenuate the switching noise by their required size and location cannot be effective at attenuating the EMI frequencies above 2 MHz. The most effective (and classic) method to successfully attenuate EMI above 2 MHz is at the circuit interface to the outside world. The filters installed at the interface are classically called the EMI filter. These EMI filters usually include both common mode and differential mode components. The EMI filter must be enclosed within a Faraday shield as described in chapter 6. The size of the EMI filter components and the package in which they are mounted are specifically designed for high-frequency filtering. The common mode elements are baluns (common mode inductor) and feed-through capacitors. The feed-through capacitor case is one of the capacitor electrodes and is mounted and bonded to the chassis wall at the interface. The differential capacitors have high resonant frequencies and are packaged for the best performance possible. Differential inductors or ferrite beads are often used, but many designs use the leakage inductance of the balun for differential inductance. The beads are very lossy above a few megahertz. The beads dissipate the high-frequency energy as heat.

Attempts to pass EMI requirements with these EMI interface filters installed on circuit boards without low-impedance paths to the chassis will generally fail. And attempts to pass EMI requirements without the continuous Faraday shield bonded to the chassis and enclosing the filter will also generally fail. Successful filter design depends upon the use of the proper components, a good circuit layout, and properly designed (mechanically and electrically) packages.

10.3 EMI SIMULATION AND LABORATORY EMI TEST SETUP

A typical system with power conversion is presented in Fig. 10.3. The example system consists of a power generator, power conditioner, and a load (a computer) mounted in a vehicle. The power conditioner has both differential and common mode filters designed to control conducted emissions. The filters and their purposes are described in detail in chapter 6. We have both common mode and differential mode noise created within the power conditioner. A computer also has internal EMI noise generators, but they will be treated as insignificant for this discussion. The computer will be treated as a resistive load. Mil-Std-461 contains conducted emissions testing such as CE03. It specifies that the test will be set up as in the companion document Mil-Std-462. The CE03 test setup requires 10–μF feed-through capacitors on each power line (see Fig. 10.4). The 10–μF capacitors are a virtual short circuit to chassis ground for high frequencies. Mil-Std-461 CE03 limits apply to frequencies from 15 kHz to 50 MHz. The reactance of a 10–μF capacitor at 15 kHz is:

$$X_c = 1/(2\pi fC) = 1/(2\pi \cdot 15e^3 \cdot 10e^{-6}) = 1.06\ \Omega \qquad (10.1)$$

The reactance of a 10–μF capacitor at 50 MHz is:

$$X_c = 1/(2\pi fC) = 1/(2\pi \cdot 50e^6 \cdot 10e^{-6}) = 0.00032\ \Omega \qquad (10.2)$$

A model of the CE03 test setup is presented in Fig. 10.5. The dc portion and low frequency components flow through the load R_L. The higher-frequency switching components flow through the 10–μF capacitors and into the chassis ground because the 10–μF

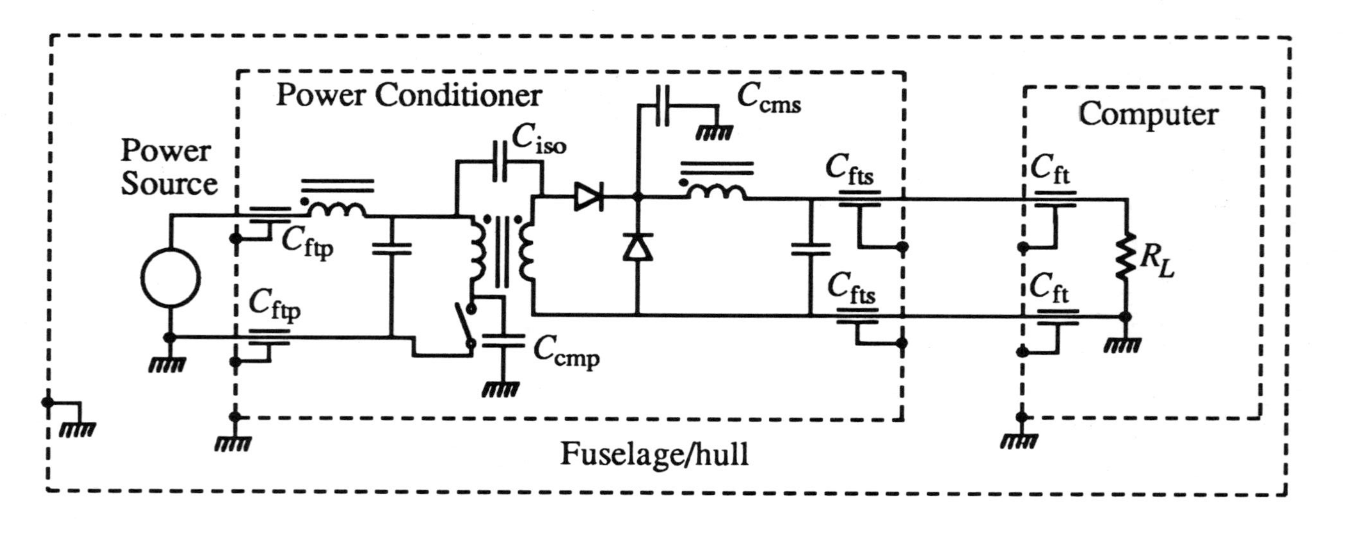

Figure 10.3 Typical system with power conversion.

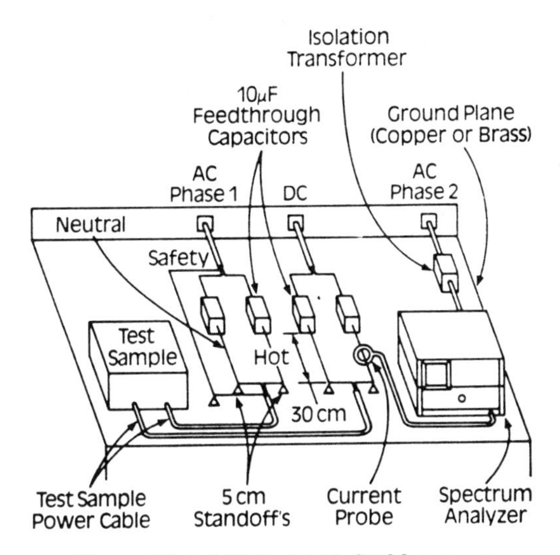

Figure 10.4 Mil-Std 462 CE03 test setup.

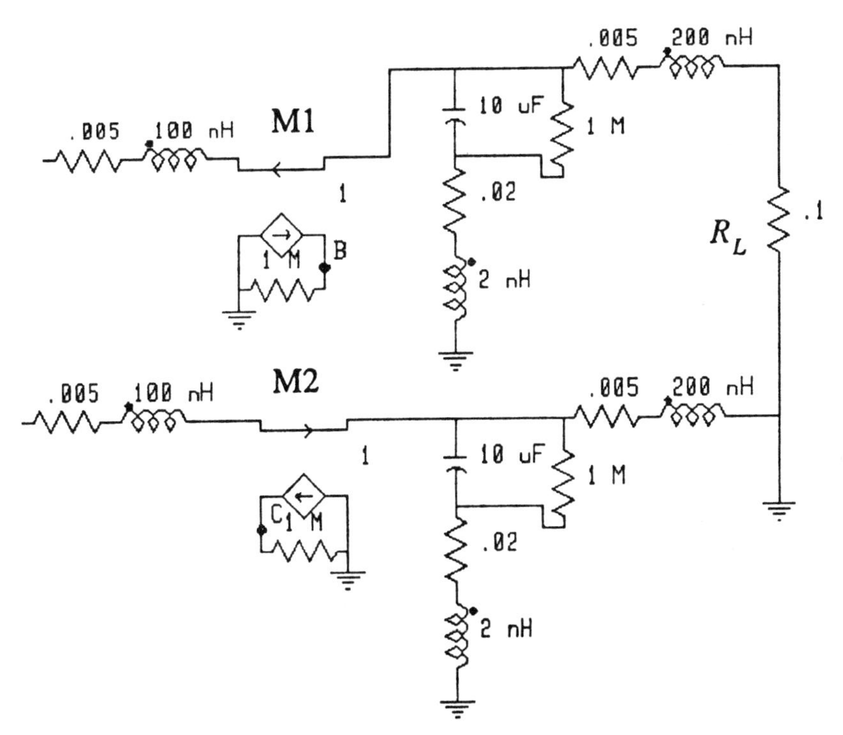

Figure 10.5 Model of Mil-Std 462 CE03 test setup.

capacitors provide a lower-impedance path than most loads. The EMI current is measured at circuit branches M_1 and M_2. The 100- and 200-nanohenry inductors model the inductance of the circuit loops created by the interconnecting wires shown in the Mil-Std-462 CE03 test setup. The 0.005 Ω resistors model the resistance of the interconnecting wires. The 10–μF feedthrough capacitors are modeled as previously shown in chapter 3.

10.4 SMPS EMI DESIGN EXAMPLE

The problem in this example is that the aviation control computer with a switching power supply fails the CE03 EMI test. The SMPS is specified to meet the requirements of Mil-Std-461B, Part 2. The spectrum analyzer plot of the CE03 narrowband measurement of the 28-V dc return lines is presented as Fig. 10.6. The 28-V dc return fails at the power supply switching frequency of 100 kHz by more than 45 dB. The upper harmonics are about 20 dB in excess of the limit. There is a peak at 4 MHz that exceeds the limits by approximately 30 dB.

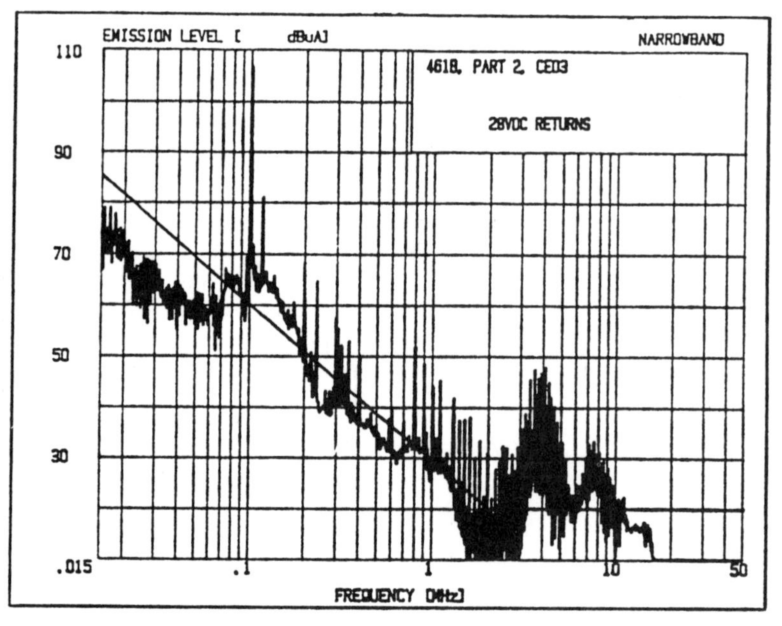

Fig 10.6 Spectrum analyzer plot of the CE03 narrowband measurement.

To solve the problem the circuit and layout were inspected. Two major problems were found with the filter circuit board layout shown in Fig. 10.7:

1. Excessive input to output capacitance on filter inductors
2. High impedance connection to chassis ground

I/O Capacitance

The darkened trace on the printed wiring board (PWB) has an identical trace on the adjacent layer. The filter input and output are each connected to one of these traces. With the area of the trace, the distance between layers, and the epoxy circuit board

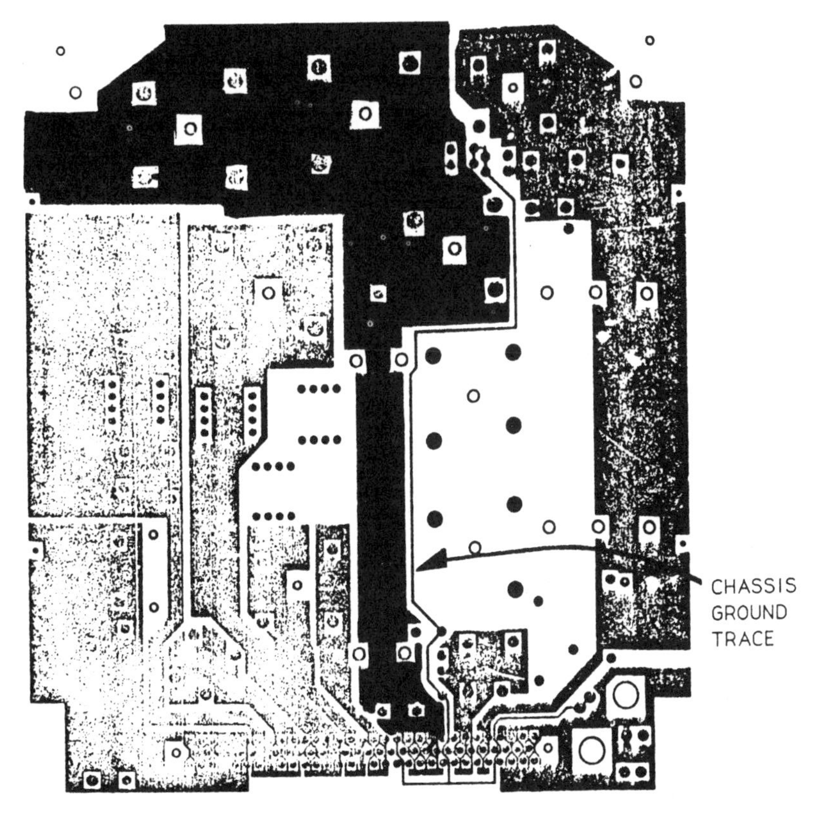

Figure 10.7 Filter circuit PWB layout.

relative permittivity of 4.7, the capacitance of these two traces was calculated to be 807 pF. An identical circuit board, without components installed, was measured to be 650 pF. The calculation is as follows:

$$C = \sigma A/d \tag{10.3}$$

$$C = \sigma_o \sigma_r A/d = 806.8 \text{ pF} \tag{10.4}$$

C = the capacitance in farads
σ_o = 8.855 pF/M
σ_r = 4.7 (relative permittivity epoxy circuit board)
A = 4.42 M^2 (area of plates)
d = 228 μM (distance between the plates)

High Impedance Connection to Chassis

The chassis ground connection trace is labeled in Fig. 10.6. The trace impedance is too high to be an effective connection to the chassis. The high-frequency ac resistance will be very high.

10.5 MODEL THE PROBLEM

A model of the filter and parasitic effects of the printed wiring board (PWB) layout was developed and is presented in Fig. 10.8.

The model goal is to simulate the performance of the components in the schematic shown in Fig. 10.9. The component labeled I_{in} represents the SPICE EMI spectrum generator shown in Fig. 10.10. The spectrum generator, I_{in}, produces the frequency domain spectrum of the time domain current pulses caused by the switch in the SMPS. An expanded view of the filter is presented in Fig. 10.11. The ground trace impedance and the impedance of the wire from the circuit board chassis capacitors to the chassis ground is modeled by the 300-nH inductance element. The input to output capacitance is modeled by the 2000-pF capacitors around the filter inductors.

The resulting simulation data presented in Fig. 10.12A can be compared to the measured laboratory data in Fig. 10.6. The curve labeled B is the simulated EMI measurement data from one of the current monitors. Since this is an input filter, it is appropriate to use the switching current waveform. For output EMI analysis it is

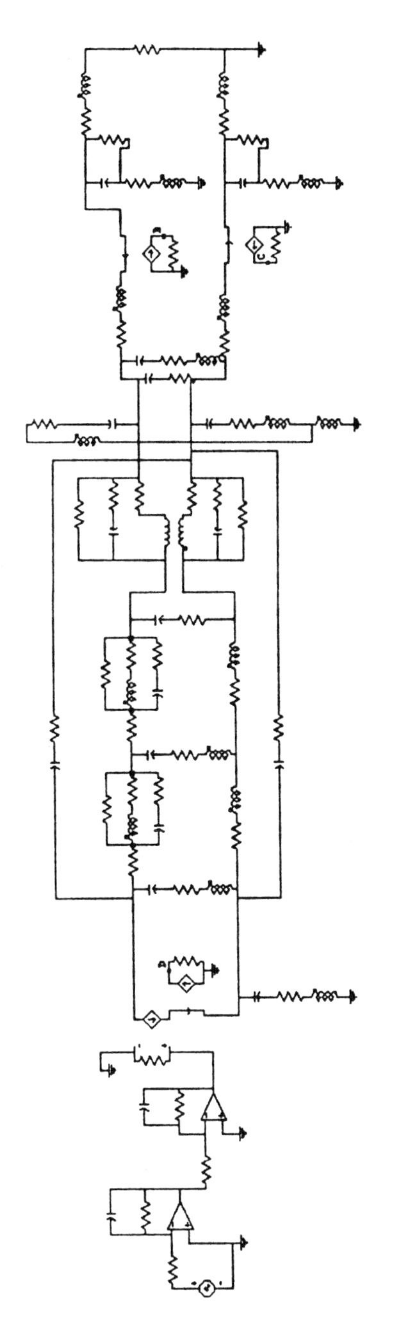

Figure 10.8 Model of the spectrum generator, filter, parasitic effects of the printed wiring board (PWB) layout, and Mil-Std 462 test setup for CE03.

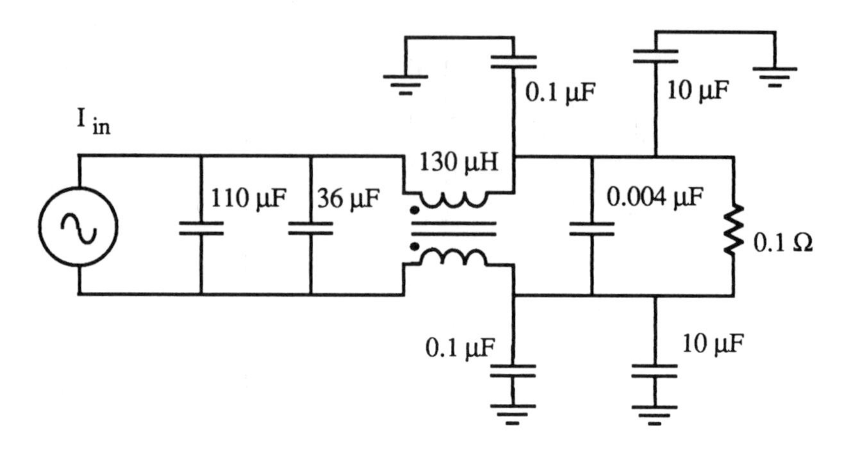

Figure 10.9 Schematic of the filter and Mil-Std 462 (CE03) test setup.

often appropriate to use the output voltage waveform. The voltage at the node of the input filter and switched transformer primary should be a constant (dc).

There are actually no harmonics created below the fundamental frequency in a stable, constant frequency SMPS. Although the analysis can be performed at frequencies lower than the fundamental, the data are meaningless.

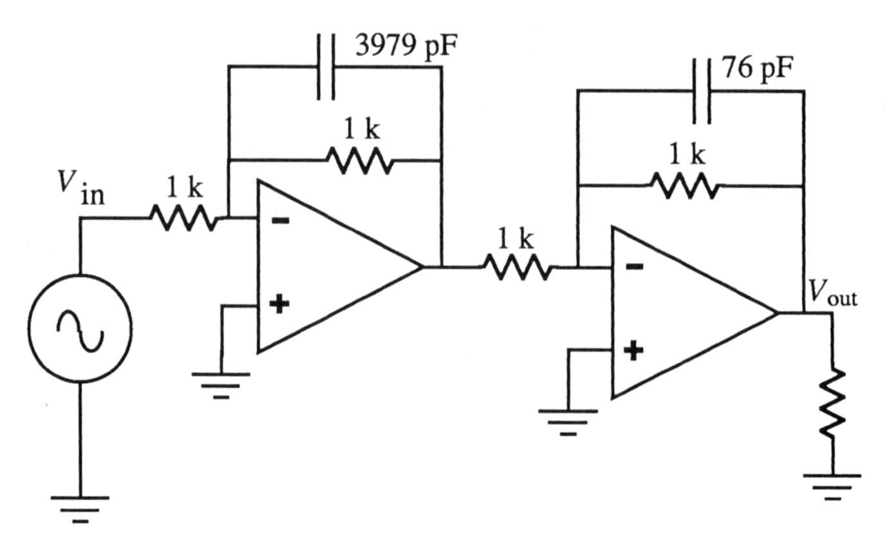

Figure 10.10 Spectrum generator model (example SMPS).

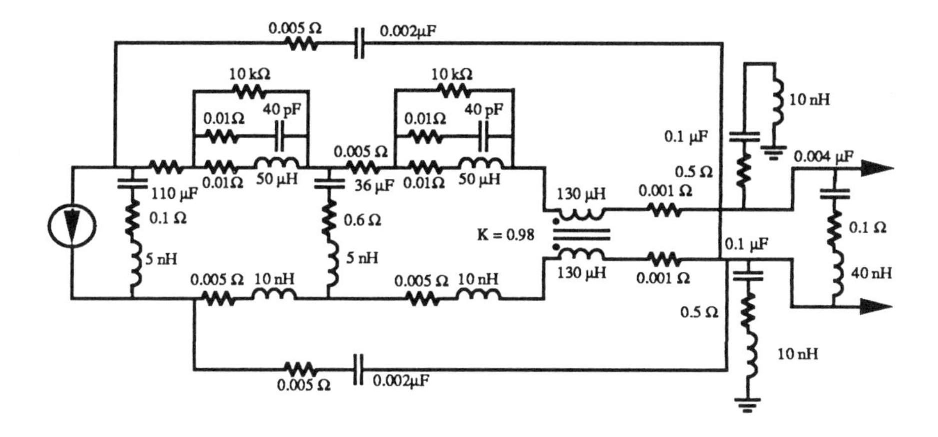

Figure 10.11 Filter model (example SMPS).

The simulation exceeds the CE03 limits in three places (typical for SMPS). Another simulation was performed with the 0.01–μF and 0.005-Ω elements (input to output capacitance) and the excessively long connection to ground removed (300-nH inductance elements change to 5-nH). The conducted emissions of the new model, with parasitics removed, are less than the CE03 limits and now pass the test requirements (see Fig. 10.12B).

The plot in Fig. 10.13A shows the performance with the input to output capacitance eliminated. In this case, if all of the input to output capacitance is eliminated, the circuit passes the CE03 test. Of course, there is always some input to output capacitance. The plot in Fig. 10.13B shows the performance with the input to output capacitance present but the excessive ground path inductance reduced. The performance is improved over the original model (Fig. 10.8) but still fails at the fundamental frequency. This is no surprise because the fundamental frequency component is a differential mode problem and the excessive ground inductance is a common mode problem.

10.6 SIMULATION PROBLEMS

The simulation as presented generally responded to changes in the model as we would expect. When a closer inspection of the simulation is performed, some obvious shortcomings become apparent.

We would not expect that the reduction of input to output capacitance would significantly change the performance of the filter at 100 kHz. If the input to output capacitance or the balun inductance is changed in value, the resonant peak (Fig. 10.12A) formerly at 100 kHz changes frequency. The fundamental switching frequency peak is generally not caused by a resonant *LC* tank. The fundamental switching frequency is generally the largest energy component of the waveform. Sometimes an *LC* tank resonates in the range of the switching frequency, which causes an increase in the peak amplitude. If the *LC* tank frequency was not exactly 100 kHz, there would be another peak close to the switching frequency.

There are three significant problems with this model even though initially the comparison to the measured data looked good:

1. Amplitude at switching frequency (100 kHz) is too small.
2. This EMI analysis provides only a worst case envelope, which should not look exactly like the measured data.
3. The measured data is for the return simulation current monitor and the analysis presented so far is for the power or signal current monitor.

No matter how tempting it is to accept results that look good, tests must be performed to verify the validity of the model to provide a confidence level that will make the simulation usable for design purposes.

The SPICE EMI generator presented generates the envelope of the real world (laboratory measurement) EMI spectrum. In the real world, frequency components between the harmonics do not exist. The SPICE EMI generator generates continuous data between harmonics. This is not a problem if we keep these shortcomings in mind when observing the simulation data. The worst case spectrum envelope is accurate and is our concern when trying to reduce emissions to below the required limits. The envelope drawn on the measured data of Fig 10.14 would be the best that we could expect from the simulation data. The envelope drawn on the measured data in Fig 10.15 would be more typical of the simulation data obtained from the SPICE EMI analysis method presented.

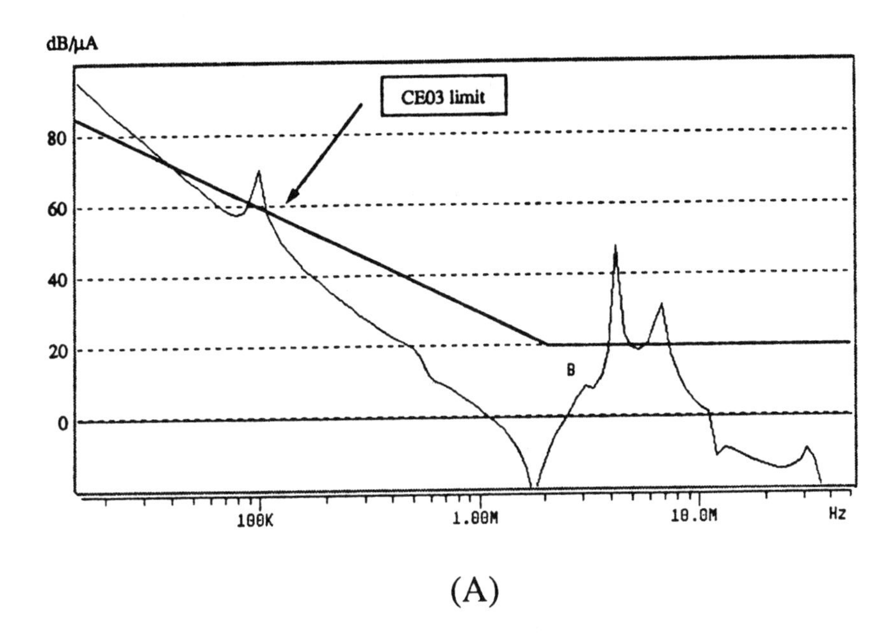

(A)

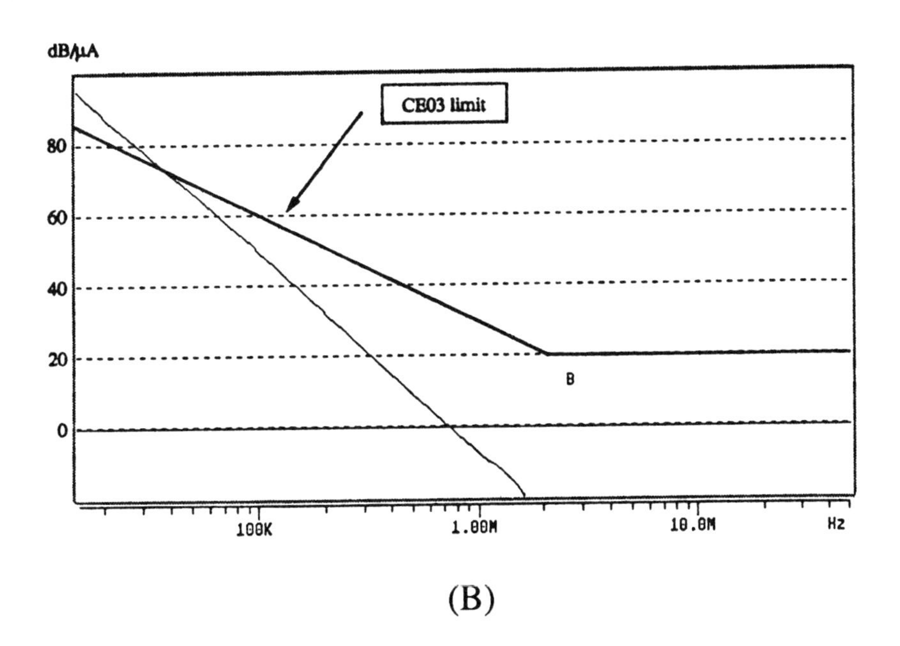

(B)

Figure 10.12 Simulation data for model with (A) as-designed parasitics and (B) without input to output capacitance and high-impedance ground connection.

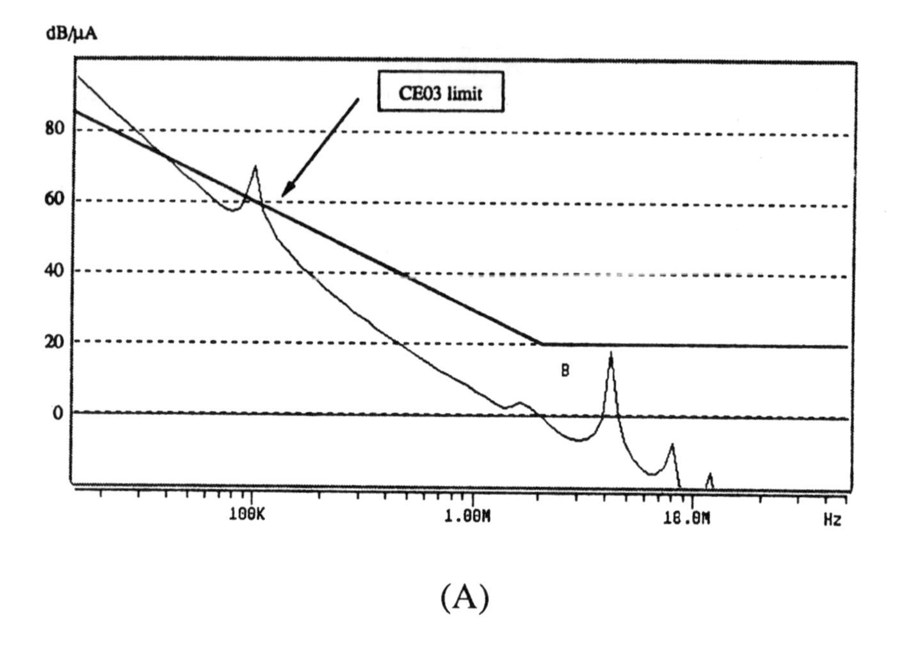

(A)

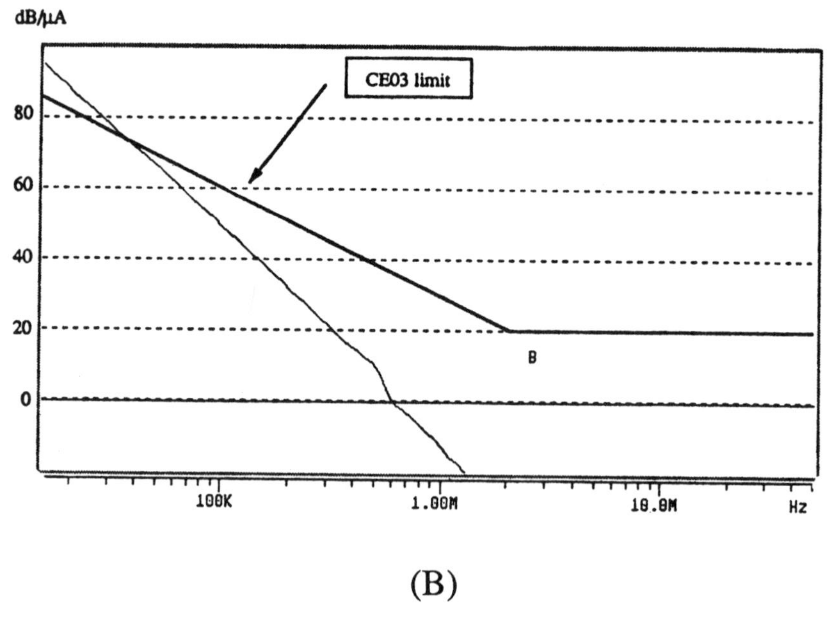

(B)

Figure 10.13 Simulation data for model with: (A) input to output capacitance only and (B) high-impedance ground connection only.

The points labeled A, B, C, D, and E are the frequencies of concern for meeting the required conducted emissions limits. The peaks labeled B, C, D, and E are usually at the frequency of the dominant *LC* tanks of the circuit and layout. The valleys between the points are frequencies where parasitic or unintentional zeros cause the filter attenuation to decrease instead of increase with frequency as intended.

10.7 BACK TO FUNDAMENTAL MODEL

At this point we will regress the model development and try to fix the shortcomings so as to make this model more useful. In order to examine this model in finer detail, we reduce it down to a simple

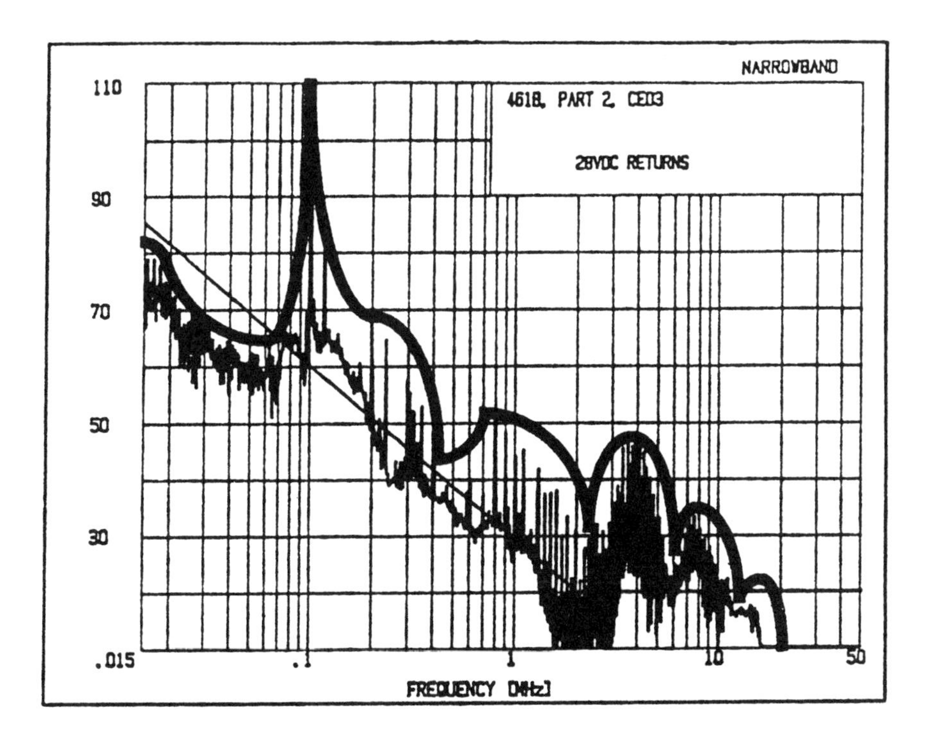

Figure 10.14 Spectrum analyzer plot with desired envelope in heavy line.

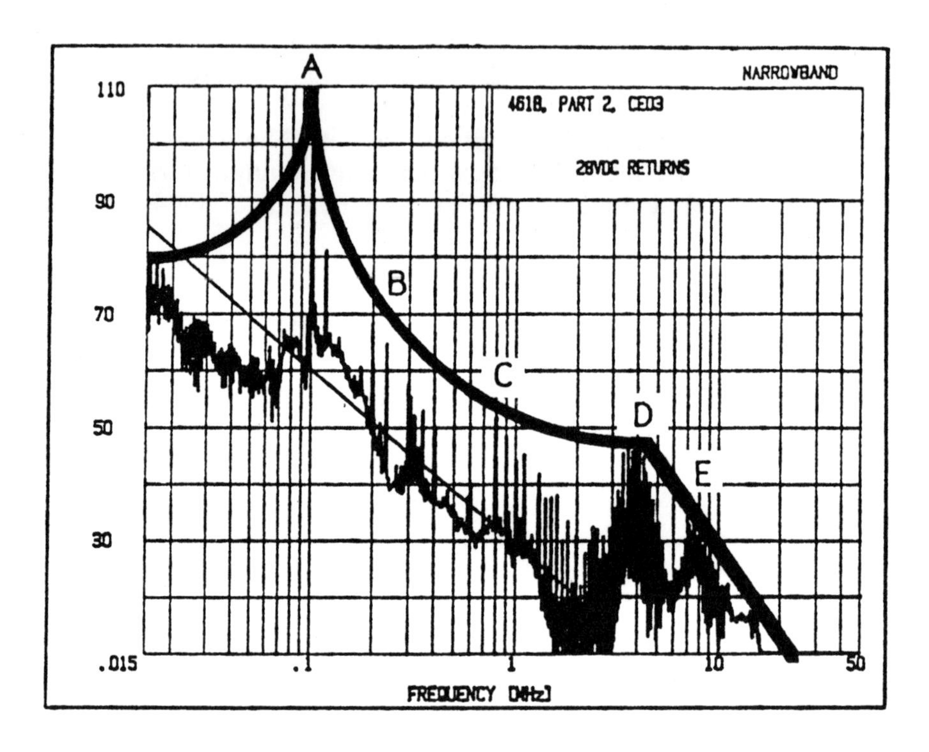

Figure 10.15 Spectrum analyzer plot with simple envelope of worst case spectral peaks.

and fundamental version (Fig. 10.16). The filter part of the model now looks very much like a schematic representation. Simulation results using this model show that the attenuation at 100 kHz is 123.4 dB. A simple hand calculation approximation for the attenuation of the two-section *LC* filter, at frequencies away from resonance, is as follows:

$$\text{Attn} \approx (X_{c1}X_{c2})/(\omega^2 L_1 L_2) \tag{10.5}$$

$$= (0.01447 \cdot 0.0442)/987 \tag{10.6}$$

$$\approx 0.648\ \text{e}^{-6} \quad \approx 123\ \text{dB} \tag{10.7}$$

There is an insignificant difference between the approximate filter attenuation and the simulated data measured at the current probes.

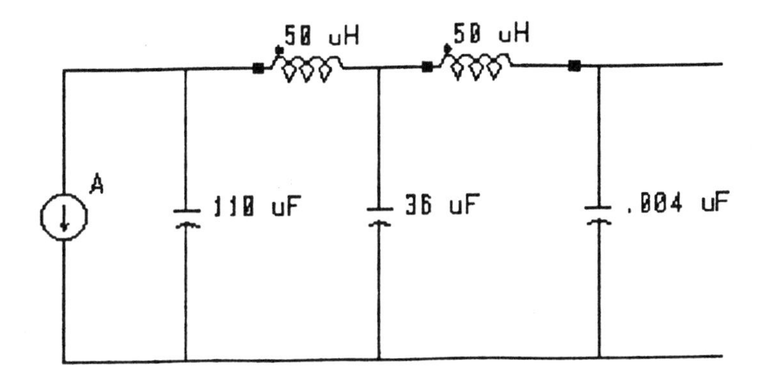

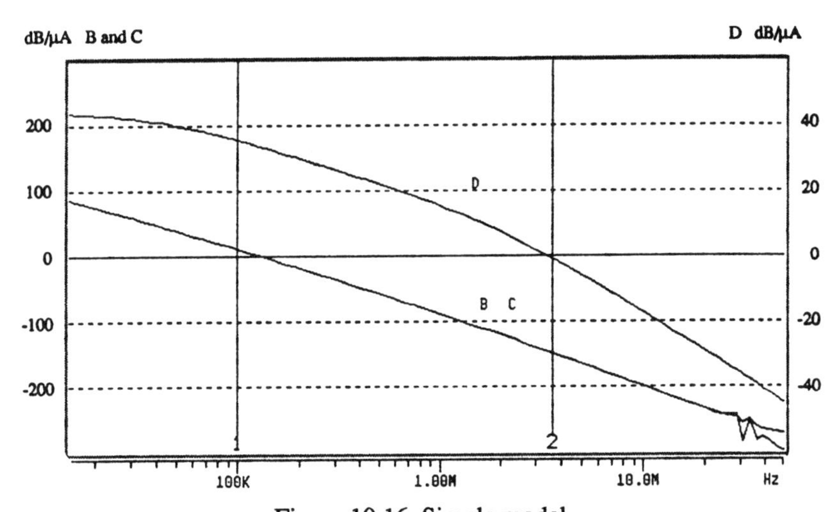

Figure 10.16 Simple model.

The simulation results do not look like the laboratory measurement EMI plot of Fig. 10.6. To create a useful model we must add elements to the model so that a more accurate simulation is achieved.

The model in Fig. 10.17 includes more of the parasitic elements of the components. Although this simulation is not accurate, it is better than the previous attempt. A hand calculation approximation for the filter attenuation is now:

$$\text{Attn} \approx (R_1 R_2)/(\omega^2 L_1 L_2) \tag{10.8}$$

$$\text{Attn} \quad = (0.065 \bullet 0.311187) \qquad (10.9)$$

$$\approx 19.7\ e^{-6} \qquad \approx 94\ \text{dB} \qquad (10.10)$$

The filter performance of the main filter capacitors is now dominated by the ESRs over the values of the reactance of the capacitors. The performance of the filter capacitors in this EMI design example (and most cases with outputs less than 300 V dc) are dominated by the parasitics of the components. The simulation results with some component parasitics modeled shows that the attenuation at 100 kHz is now 93.4 dB. We are achieving the expected filter attenuation in the simulation at the fundamental switching frequency of 100 kHz. The high-frequency performance still does not compare favorably with the measured data in Fig. 10.6.

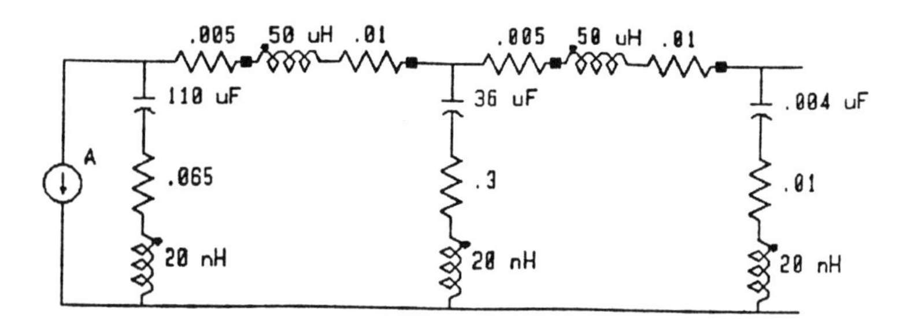

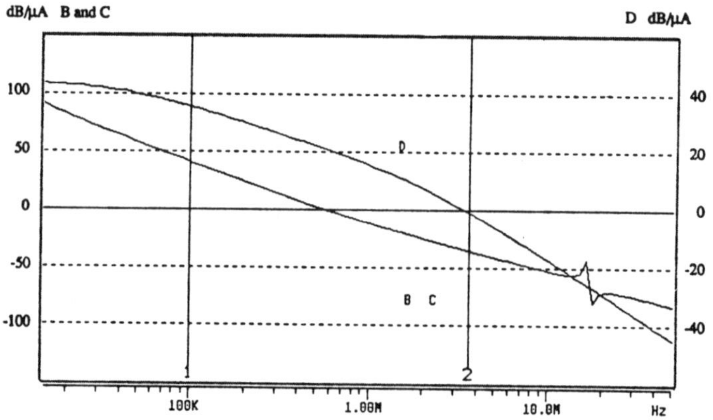

Fig 10.17 Model with parasitic elements added to components.

Parasitic capacitances (40 pF) are now added to the 50-μH inductor models. The new simulation results are presented in Fig. 10.18. The filter attenuation at the switching frequency (100kHz) is still accurate at about 94 dB. At the resonant frequency of the 50-μH inductors and 40-pF capacitances (3.59 MHz), the impedance is at a minimum. At frequencies higher than 3.59 MHz the attenuation reduces with increasing frequency. The filter performance is reduced. The good news is that the simulation performance data is "looking" more like the measured data.

So far we have only modeled the differential aspects of the problem at hand. We know that the common mode paths greatly affect EMI performance so we now add these elements to the model (see Fig. 10.19). The 0.1 microfarad chassis EMI filter ca-

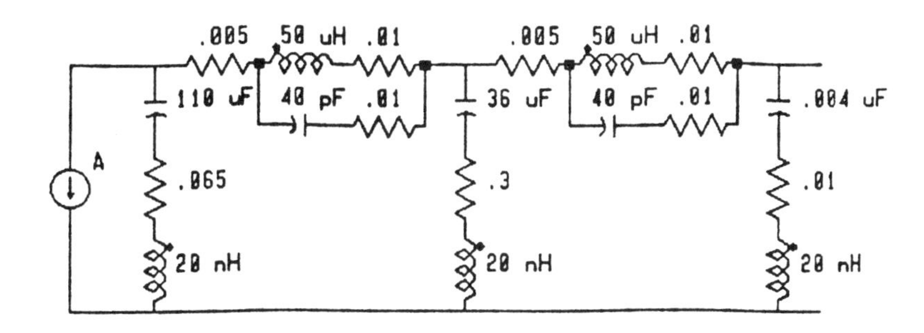

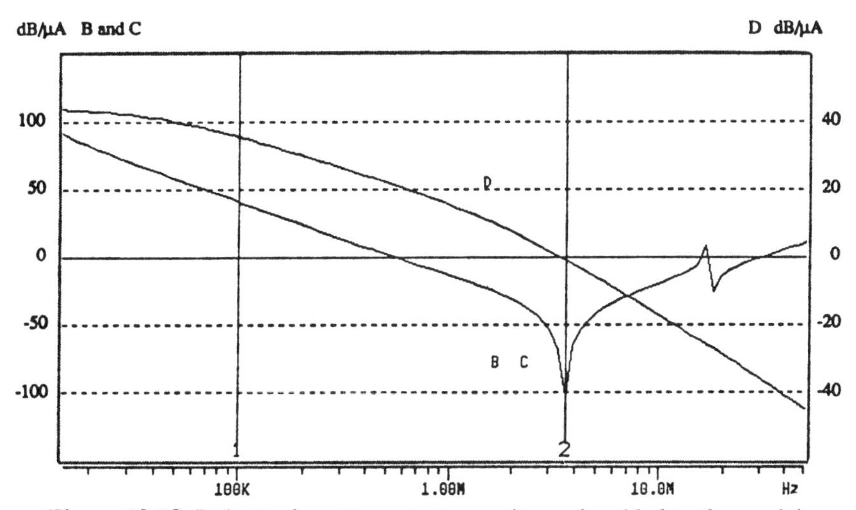

Figure 10.18 Inductor input to output capacitance is added to the model.

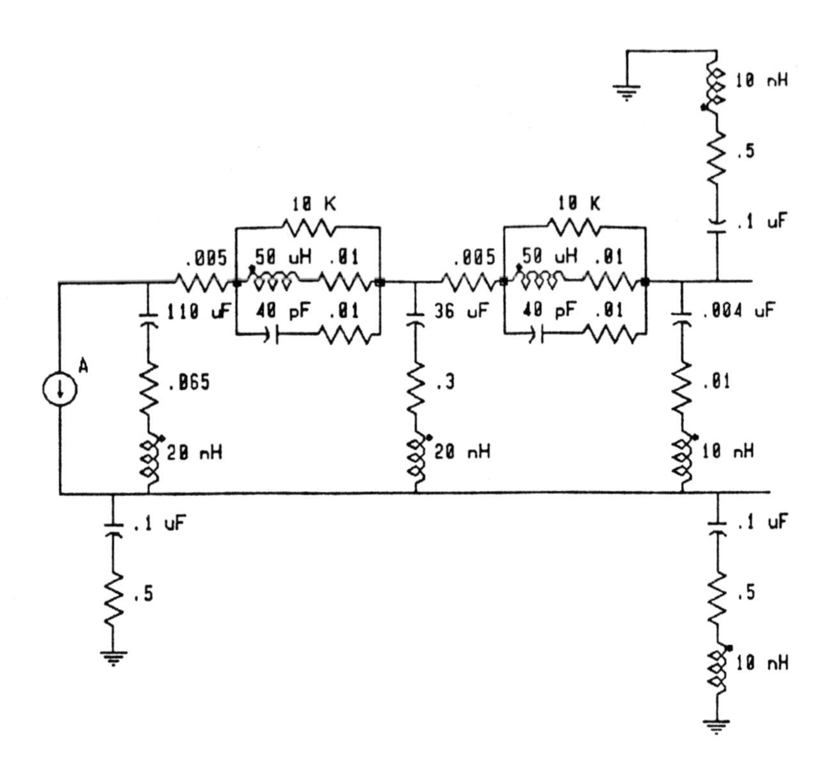

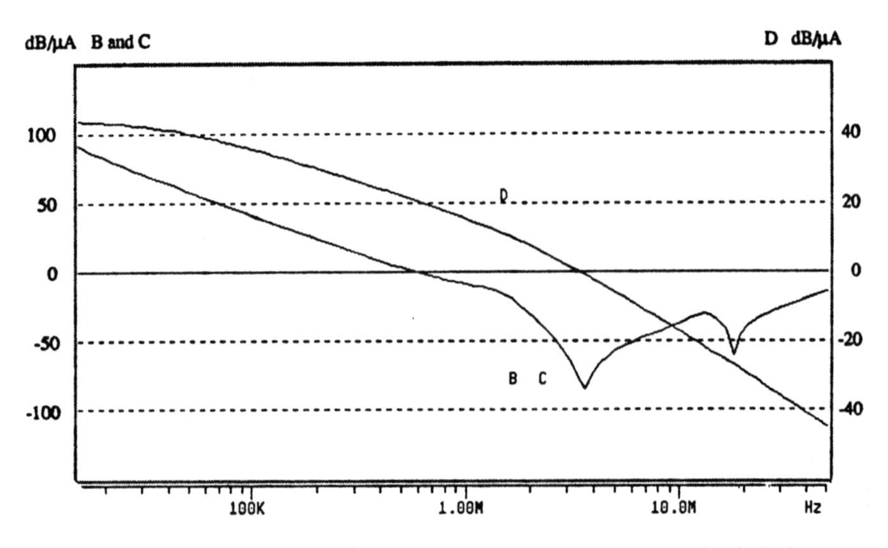

Figure 10.19 Model with the common mode components included.

pacitors are now included in the model. A 200-pF capacitance modeling the common mode capacitance of the switching electrode (FET drain) is also included. The filter performance is not, at this point, significantly affected by the addition of the common mode elements. The return line in this model is a short circuit from the EMI source to the test setup. This of course is not a good model for a return line. Also there is very little driving the common mode currents in the common mode path. There is a great temptation at this point to drive the common mode circuit with a common mode EMI spectrum in addition to the differential EMI spectrum generator as shown in Fig. 10.20. The problem here is that we are "double banging" the EMI source of energy.

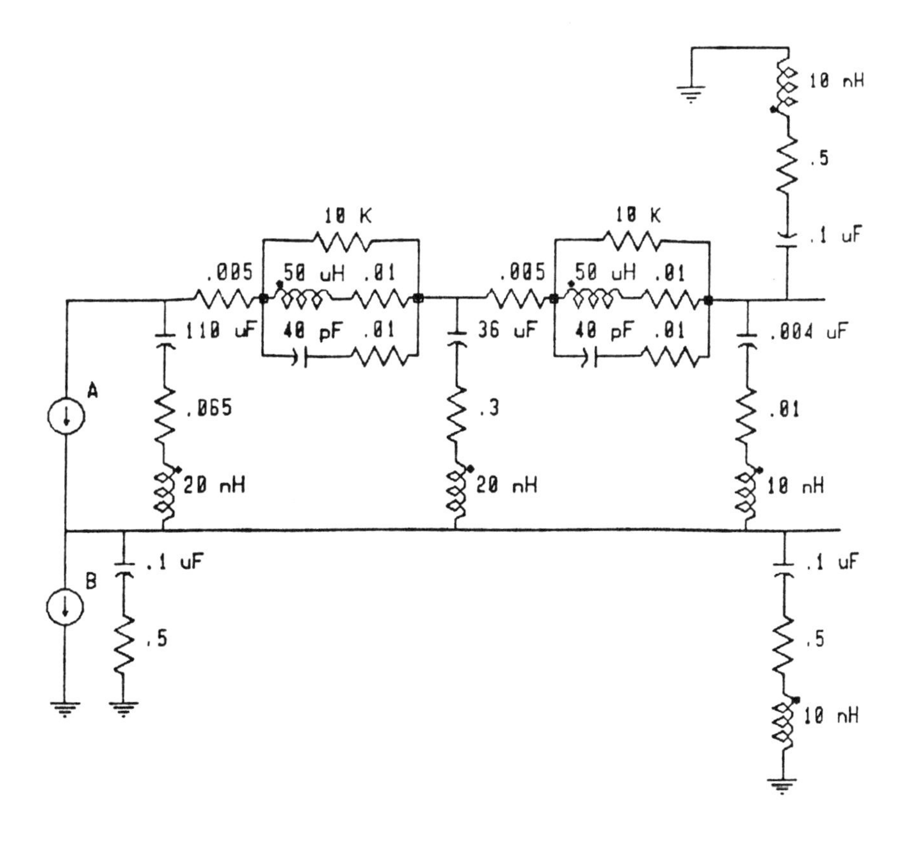

Figure 10.20 Tempted addition of common mode spectrum generator is inappropriate.

The addition of the parasitic elements of the circuit layout is needed for an accurate EMI simulation. The circuit board layout affects the conducted emissions of the circuit. With the circuit layout parasitics included as in Fig. 10.11, the differential spectrum generator drives currents through the power and return lines, which develops voltages in the common mode circuit. The resulting common mode currents pass through the measured common mode capacitance and the rest of the common mode filter elements. The resulting simulations are not perfect copies of the actual circuits but are close enough to be very useful in the design process.

The limitations of these models and simulations are detailed in chapter 14. If the limitations are always kept in mind when doing simulation, the shortcomings can usually be minimized so as to get useful results. Most of the elements of the models should and can be measured. The very few elements that cannot be measured accurately need to be tested in the model and compared to measured data to improve the accuracy. Fudging numbers is not acceptable, but at this level of modeling a little bit of carefully checked empirical data can be used to obtain useful results in the simulation process.

Information about the model can be found by changing a single element value and observing the difference in the resulting data. This can be a very useful tool to achieve a better understanding of the performance of the model, which can be extrapolated to the real circuit.

Laboratory experiments can then be performed to gain confidence in the information that comes from the simulation. In the simulation presented, there are so many elements that the changing of each element and observing the results would be an extremely time-consuming undertaking. We need to identify the major players (elements) at the frequencies of interest (15 kHz to 50 MHz) for reduction of design effort and for better understanding of the actual operation of the circuit.

10.8 IDENTIFY THE PLAYERS

A schematic of the model is presented in Fig. 10.21 with the inductor, and capacitor reactances at 100 kHz and 50 MHz listed by each reactive element. The resistances listed are those used in the model. For simplification at this point we don't worry about the fact that reactive components are orthogonal to resistive components. We will treat reactance and resistance equivalently as resistances, remembering that we would actually have to use complex algebra for precise calculations. We often neglect the phase angle caused by the small resistance of a capacitor in much the same way. Often we don't really care if the phase angle is exactly 90°; if the ESR is small compared to the reactance, it is neglected.

We can now create two simplified equivalent circuits at 100 kHz and 50 MHz by neglecting relatively small impedances in series branches and neglecting relatively large impedances in parallel branches. The resulting equivalent models at 100 kHz and 50 MHz (Fig. 10.22) consist of the elements that are the players in the circuit.

The resulting simplified circuit at 100 kHz (Fig 10.22A) consists of the L and R elements that were used in the simple hand calculation. This is no accident and should be expected. Remember that at frequencies lower than 2 MHz, the EMI problems

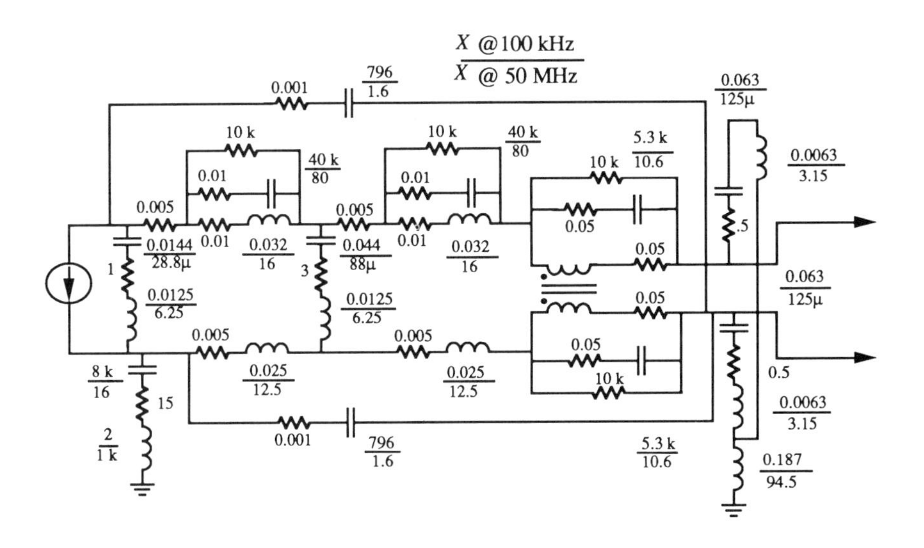

Figure 10.21 Model with reactances at 100 kHz and 50 MHz labeled.

are principally differential. The common mode elements show that the chassis ground path at 100 kHz is of relatively low impedance. The common mode elements do not significantly affect the filter attenuation at 100 kHz.The resulting simplified circuit at 50 MHz (Fig. 10.22B) is more complicated than for 100 kHz. This is not surprising since the number of features in the data curves are more complicated in the higher-frequency range. Some of the parasitic layout elements of the model along with component parasitics are

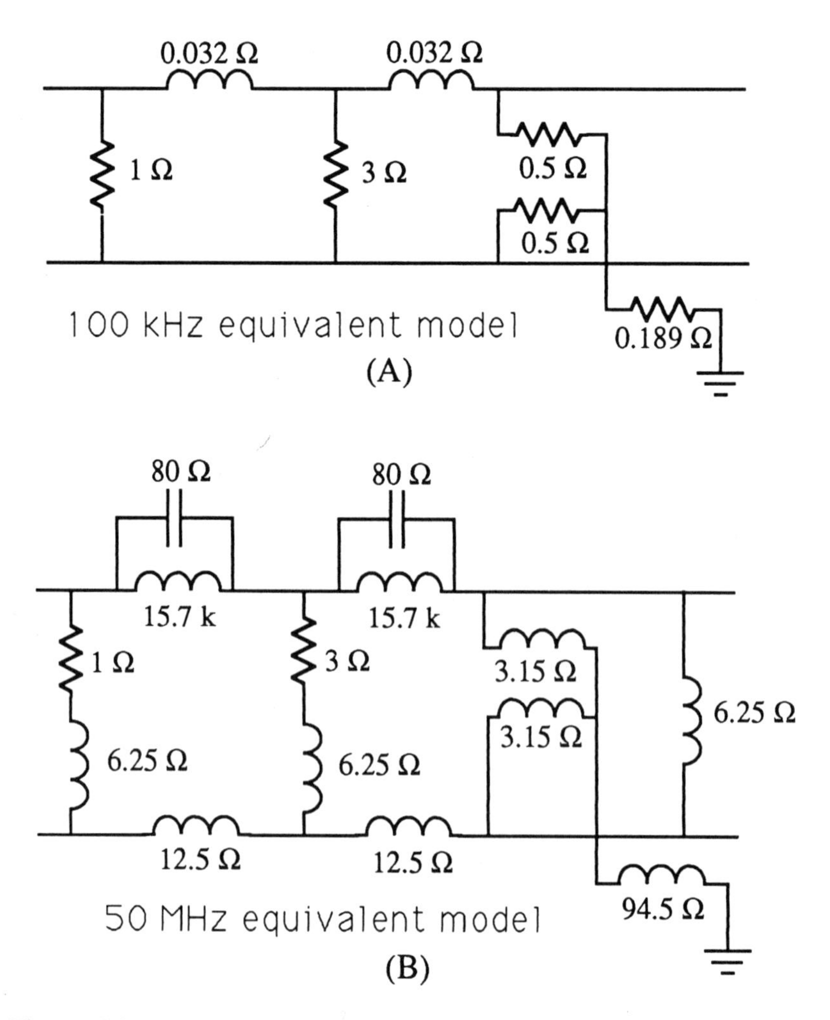

Figure 10.22 Reduced equivalent models at (A) 100 kHz and (B) 50 MHz.

now players in the filter attenuation game. The inductive reactances of the capacitor leads and layout are now comparable to the ESRs and are players at 50 MHz. The reactance of the input to output capacitance of the 50-μH inductors is now comparable to the reactance of the 50-μH inductors. A simple "by hand calculation" approximation at 50 MHz would now result in a voltage divider ladder consisting of the reactance of the input to output capacitance in place of the inductor reactance in the denominator and the reactance of the inductance of the capacitors in place of the ESRs. The filter inductors and capacitor ESRs dominate the filter performance at 100 kHz. The inductor input to output capacitance and capacitor equivalent series inductance differentially dominate the filter performance at 50 MHz.

The high impedance path to chassis from the chassis capacitors is evident in this simplified model. The reactance of the path to chassis is 94.5 Ω. The ability of the chassis capacitors to attenuate high-frequency common mode harmonics is greatly reduced by the poor layout.

Another informative view of the circuit can be created by using the model schematic and labeling the possible *LC* tank combinations with the *LC* characteristic (or driving) impedance and the resonant frequency of each tank as is done in Fig. 10.23. In any particular frequency range the *LC* tank of lowest impedance in that frequency range will dominate performance. This view enables the designer to find the dominate *LC* tanks and the accompanying resonant effects relative to the circuit.

The conducted emissions at the switching frequency are more than the original simulation predicts. As a result of inspection of Fig. 10.22A (equivalent model at 100 kHz) we can now predict that the ESRs of the original 110– and 36–μF capacitor models are not appropriate. With the ESRs modeled as 2 and 4 Ω respectively, the resulting simulation data are presented in Fig. 10.24. This data compares well to the measured spectrum analyzer data of Fig. 10.6. The ESRs can be changed to obtain any attenuation desired at 100 kHz in the simulation without affecting the rest of the data curve significantly. This new model, when exorcized, will give usable results. With the input to output capacitance of the layout removed and the impedance path to the chassis for the chassis capacitors reduced, the simulation results as shown in Fig. 10.25. The simulation now easily passes the CE03 conducted emissions test.

A layout parasitic was ignored for this example but must be mentioned here for completeness. In some cases, the capacitance

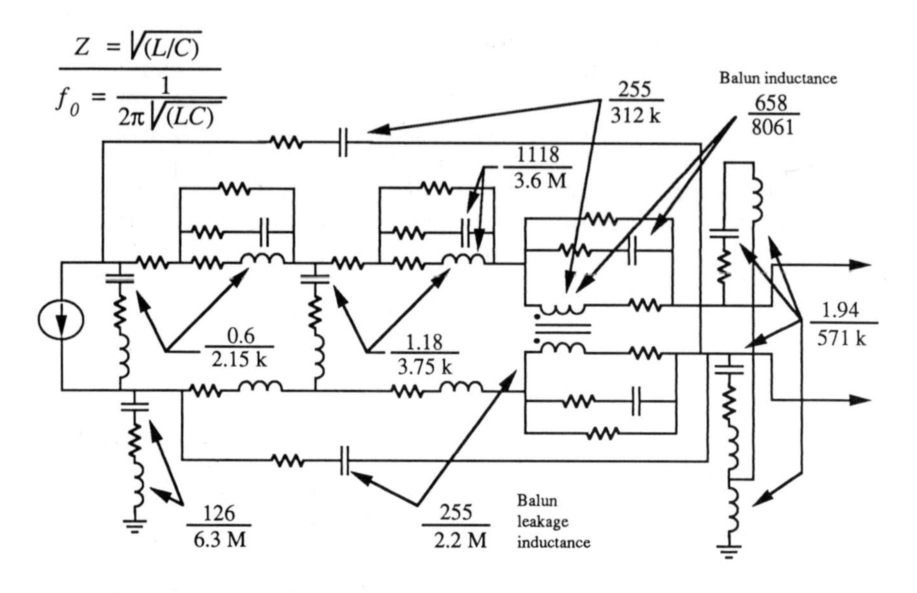

Figure 10.23 Model schematic with LC tank characteristic impedance and resonant frequencies labeled.

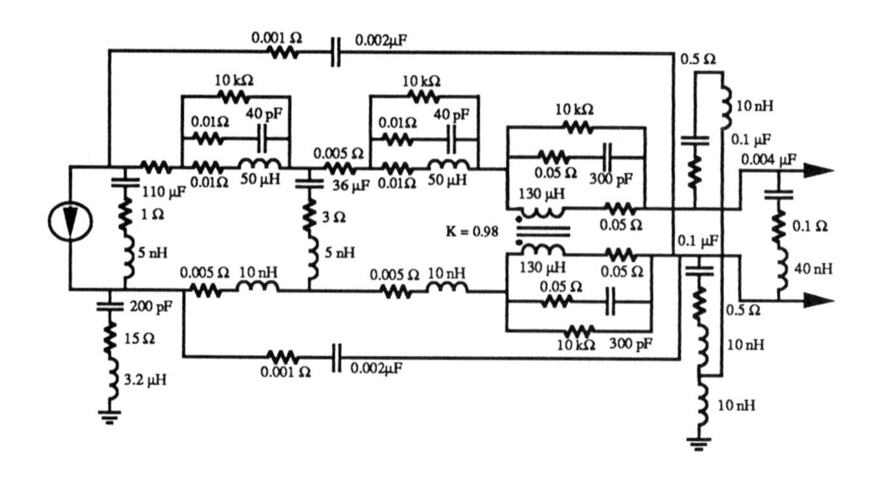

Figure 10.24 (A) Relatively accurate model.

from adjacent power and return lines is significant to circuit EMI performance and must be part of the model. In this simulation, the inductance and resistance of the PWB traces were included in the model, and at 50 MHz the inductance of the traces was a player. The model in Fig. 10.26A includes many 20-pF capacitances with series 5 Ω resistances to model the trace capacitances formed by

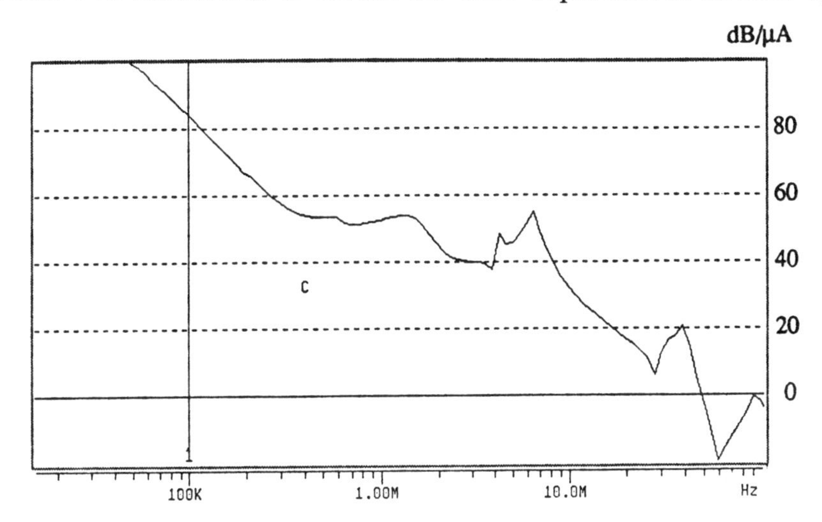

Figure 10.24 (B) Relatively accurate model simulation data.

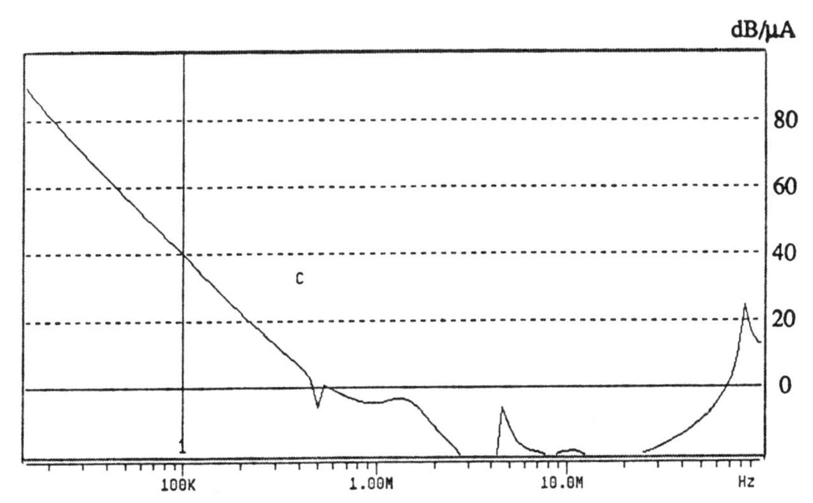

Figure 10.25 Simulation results with input to output capacitance and ground path impedance reduced.

adjacent traces with the glass epoxy acting as a dielectric between the traces. The additional elements make the model appear a little messy and were excluded for clarity. The simulation data with and without the trace capacitances modeled can be compared in Fig. 10.26 B and C, respectively. The difference between the two simulation data plots is insignificant. The only slight differences are apparent in the frequency range around 30 MHz. A review of Fig. 10.2 (typical EMI problem areas) and consideration of the analysis performed in this chapter give strong understanding of the conducted emissions problems in SMPS design.

10.9 OTHER TYPES OF EMI MODELING FOR SMPS

Output EMI simulation

EMI spectrum generators can be applied to the output filters so as to simulate the EMI threat to loads connected to the power supply. The output lines are the path for the conducted EMI emissions from the internal power supply switching. The victim in this case is the computer, but it is modeled and tested as a resistive load.

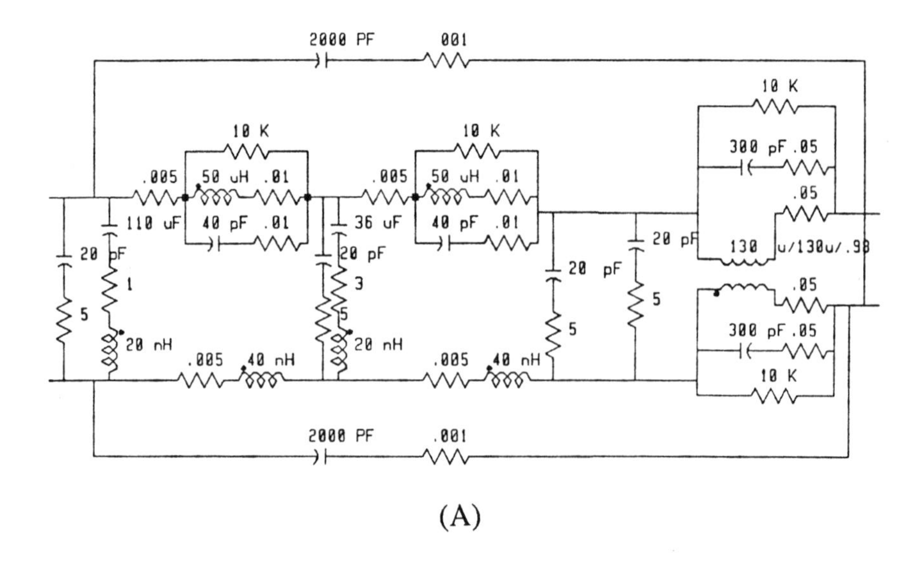

(A)

Figure 10.26 (A) Simulation model with PWB capacitances.

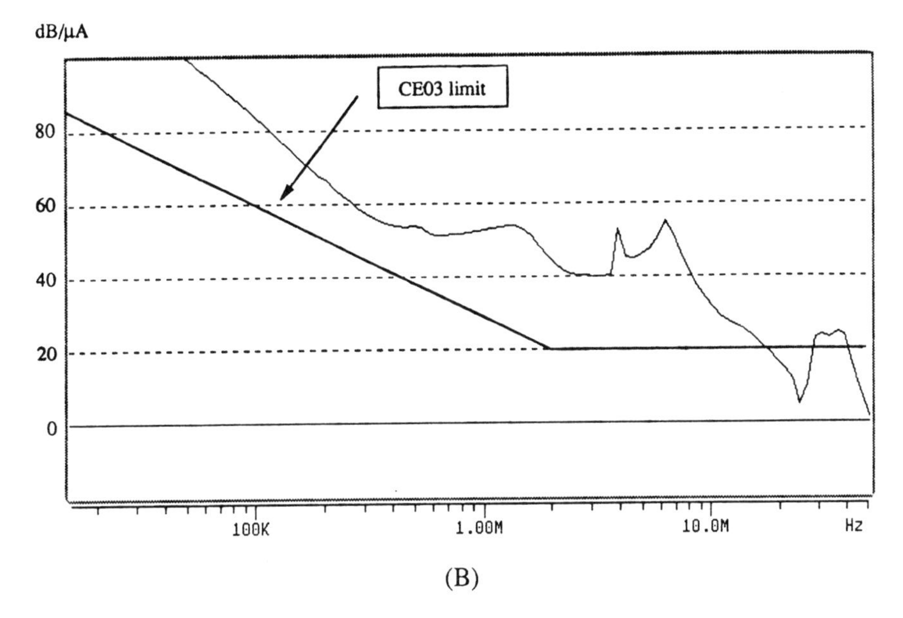

(B)

Figure 10.26 (B) Simulation data for model with PWB capacitance included.

The spectrum generator output is the frequency domain representation of the waveform at the cathode of the output rectifier diode cathodes. The capacitor C_{cm} and resistor R_{cm} model the common mode capacitance created by the cathode connected diode cases and the chassis connected heat sink separated by a thin insulator. An output EMI simulation is presented in chapter 11.

Power supply output EMI simulation is similar to power supply input EMI simulation with the major difference that the EMI threat is a current instead of a voltage waveform (as in output EMI simulation). The spectrum generator output is the frequency domain characterization of a current waveform.

The input current waveform is often a trapezoid instead of a rectangle as in the output waveform. The frequency domain spectrum for the trapezoid is the same as the rectangle except that the amplitude, A, is the average of amplitude during the ON time of the power switch.

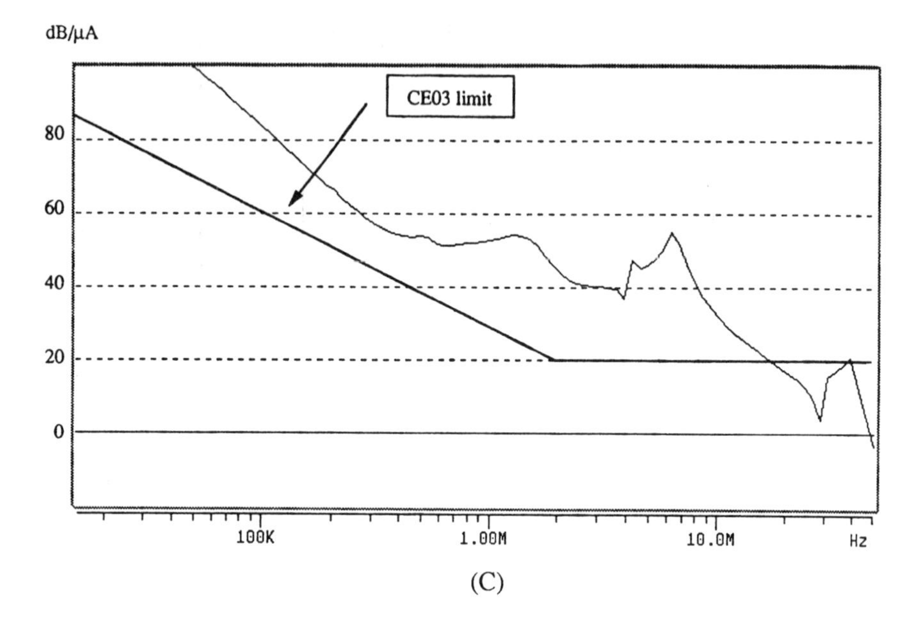

(C)

Figure 10.26 (C) Simulation data for model without PWB capacitance.

AC/DC Power Supplies

The analysis can be generally done in the same manner for ac or for dc circuits. The input rectifier diodes in 60- or 400-Hz ac input circuits change conduction states, which causes the natural response of the resonant frequencies of the circuit and layout to be stimulated. When the diodes switch conduction states, a step change in current is introduced. The repetition rate of this "diode switching EMI" is very slow when compared to the switching frequency of a typical power supply and/or the frequency range of CE03. Because of this low repetition rate, the diode switching EMI is generally broadband. The analysis described previously only addresses narrowband conducted emissions. When good layout and design practices are used in the ac/dc rectifier filter portion of the ac/dc power supply, the broadband EMI will generally not exceed the CE03 broadband limits. The fundamental power supply switching EMI is narrowband noise because of its repetitive nature, as previously mentioned. The diode switching EMI is an important concern for Mil-Std-461 CE01.

CE01 Low-frequency Conducted Emissions

The frequency span for which Mil-Std-461 CE01 applies is 30 Hz to 15 kHz. The CE01 requirements are generally applied to ac input lines, but they can also be applied to dc and interconnecting control leads as well. The fundamental line frequency component amplitude is usually not limited and is generally exempt from the requirement. The harmonics of the line power waveform must not exceed the limits. CE01 is actually a harmonic distortion requirement. A CE01 simulation is included in chapter 14.

The intention of the requirement is that the power to electronic equipment will be transmitted at the fundamental frequency. Harmonic distortion of line power waveforms can cause the flattening of the sinusoidal waveform peaks and poor power factors. The power factor can be compensated for pure sinusoids, but the harmonics components' power factor will not be improved. Some types of electronic equipment depend upon the integrity of the waveform. An example is aircraft generators that peak detect to regulate the ac line power outputs.

10.10 CONCLUSION

The maximum amplitude limits for Mil-Std-461 CE03 are plotted in these simulations. The amplitude of the EMI currents must not exceed the limits. The same type of simulation could just as easily be done for FCC, VDE, or any other standards. The model of the test setup and the plotted limits can be changed appropriately for any standard. This described simulation process can be used in any design requirements environment and can be successful in predicting the ability of an electronic circuit design to meet its EMI specification requirements. An electronic circuit assembly can be designed (early in the design process) to meet the conducted emissions requirements. Designing to meet conducted emissions requirements and effective packaging methods are the first steps in meeting all of the EMI requirements in any electronic assembly in which the greatest EMI threat is internal.

QUESTIONS

10.1 What is the major internal source of EMI in a switching power supply? Why?

10.2 Why are the typical differential problems below 2 MHz and the typical common mode problems above 2 MHz?

11

TRANSISTOR AND DIODE PACKAGING PROBLEM FOR EMI

This chapter presents one of the main problems causing conducted EMI in transistor and diode packages and their mounting. This is a good example of the whole process of analysis, verification, and correlation. The results of analysis and tests are used to greatly improve the EMC aspects of design early in the design cycle. It was found that traditional methods of transistor and diode package mounting were creating a great deal of EMI. This EMI can be greatly reduced by simple design changes, resulting in the use of smaller filters which facilitate better overall designs. The design in this example did not meet EMI requirements with the TO-3 transistor package and DO-4 diode package; by changing the package of the power transistors and rectifier diodes, the design was improved to meet the requirements.

This chapter will define the problem of the common mode current paths, develop ways to quantify the problem, and show ways to reduce the EMI currents and resulting voltages in order to design to meet military EMI standards.

11.1 NEW SEMICONDUCTOR DEVICE PACKAGES

The new TO-247 plastic and TO-254 hermetic packages offer many great advantages over the old packaging technologies. Mechanically they are much easier to install. The transistor is electrically isolated from the case so that it may be mounted directly to a heat sink without the need for an insulator. In the old TO-3 packages the transistor chip was mounted directly to the case, so an insulator was required for mounting to heat sinks. The transistor chip in the new packages is fabricated on an electrically insulating and thermally conductive substrate. The chip to package capacitance has been measured to be approximately 5 pF on both the TO-247 and TO-254 packages. The common mode capacitance caused by the TO-3 transistor packaging method is approximately 130 pF.

For all new designs the new TO-247 and TO-254 packages are highly recommended. The advantages are greatly reduced manufacturing costs because of the ease of installation, the opportunity for better circuit layout because all of the electrodes come out on the same (and convenient) side of the part, and the ability to reduce the common mode current injection into the chassis.

11.2 COMMON MODE SHORTING SCREENS

In Mark J. Nave's article *Switched Mode Power Supply Noise: Common Mode Emissions* (*PCIM*, May 1986) a method is presented to improve the mounting for TO-3 packages by the use of a shorting screen. Shorting screens are now available from the Chomerics company. The screen is two TO-3 insulators laminated with a copper sheet in between them with an external contact to connect the copper screen to circuit ground. This is acceptable for some existing designs on retrofit or new designs in which only TO-3 packages can be used, but now the common mode current is conducted into the circuit ground. If the circuit ground layout is done so that this extra current won't cause voltage potentials causing an upset, this Band-aid fix can be acceptable.

11.3 TYPICAL SYSTEM WITH POWER CONVERSION

In chapter 10 a block diagram (Fig. 10.3) was shown of a typical system that includes a motor generator, power conversion equipment, and a computer. The processes that occur in this system are basic to all systems, from a complete naval ship with many communications, navigation, and ordnance systems to a single circuit board in a subassembly. The basic elements (*LC* input filter, power transformer, switch, rectifiers, output *LC* filter, and feed-through capacitors) of a generic power supply are shown. The fundamental problems are basically the same for all topologies. Within computers, with fast rise times and low-impedance clock circuits, common mode current injection is a problem that is similar to the system-level problem.

C_{cmp} is the common mode capacitance between the switch element (the drain or collector of a transistor) and the chassis. Often this is the case of a TO-3 package to a heat sink. Usually a thin electrical insulator of a few mils thickness is installed between the case and heat sink to electrically isolate them but to allow the heat to conduct from the transistor to the heat sink. This mounting creates a capacitor that is a low impedance to the fast voltage rise and fall times.

C_{cms} is the common mode capacitance on the secondary side of the power and isolation transformer. This capacitance is created typically by mounting a DO-4 or DO-5 or similar stud mount type diodes for rectifiers. Once again there is usually a thin electrical insulator mounted between the diode body and the heat sink that creates a common mode capacitor to the chassis.

C_{iso} is a lumped model of the primary to secondary winding capacitance. The tighter the winding coupling from input to output to reduce leakage inductance, the more input to output capacitance is created. This capacitance (C_{iso}) reduces the input to output isolation at high frequencies.

C_{ftp} and C_{fts} are the feed-through capacitors on the primary and secondary sides, respectively.

11.4 COMMON MODE CURRENT PATHS

A typical waveform for the drain or collector of a switcher is shown in Fig. 11.1. A simplified model of a primary side common mode current path is presented in Fig. 11.2. During the OFF time the common mode capacitor is charged through the chassis ground system, the source generator, the winding capacitance of L_p, and the primary of the power transformer. When the switch is turned ON, the common mode capacitor C_{cmp} is discharged through the power switch, the circuit ground return, and the chassis ground system. This injected current at the rate of $i = C(\delta v/\delta t)$ causes voltage gradients on the grounds. The grounds are supposed to be zero potential planes. The capacitance $C_{cmp} = \sigma A/d$, where σ is the permittivity of free space ($8.855e^{-12}$), A is the area of the electrodes in square meters, and d is the distance between the electrodes in meters. For a TO-3 package (area approximately 1.5 in^2) mounted on a heat sink with an insulator 0.006-in thick and dielectric constant of 2.3, the common mode capacitance created is approximately 130 pF. For a 100-ns rise time on a 280 V input

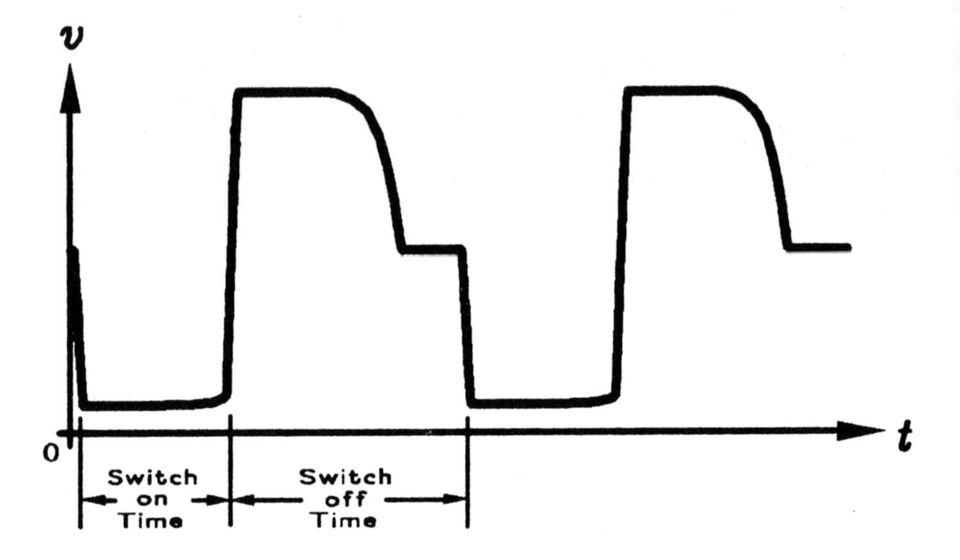

Figure 11.1 Typical switching waveform.

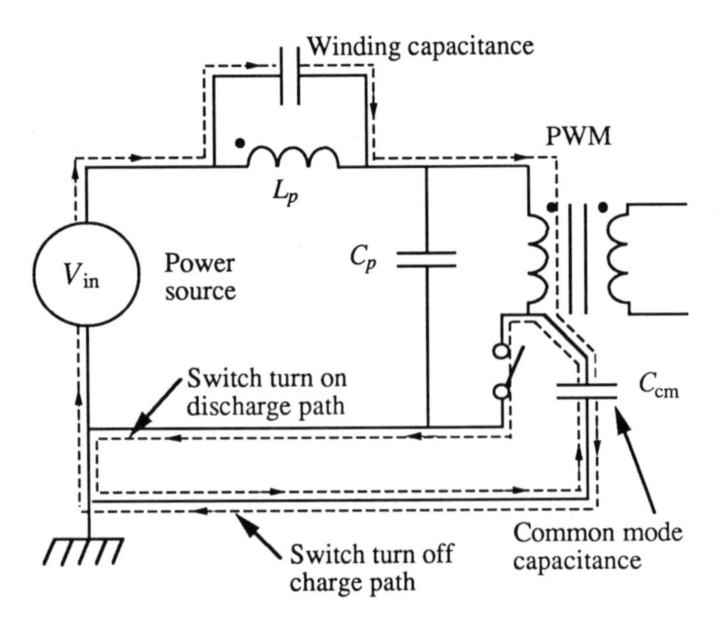

Figure 11.2 Common mode current paths.

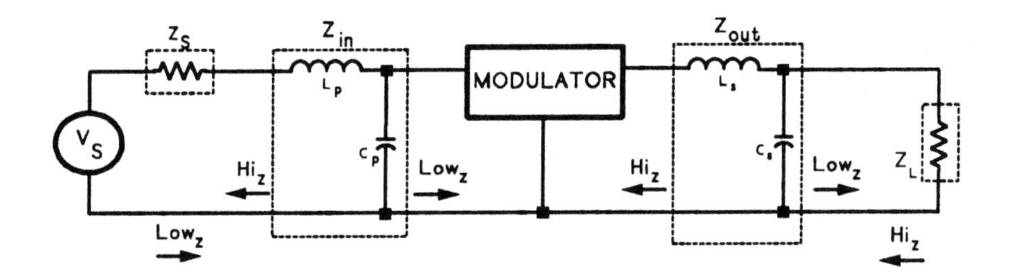

Figure 11.3 Differential mode.

convertor, a 130 pF capacitance gives a peak common mode EMI current of 364 mA or 111 dB above 1 μA.

A basic block diagram for a differential model of a power source, switching power convertor, and load is presented in Fig. 11.3. The modulator consists of a pulse width modulated switch (PWM), control circuit, and input and output LC filters. The functions of the input *LC* are smoothing or averaging input voltage, providing hold-up time, differential EMI filtering, and local energy storage. The output filter is for averaging the PWM waveform and attenuating the differential EMI to an acceptable level.

Common mode filters are added to Fig. 11.3 and are shown in Fig. 11.4. Common mode currents flow in the same direction through both the power and return (circuit ground) lines and flow back through the chassis ground. This leads to the common mode

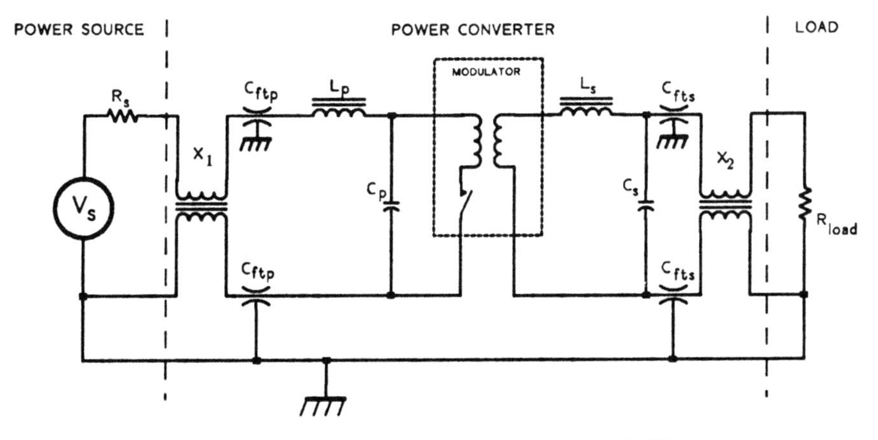

Figure 11.4 Generic differential and common mode filter system.

model presented in Fig. 11.5 where the differential elements are removed and the common mode capacitances of circuit packaging are included. The EMI common mode sources (electrodes that change voltage at fast rates causing $i_{cm} = C(\delta v/\delta t)$) are shown as voltage sources. The transistor switch drain drives current through C_{cmp} with the series combination of C_{iso} and C_{cms} in parallel with C_{cmp} and the rectifiers cathode drives C_{cms} with the series combination of C_{iso} and C_{cmp} in parallel with C_{cms}. Electrically conductive thermal planes grounded to the chassis and mounted to printed circuit boards can create common mode capacitances from all of the traces to the chassis ground. For the removal of heat this works well, but large currents can be injected into the chassis, causing failures in EMI testing and circuit upsets in actual operation of the unit under consideration. Common mode current injection can also be caused by mounting printed circuit boards too close to the chassis. The feed-through capacitors act to bypass common mode current by creating a low-impedance circulation path within the convertor. The balun acts to increase the impedance of the common mode path going outside of the convertor. The diagram in Fig. 11.6 shows that the common mode current problem is the same for output circuitry as for input circuitry. We are willing to circulate a little more common mode current locally within the power supply chassis to avoid the flow of common mode current outside of the chassis. The new TO-247 and TO-254 hermetic packages and TO-3 packages are diagrammed in Fig. 11.7.

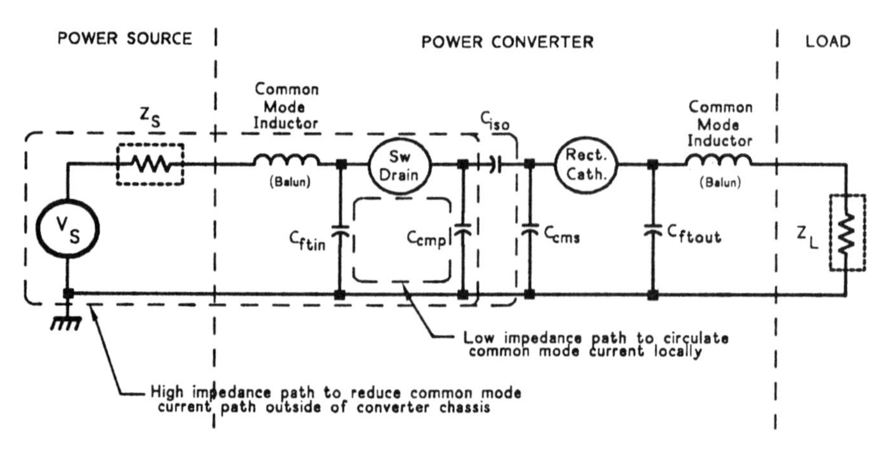

Figure 11.5 Common mode current path of input circuitry.

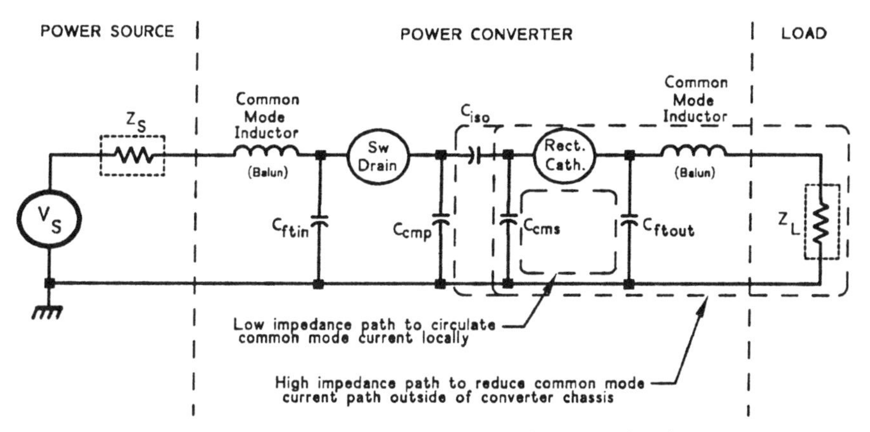

Figure 11.6 Common mode path of output circuitry.

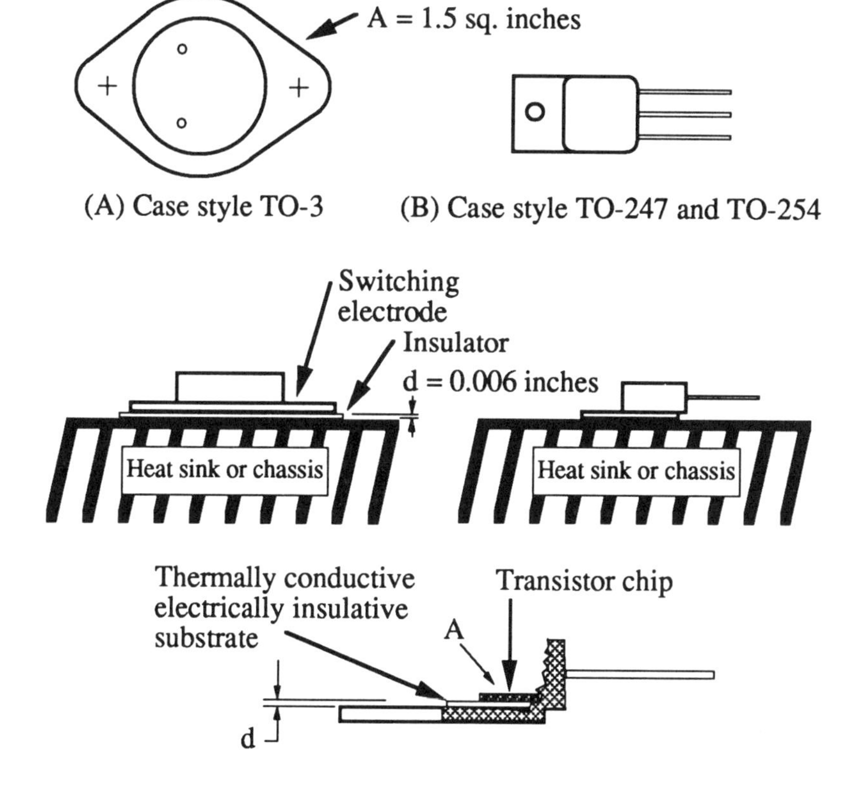

Figure 11.7 TO-3 versus TO-247 and TO-254 package comparison.

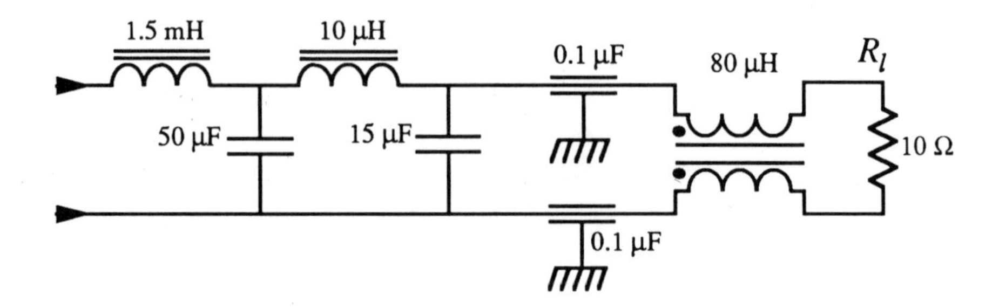

Figure 11.8 Typical differential and common mode output filter.

The schematic in Fig. 11.8 is a typical differential and common mode output filter and load. A model for a simulation to compare the EMI performance of the TO-3, DO-4, TO-247, and TO-254 packages is presented in Fig. 11.9. The model includes an EMI spectrum generator, model of the filter in Fig. 11.8, model of the Mil-Std-462 CE03 test setup, and load. The simulation results are presented in Figs. 11.10 and 11.11.

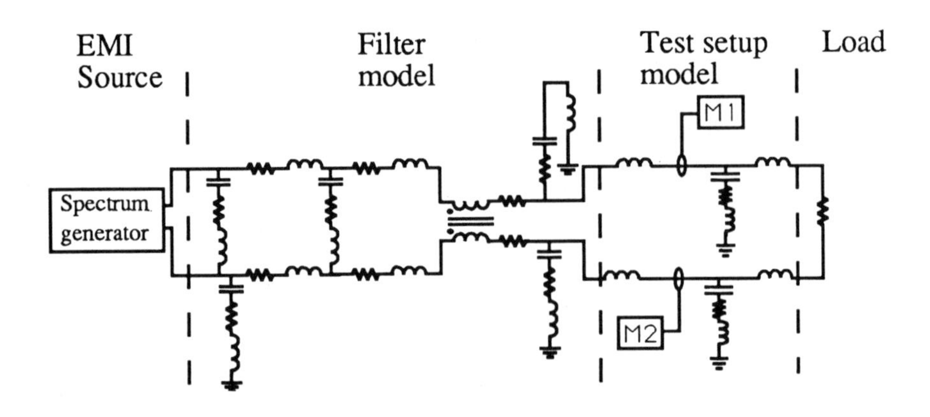

Figure 11.9 EMI test configuration model.

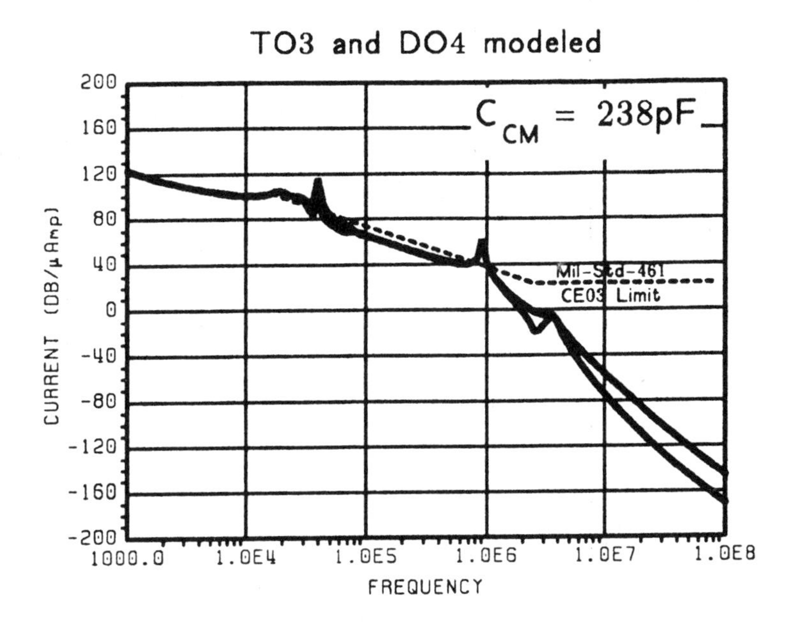

Figure 11.10 Similation data for design with excessive packaging parasitics.

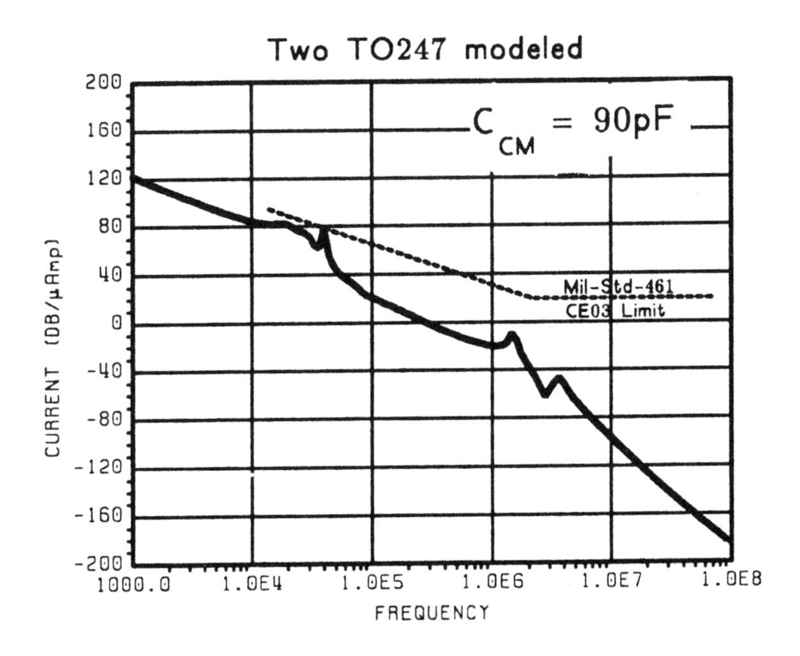

Figure 11.11 Similation data for design without excessive packaging parasitics.

11.5 CONDUCTED EMISSIONS REDUCTION BY CHOICE OF PACKAGE

The graphs in Figs. 11.10 and 11.11 compare the TO-3 package EMI performance to the improvement in performance using the new TO-247 or TO-254 packages. The simulated data for the circuit modeled with TO-3 and DO-4 packages fails Mil-Std-461 (Fig. 11.10). The same circuit, transistor, and diode (but modeled with TO-247 packages), now easily passes the EMI test (Fig. 11.11).

QUESTIONS

11.1 What are the physical aspects of component packages that are important to consider when designing a circuit that must meet EMI standards?

11.2 What are the important physical features of diode and transistor packages that affect EMC performance?

11.3 Why do we prefer to make the common mode conduction paths as small (localized) as possible?

11.4 In a power supply, what are the differences in impedance matching for power transfer from input to output and the desired matching for EMI attenuation?

12

CIRCUIT EXAMPLES

12.1 EXAMPLE 1

Problem

Bus voltage on aircraft exceeds specified maximum voltage limit.

During flight testing, high main bus voltages were observed. Amplitudes as high as 122 V were recorded by the flight engineer on one of the power buses. The maximum specified voltage is only 120 V maximum. When the radar was powered but not operating, the effect was observed. By isolating loads, it was found that the radar system contributes to this high voltage by as much as 2 V. The radar system consumes only 1240 W.

Solution

The 400 Hz aircraft generator feedback scheme regulates peak voltage. The peak voltage must be proportional to the RMS voltage. The success of this scheme depends upon low harmonic distortion. The voltage peaks of the aircraft's 400-Hz generator are flattened by high-current pulses going into the radar. The aircraft generator feedback compensates for this by increasing the voltage peak to the nominal value. The RMS voltage of the 400-Hz sinusoidal voltage with a flattened peak becomes higher than designed for. The input current to the radar was measured with an oscilloscope. Current peaks of 41 A were measured with very low conduction angles.

Three possible solutions were considered:

1. Change the feedback control of aircraft generator to regulate proportionally to RMS voltage rather than peak voltage.

2. Add an inductor to the radar to spread the input conduction angle.
3. Use a notch filter to remove the third-harmonic. Third harmonic distortion causes the flat top on the voltage waveform.

Changing the feedback control of an aircraft generator would be expensive. Also many of the aircraft of this fleet do not use the offending type of radar. This solution was abandoned.

Solutions 2 and 3 are filters designed to change the spectrum of the power waveform. Simulation was performed to get insight into the effect of the small conduction angle on the spectrum of the power conversion waveform. Two simple high peak current waveforms (Fig. 12.1) were run for comparison on the FFT program. The average currents of each waveform are equal, but one waveform has twice the peak current and half of the conduction angle of the other waveform. The linear versus linear frequency spectrum plots of Fig. 12.1 are presented in Figs. 12.2 and 12.3. The harmonic amplitudes of the small conduction angle waveform are much larger than for the larger conduction angle waveform. The third harmonic in Fig. 12.3 is much smaller than the fundamental. The fifth harmonic is almost nonexistent. Log versus log frequency domain plots are shown in Figs. 12.4 and 12.5. The envelope of harmonic amplitudes drops at 20 dB per decade for both plots, but the plot with the larger conduction angle (Fig. 12.5) turns over at a much lower frequency than does the one in Fig. 12.4.

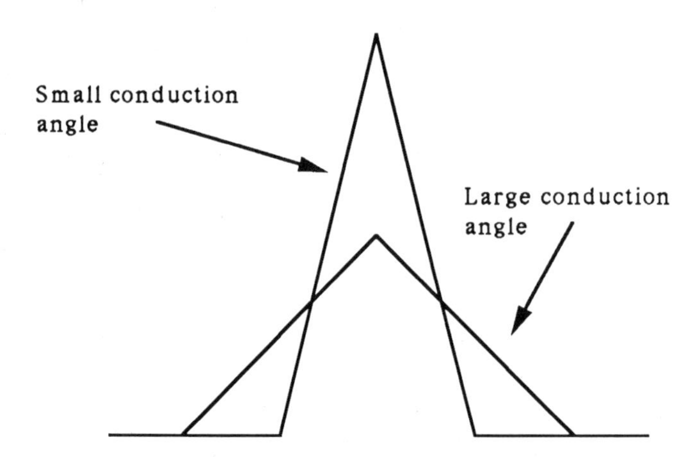

Figure 12.1 Comparison of large conduction angle to small conduction angle.

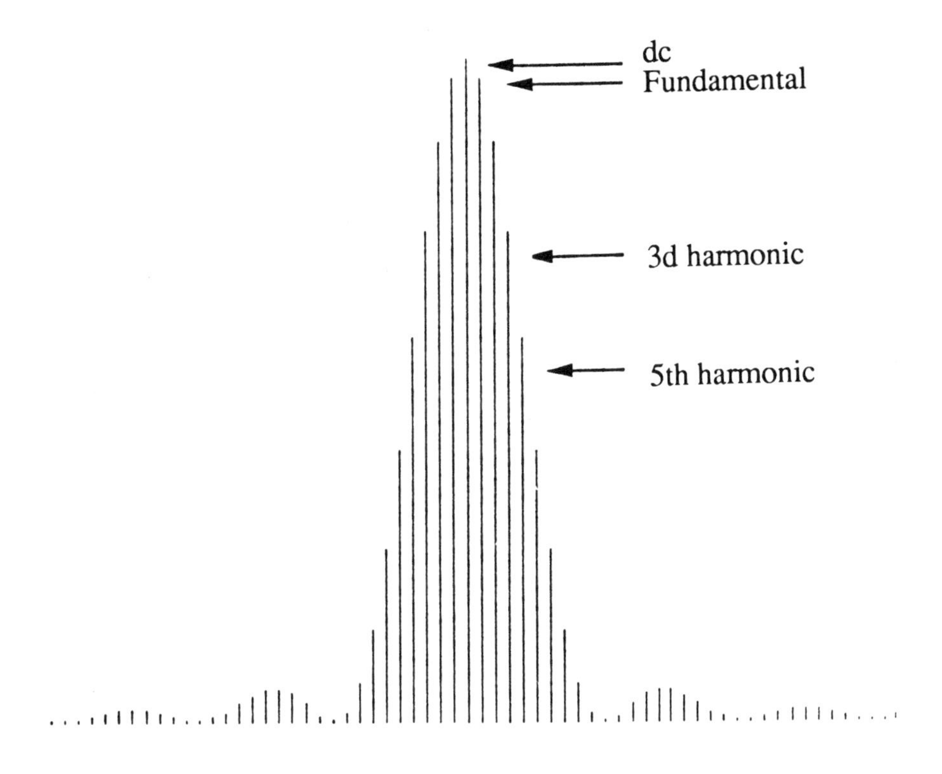

Figure 12.2 Small conduction angle.

In Solution 2 an inductor added to the radar unit would spread the conduction angle. The radar unit is a purchased item. Modification of a standard radar unit that functions well in other applications would be very expensive and would also require recertification. Also the addition of an inductor to the radar unit would lower the dc voltage so that the whole power input portion of the radar would have to be redesigned for lower dc voltage input. This solution is abandoned for these reasons.

Solution 3 (the only one left) is now looking more promising. We proposed to reduce the third harmonic with a notch filter. The first step to verify the proposed modification is to simulate the problem circuit.

A model of the existing aircraft generator and radar input is presented in Fig. 12.6. The ECA program to simulate the circuit operation is shown in Fig. 12.7. The input current waveform (Fig. 12.8) of the simulation is 41 A current pulses, the same as was mea-

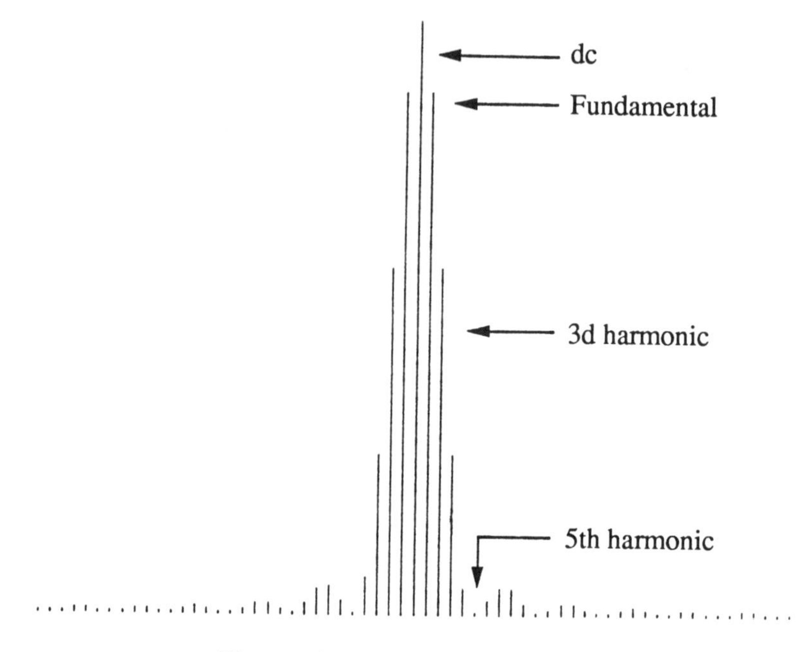

Figure 12.3 Large conduction angle.

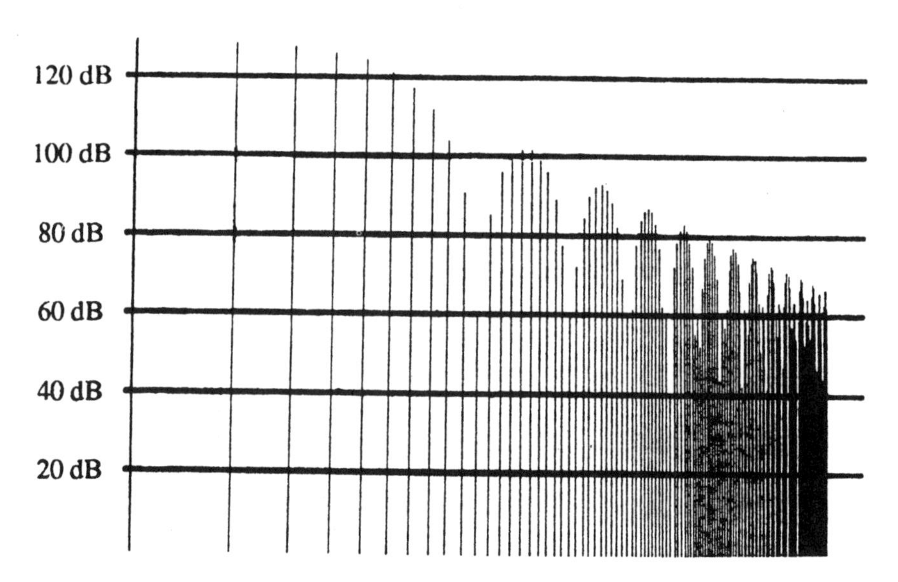

Figure 12.4 Small conduction angle (plotted log versus log scale).

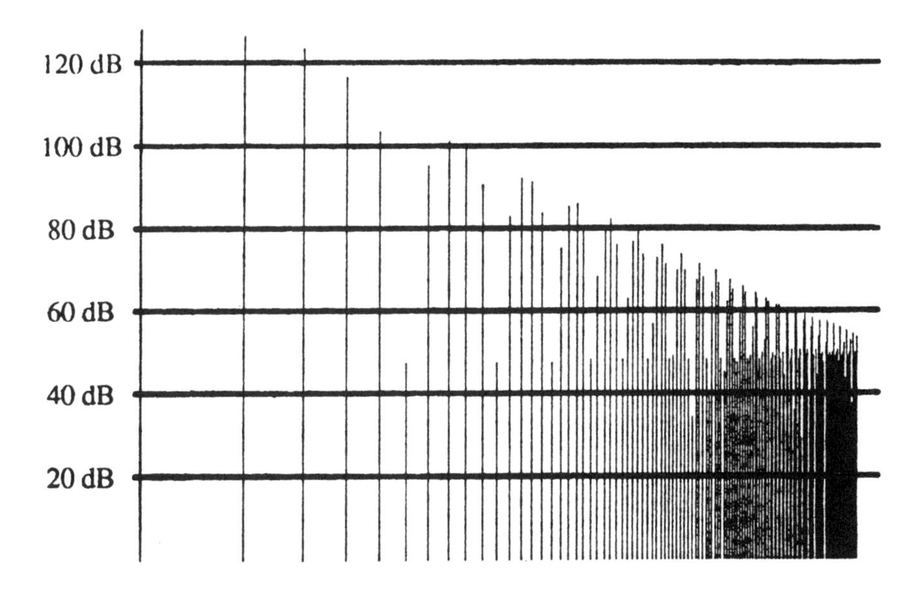

Figure 12.5 Large conduction angle (plotted log versus log scale).

sured in the aircraft with a high bus voltage. The conduction period is short because the input to the weather radar is a large capacitance. The rectifier diodes are forward biased for a short period of time. When the diodes are forward biased, there is a very low-impedance path to the capacitors. If the input of the radar unit were an *LC* filter, the inductor would provide a high input impedance and store and return energy each cycle to spread the conduction angle over a longer period of time. The simulation input voltage peaks are flattened (Fig. 12.9) and appear similar to the aircraft bus voltage when the radar is operating.

The third harmonic shown in Fig. 12.10 tends to flatten the peaks of the fundamental. The fifth harmonic in Fig. 12.11 reinforces the peaks of the fundamental. A filter designed to remove the third harmonic should reduce the amount of flattening of the input voltage waveform. The filter *Q* should be fairly high so as to reduce the circulating current at higher harmonics. Circulating currents in the notch filter cause unnecessary dissipation.

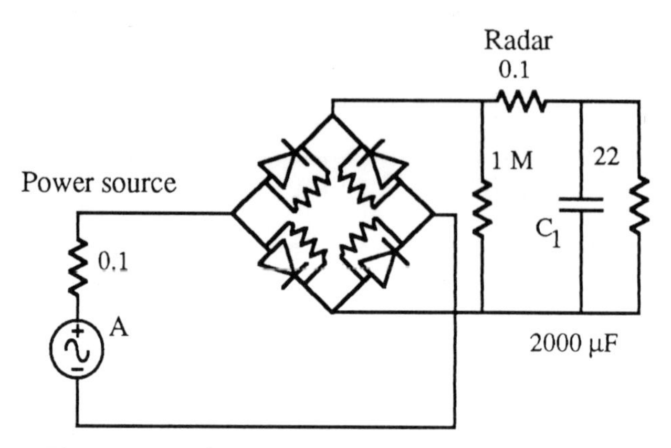

Figure 12.6 Model of radar unit power input circuit.

The notch filter in Fig. 12.12 is designed to resonate at 1200 Hz (third harmonic in 400-Hz system). The insertion loss curve for the notch filter is shown in Fig. 12.13. At resonance, the filter has its maximum impedance. The insertion loss at 1200 Hz is 28 dB.

branch	label	nodes		value
1	Vin	0	0	1.
2	R8	1	0	0.2
3	L1	1	4	1.E-15
4	C2	1	3	100.E-18
5	R9	3	4	1.G
6	L2	4	5	1.E-15
7	C3	5	6	100.E-18
8	R10	6	0	1.G
9	D1	4	7	10.p
10	D2	0	7	10.p
11	D3	8	0	10.p
12	D4	8	4	10.p
13	R1	4	7	10.K
14	R2	7	0	10.K
15	R3	8	0	10.K
16	R4	8	4	10.K
17	R6	7	9	0.1
18	C1	9	8	0.002
19	R7	9	8	22.
20	R5	7	8	100.K

Figure 12.7 ECA program listing.

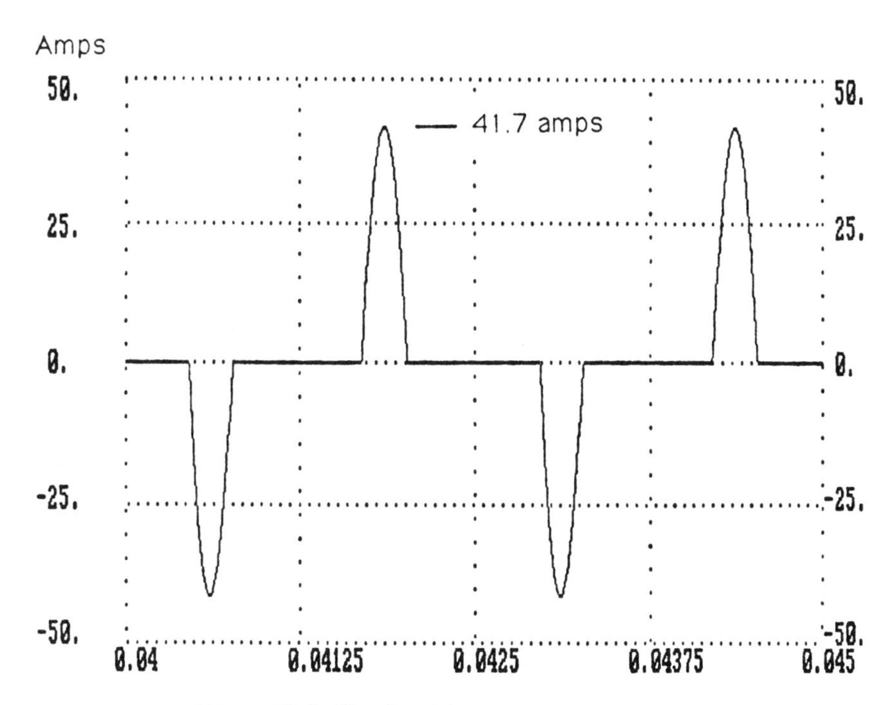

Figure 12.8 Simulated input current waveform.

The notch filter is added to the model in Fig. 12.6, creating the model shown in Fig. 12.14. The notch filter is an assembly that interfaces the radar unit to the input power. The ECA program for the simulation is presented in Fig. 12.15. With the notch filter installed, the input current waveform (Fig. 12.16) peaks are 23 A. The radar unit input current conduction angle is much larger with the notch filter installed (compare to Fig. 12.8). The radar unit input current peaks are reduced by 18 A. There is less peak flattening of the bus voltage plot in Fig. 12.17. Without the notch filter the peak voltage is only 157 V. With the notch filter installed, the voltage peak is raised to 161.58 V, much closer to the nominal 165 V. The simulated aircraft generator source current, with the notch filter installed, is shown in Fig. 12.18.

A breadboard of the notch filter was built and two tests were performed. The test setup schematic for the insertion loss test is shown in Fig. 12.19. The insertion loss test data are presented in Fig. 12.20. The data compares well to the data predicted in the notch filter insertion loss simulation (Fig. 12.13).

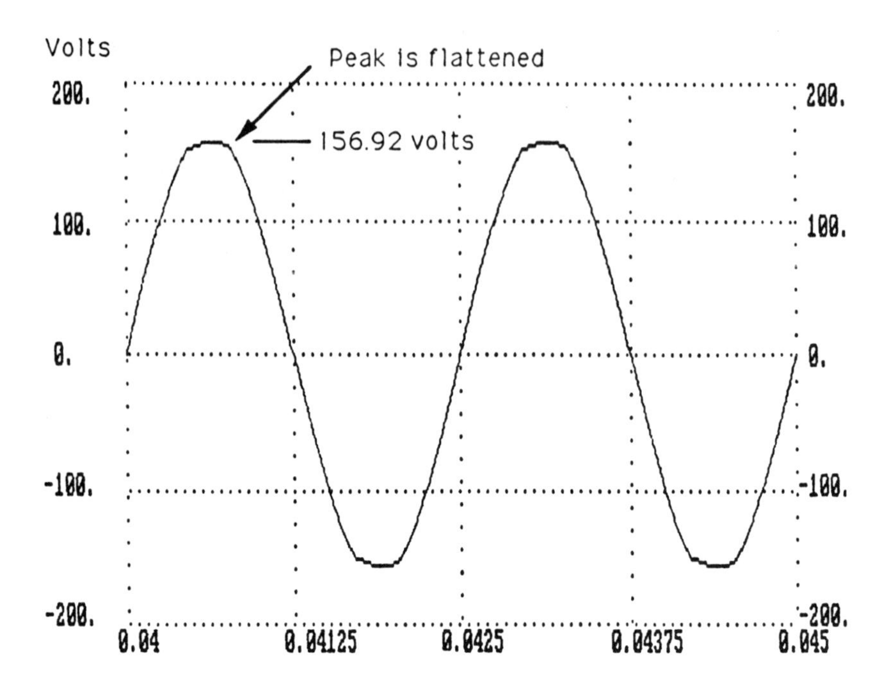

Figure 12.9 Input voltage with flattened peaks.

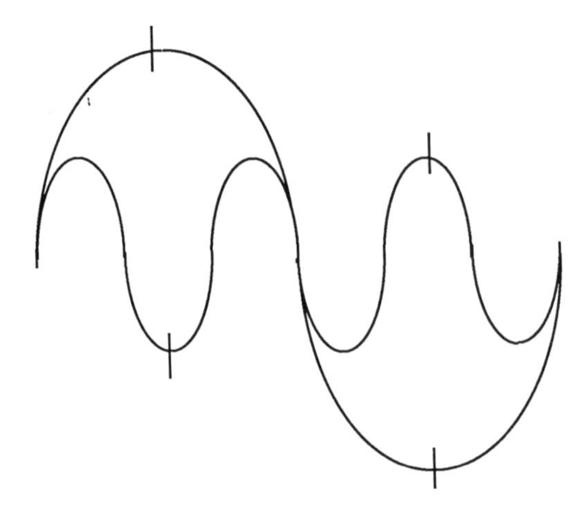

Figure 12.10 Third harmonic flattens peak of fundamental.

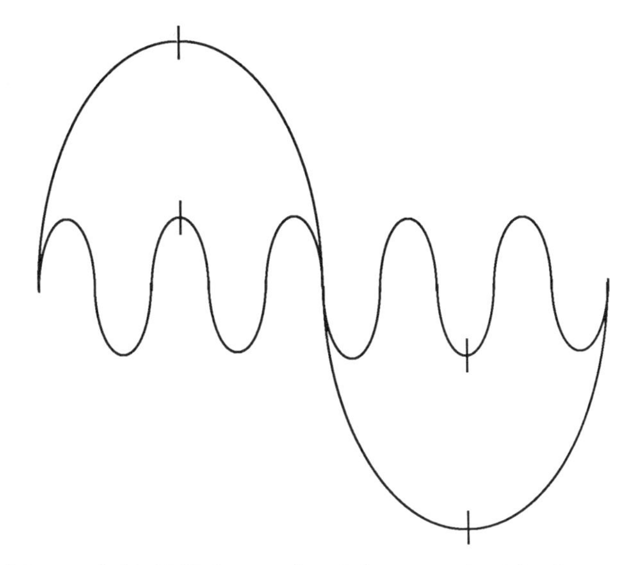

Figure 12.11 Fifth harmonic reinforces peak of fundamental.

A 400-Hz line test was also performed on the breadboard notch filter, which was connected to a load that simulates the radar unit. The test setup schematic for the 400-Hz line test is shown in Fig. 12.21. Scope pictures of the test data are presented in Fig. 12.22.

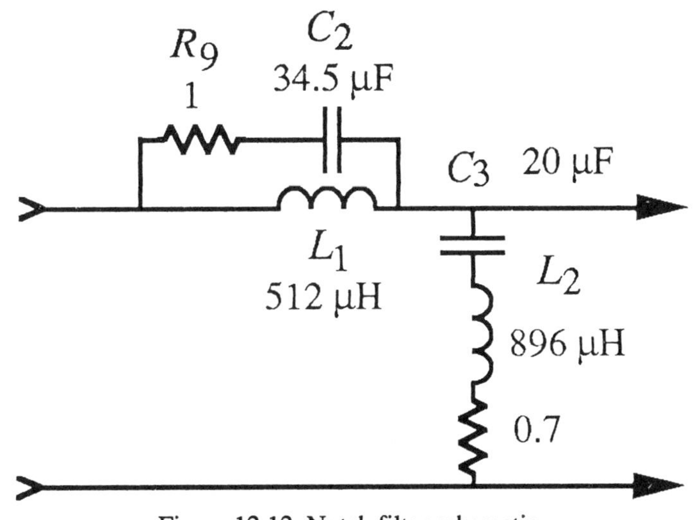

Figure 12.12 Notch filter schematic.

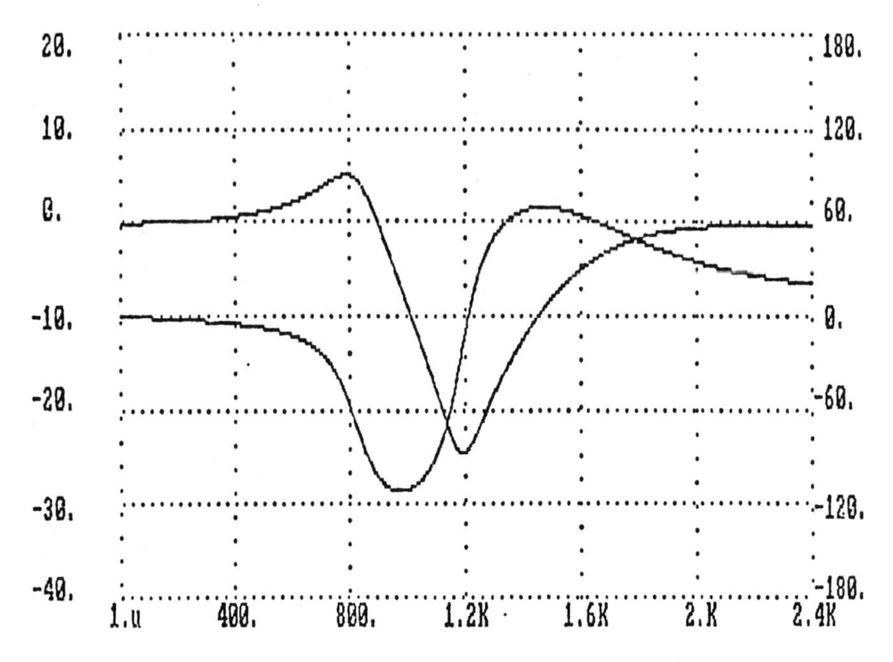

Figure 12.13 Insertion loss curve of notch filter.

The reader should compare Fig. 12.22A to Fig. 12.7, Fig. 12.22B to Fig. 12.18, and Fig. 12.22C to Fig. 12.16. The lab test data verify the accuracy of the simulation prediction.

12.2 EXAMPLE 2

Problem

During the design process of a switching power supply for a missile guidance computer, a proposal was made by a new engineer on

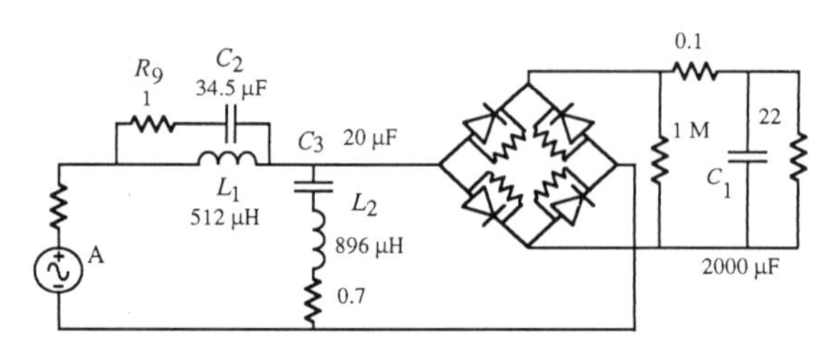

Figure 12.14 Model of the radar with the notch filter.

branch	label	nodes		value
1	Vin	0	0	1.
2	R8	1	0	0.2
3	L1	1	4	512.u
4	C2	1	3	34.5u
5	R9	3	4	1.
6	L2	4	5	896.u
7	C3	5	6	20.u
8	R10	6	0	0.7
9	D1	4	7	10.p
10	D2	0	7	10.p
11	D3	8	0	10.p
12	D4	8	4	10.p
13	R1	4	7	10.K
14	R2	7	0	10.K
15	R3	8	0	10.K
16	R4	8	4	10.K
17	R6	7	9	0.1
18	C1	9	8	0.002
19	R7	9	8	22.
20	R5	7	8	100.K

Figure 12.15 ECA program for simulation of radar with notch filter.

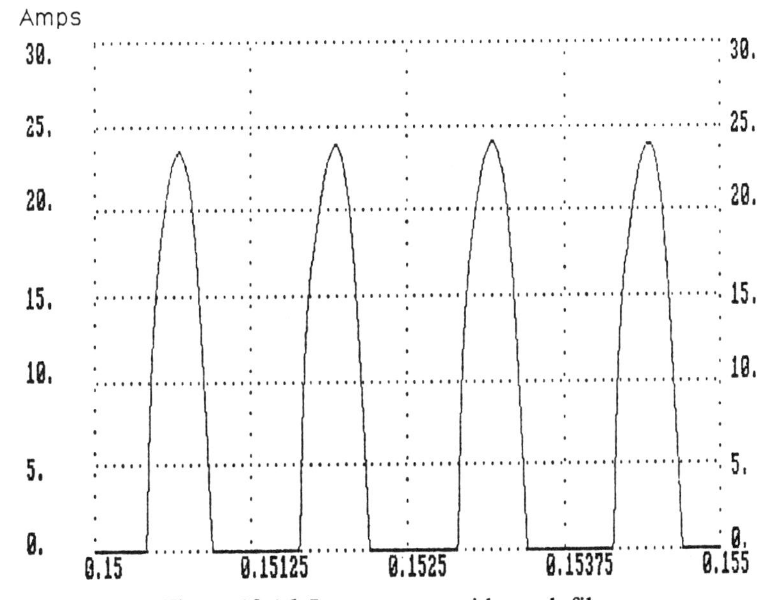

Figure 12.16 Input current with notch filter.

The input filter has reduced the flattening of the peaks.

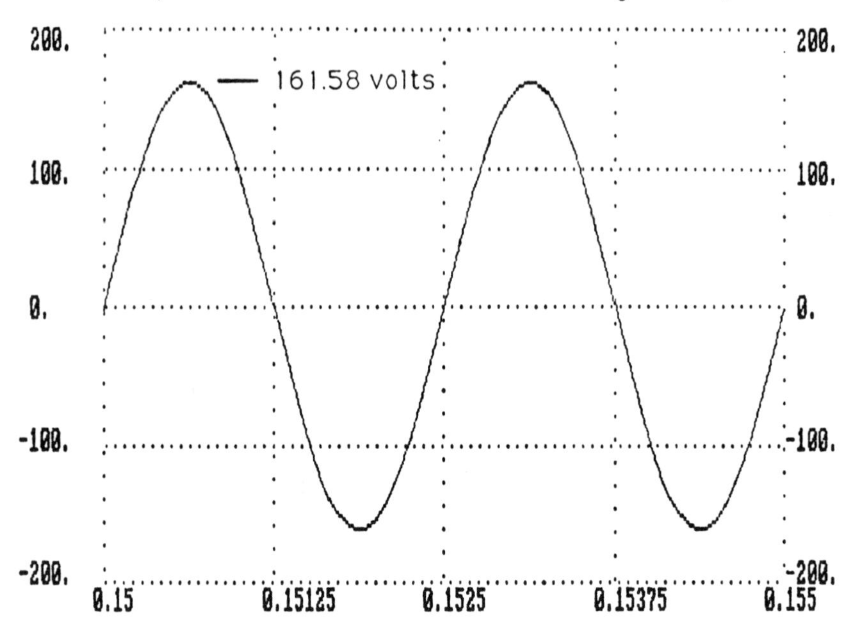

Figure 12.17 Input voltage peak flattening is less.

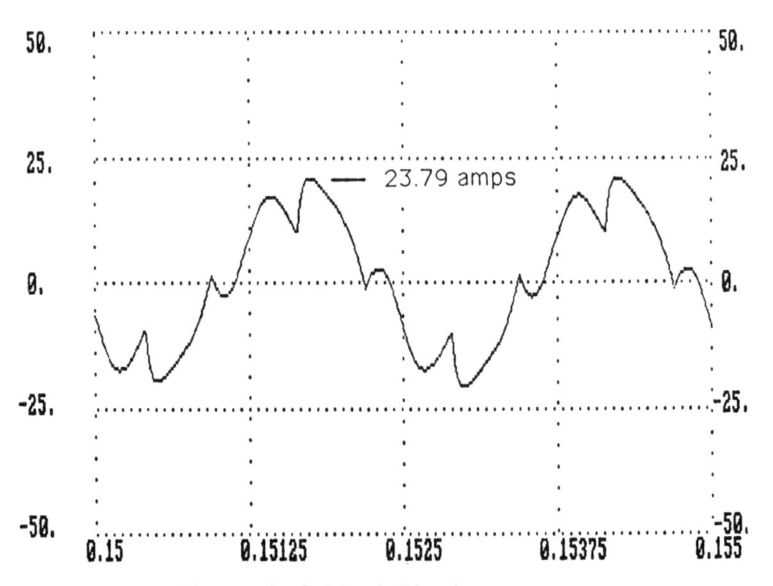

Figure 12.18 Notch filter input current.

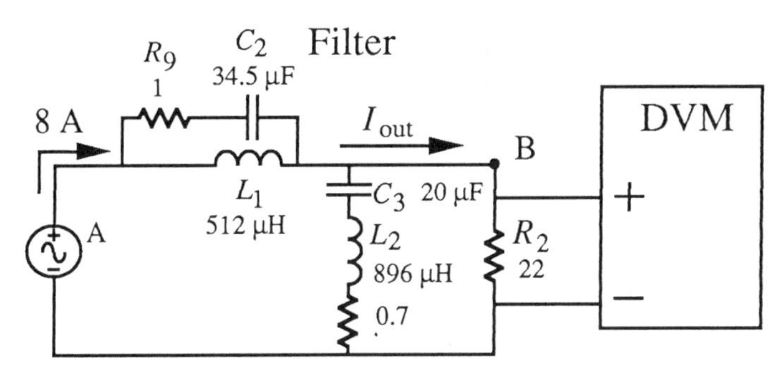

Figure 12.19 Test setup for the insertion loss test.

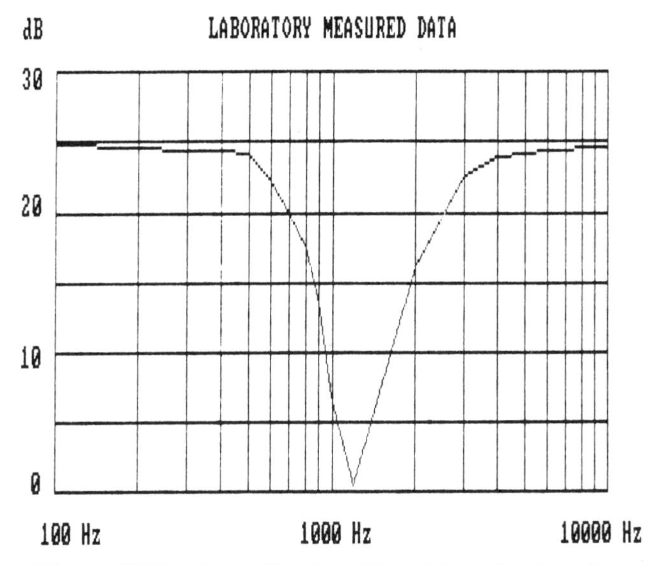

Figure 12.20 Notch filter breadboard insertion loss data.

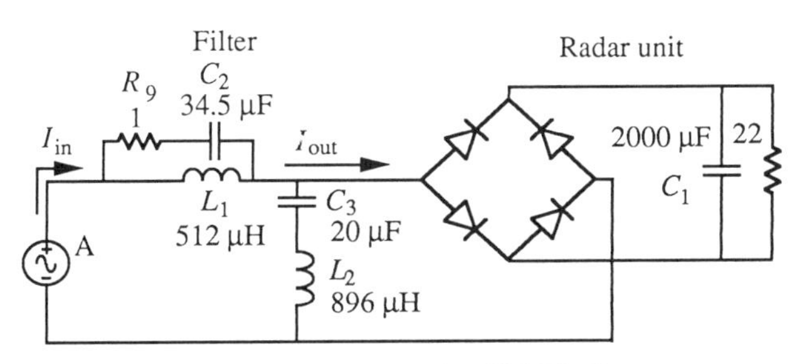

Figure 12.21 Test setup for 400-Hz line test.

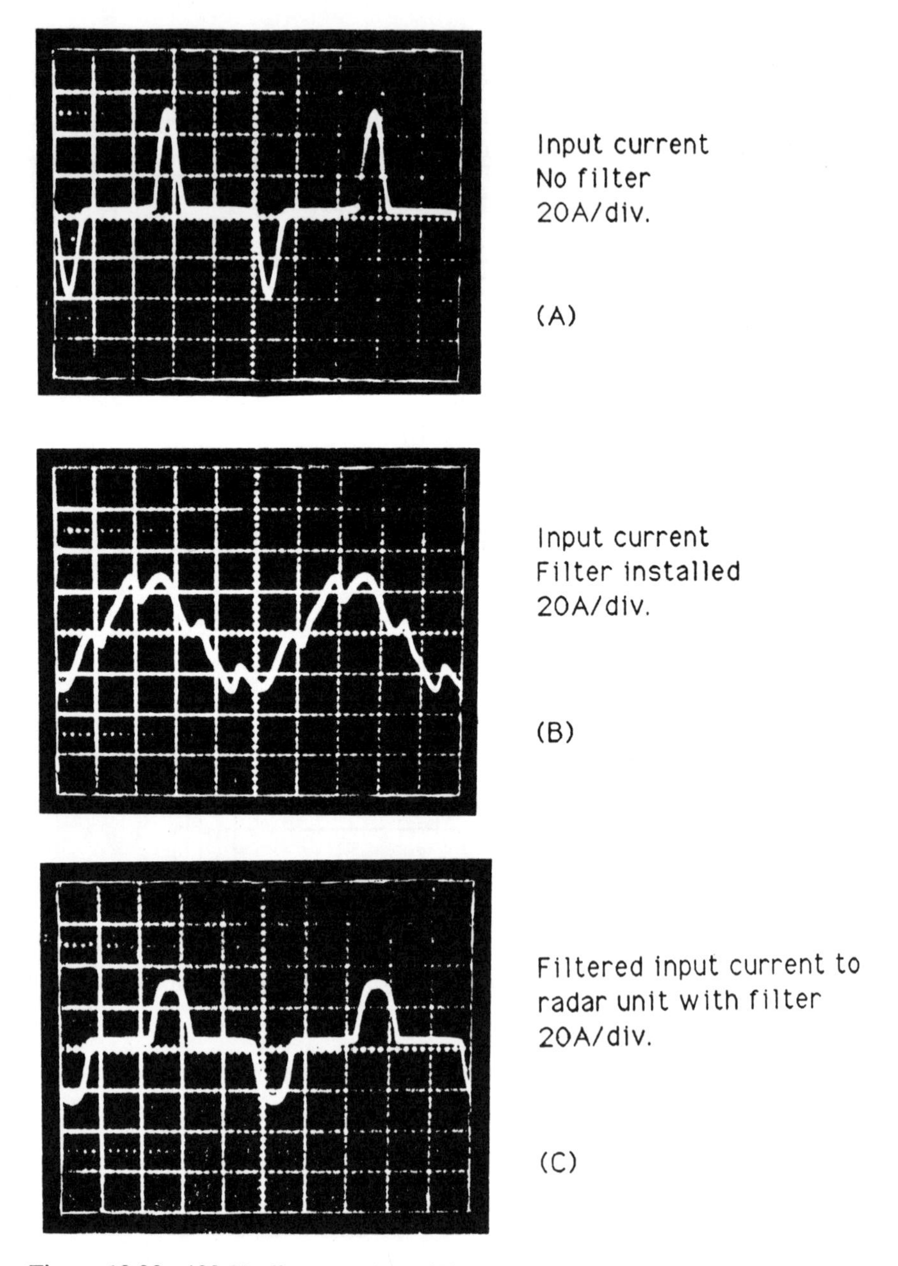

Figure 12.22 400-Hz line test data. (A) Input current, no filter, 20 A/div. (B) Input current, filter installed, 20 A/div. (C) Filtered input current to radar unit with filter 20 A/div.

the project to modify the existing input filter design. The proposal was to replace the tantalum capacitors with a combination of aluminum electrolytic and polycarbonate type capacitors. A simulation to predict if the proposed filter modification performs better than the existing filter was requested by engineering management.

Solution

The following simulations and tests were performed to solve the problem:

1. Simulate tantalum filter insertion loss test
2. Simulate aluminum electrolytic and polycarbonate filter insertion loss test
3. Lab breadboard test to verify prediction

To determine which of the two designs was better, a simulation of an insertion loss test was performed on each design. The simulation was performed using *Electronic Circuit Analysis* (ECA) on an IBM personal computer. The same inductors were used in both simulations.

The existing design used seven 15–μF tantalum electrolytic capacitors divided into five in parallel and two in parallel separated by L_2. A model of the tantalum capacitor version of the filter is shown in Fig. 12.23. The program listing for the simulation is presented in Fig. 12.24. The insertion loss simulation data is presented in Fig. 12.25. At 250 kHz (the switching frequency) the insertion loss is -100 dB. Above 3 MHz the filter attenuation decreases with increasing frequency.

The proposed new design uses a single 300-μF aluminum electrolytic and two 5–μF polycarbonate capacitors. The 300-μF and the two parallel 5–μF capacitors are separated by L_2. A model of the proposed filter is shown in Fig. 12.26. The program listing is presented in Fig. 12.27. The insertion loss simulation data is presented in Fig. 12.28. At 250 kHz, the insertion loss of the aluminum electrolytic and polycarbonate design is only -80 dB.

The existing tantalum capacitor input filter design has about 20 dB more attenuation (insertion loss) in the simulation than the proposed aluminum electrolytic and polycarbonate design. The two versions of the filter were breadboarded and tested in the labora-

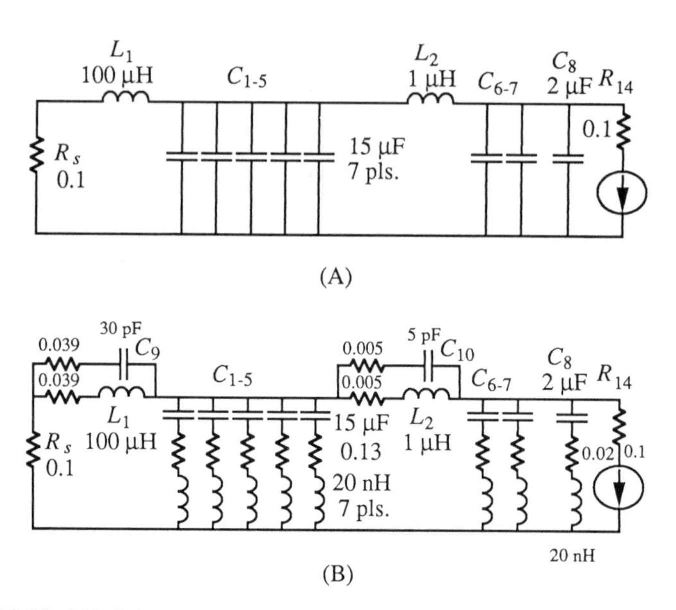

Figure 12.23 (A) Schematic and (B) model of tantalum capacitor filter version.

tory. Each filter was connected to the same breadboard switching power supply (one at a time, of course). Spectrum analysis was performed on the two filter versions operating with the breadboard power supply. The frequency domain data are presented in Fig. 12.29. The upper plot is the data from the aluminum electrolytic version. The lower plot is the data from the tantalum capacitor version. At 250 kHz, the amplitude of the aluminum electrolytic capacitor version is 62 dB/μA and the amplitude of the tantalum capacitor version is 42 dB/μA. The lab measurements show that the noise at the switching frequency of 250 kHz is 20 dB/μA less for the existing tantalum capacitor version than for the proposed aluminum electrolytic version. The results of the tests concur with the simulation prediction. The existing filter design is better than the proposed modification.

The frequency components at 65 kHz and 130 kHz in the plots are produced by an auxiliary power supply that is also part of this system. This was confirmed by shutting the auxiliary power supply off. With the auxiliary power supply off, the 65 and 130 kHz components are gone. The switching frequency of the main power supply is evident at the fundamental frequency of 250 kHz and at the harmonics at 500 kHz, 750 kHz, 1 MHz, and integer multiples of 250 kHz above.

branch	label	nodes		value
1	IIN	0	0	1.
2	R14	1	0	0.1
3	R9	2	0	0.1
4	R10	2	3	0.039
5	R11	2	4	0.039
6	C9	3	5	30.p
7	L1	4	5	100.u
8	R12	5	6	0.005
9	R13	5	7	0.005
10	C10	6	1	5.p
11	L2	7	1	1.u
12	C1	5	10	15.u
13	C2	5	12	15.u
14	C3	5	14	15.u
15	C4	5	16	15.u
16	C5	5	18	15.u
17	R1	10	11	0.13
18	R2	12	13	0.13
19	R3	14	15	0.13
20	R4	16	17	0.13
21	R5	18	19	0.13
22	L11	11	0	20.n
23	L12	13	0	20.n
24	L13	15	0	20.n
25	L14	17	0	20.n
26	L15	19	0	20.n
27	C6	1	20	15.u
28	C7	1	22	15.u
29	R6	20	21	0.13
30	R7	22	23	0.13
31	L16	21	0	20.n
32	L17	23	0	20.n
33	C8	1	24	2.u
34	R8	24	25	0.02
35	L18	25	0	20.n

Figure 12.24 ECA program listing.

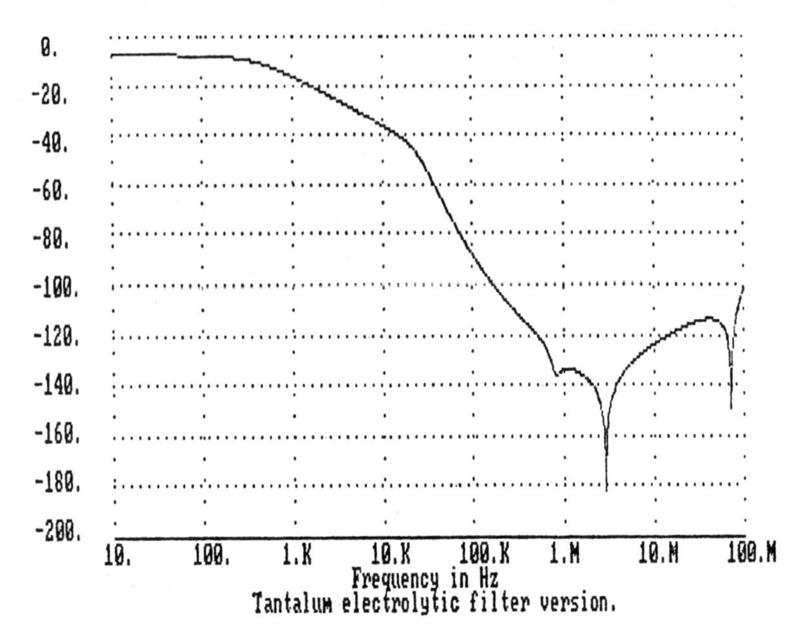

Figure 12.25 Tantalum capacitor filter version insertion loss data.

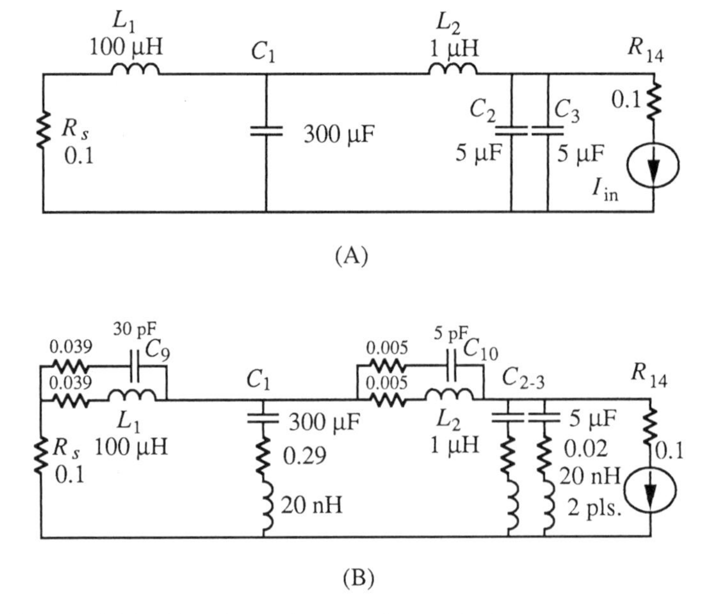

Figure 12.26 (A) Schematic and (B) Model of proposed filter version.

branch	label	nodes		value
1	IIN	0	0	1.
2	R14	1	0	0.1
3	R9	2	0	0.1
4	R10	2	3	0.039
5	R11	2	4	0.039
6	C9	3	5	30.p
7	L1	4	5	100.u
8	R12	5	6	0.005
9	R13	5	7	0.005
10	C10	6	1	5.p
11	L2	7	1	1.u
12	C1	5	10	300.u
13	C2	5	12	5.u
14	C3	5	14	5.u
15	R1	10	11	0.29
16	R2	12	13	0.02
17	R3	14	15	0.02
18	L11	11	0	20.n
19	L12	13	0	20.n
20	L13	15	0	20.n

Figure 12.27 Program listing for proposed filter version.

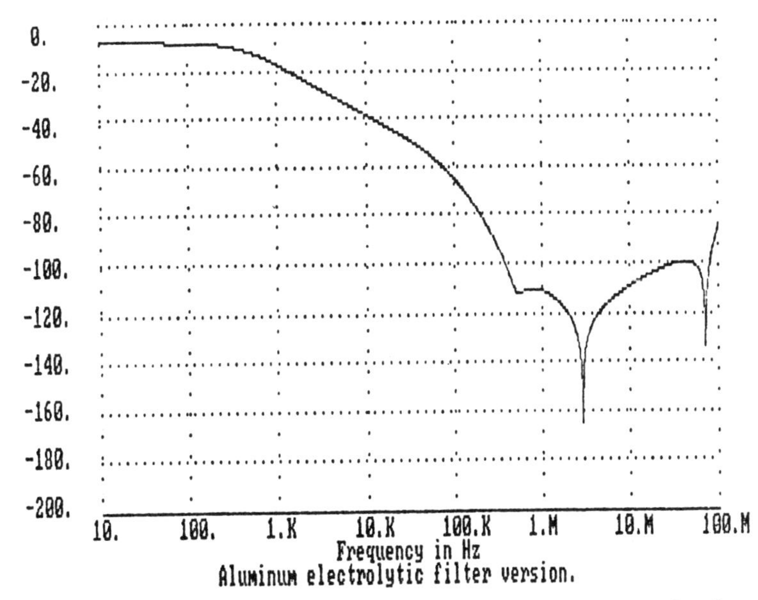

Figure 12.28 Aluminum electrolytic capacitor filter version insertion loss data.

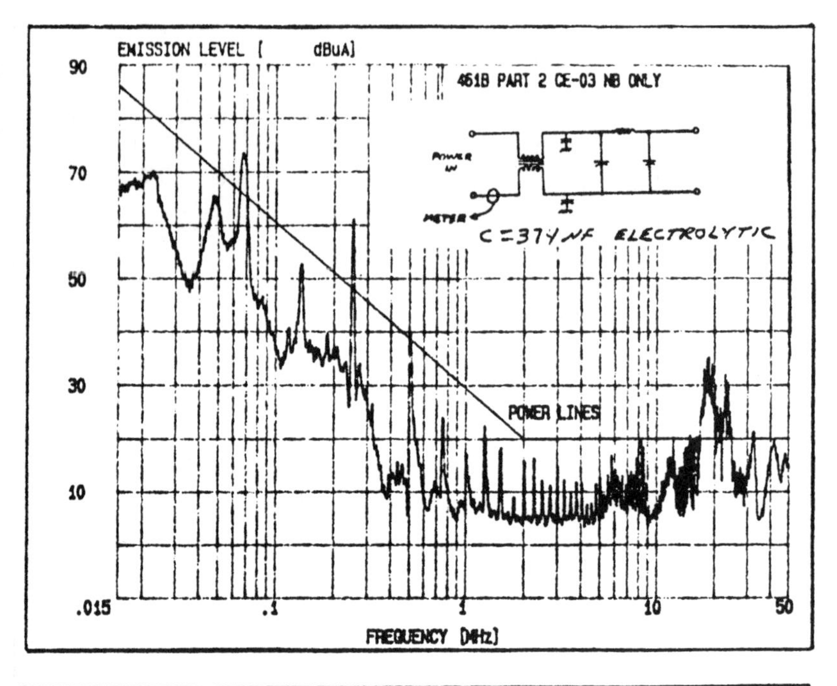

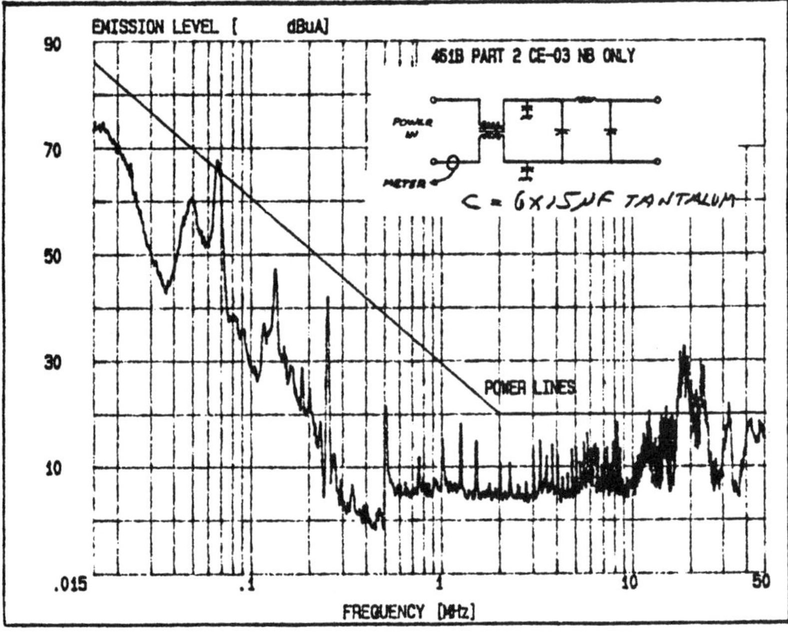

Figure 12.29 Spectrum analyzer plots of the two filter versions.

12.3 EXAMPLE 3 (FFT)

Problem

Predict, by simulation, if an ac/dc 60-Hz magnetic amplifier convertor will meet the requirements of Mil-Std-461 CE01 and CE03 conducted emissions tests.

It was determined that compliance with CE01 was the most critical input EMI test for the 60-Hz convertor. There are no relatively fast rise times in the input waveform so that the upper harmonics will not be significant in the CE03 test.

Solution

A scope picture of the input current waveform (Fig. 12.30) was digitized graphically by hand. The data were entered into the fast Fourier transform (FFT) written in BASIC. The input data were displayed by the program for verification. The displayed waveform is presented in Fig. 12.31. A plot of the frequency domain data is shown in Fig. 12.32. The CE01 requirement limits the harmonics of the waveform to 130 dBμA or less. The fifth harmonic (300 Hz) just meets the limit. A printout of the first 16 components of

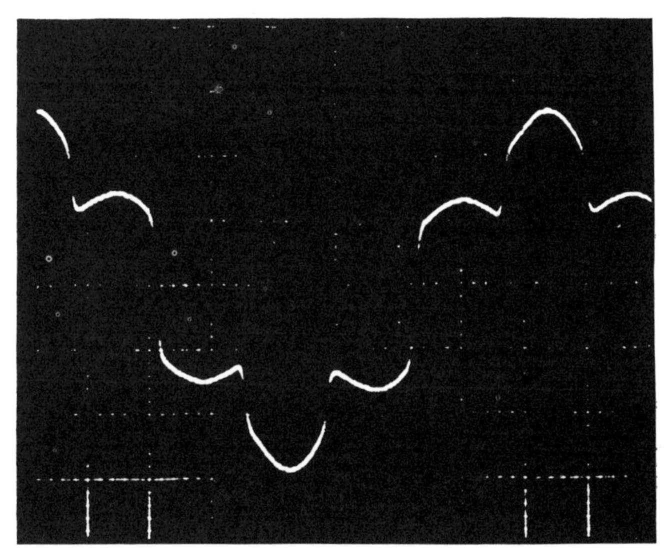

Figure 12.30 Scope picture of input current waveform.

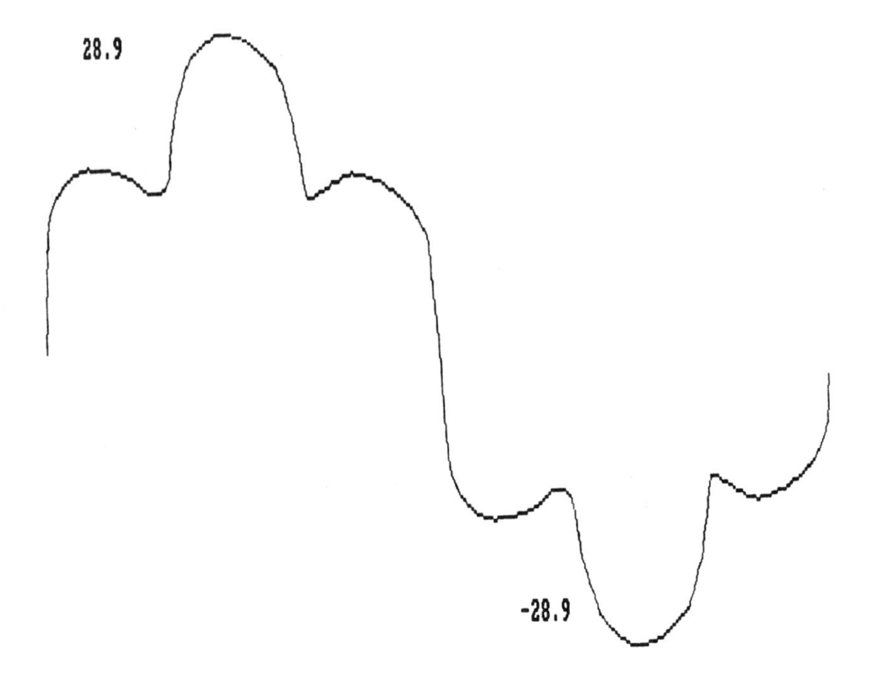

Figure 12.31 Displayed time domain waveform.

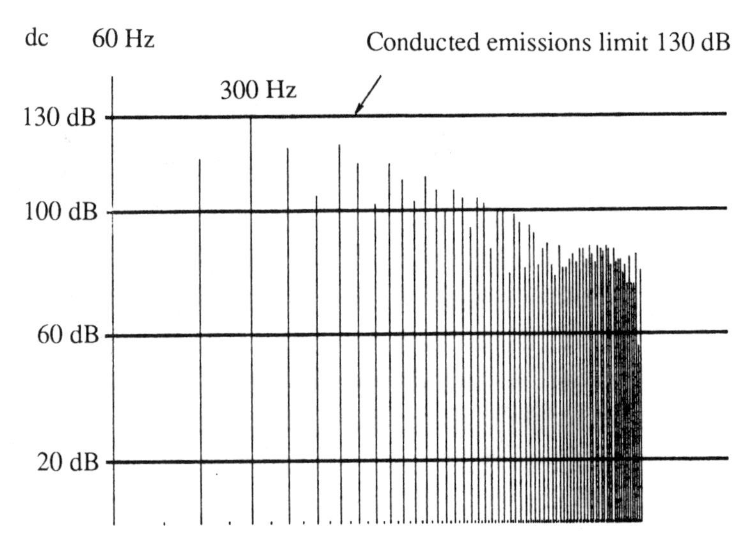

Figure 12.32 CE01 frequency domain data.

VOLTS	DB/MICRO VOLT		FREQUENCY
.000001	0	0	
13.12842	142.3643		60
.000001	0	120	
.6536976	116.3075		180
.000001	0	240	
3.194396	130.0878		300
.000001	0	360	
.957101	119.6192		420
.000001	0	480	
.1624781	104.2159		540
.000001	0	600	
1.049969	120.4235		660
.000001	0	720	
.489723	113.799		780
.000001	0	839	
.1123773	101.0136		899

Figure 12.33 Table of first 16 frequency domain components.

the frequency domain data (Fig. 12.33) lists the fifth harmonic at 130.0878 dBμA.

We would prefer to have a little more design margin for meeting this requirement. A very large filter would be necessary to be effective at the low frequency and high current of the ac/dc convertor. A large filter would not be feasible in this application. If the rise and fall times could be slowed down, there would be some design margin.

Review of the data confirmed our assumption that the limits of CE03 would be met because of the low fundamental frequency and relatively slow rise and fall times. The amplitudes of the harmonics fall rapidly with increased frequency. The CE03 limit starts at the 250th harmonic of 60-Hz power conversion.

QUESTIONS

12.1 What are the major steps in the approach to solving the problem in example 1?

12.2 If the radar unit of example 1 was designed to meet the harmonic distortion requirements of CE01, do you think the radar unit would have caused the system problem in the example? Why?

12.3 What are disadvantages of using a notch filter as a Band-aid to solving the problem in example 1?

12.4 Choosing the best component for an application is of major importance in the design process. In example 2, what should be the criteria (or parameters) for the selection of the best capacitor type for this application?

12.5 Describe the conducted emissions problem of each of the three examples in this chapter.

12.6 In example 3, why is CE01 but not CE03 a major design concern?

13

COMPUTERS AND DIGITAL LOGIC CIRCUITRY

Computers and digital logic circuits are ubiquitous in our daily lives. The EMI performance of computers is controlled by FCC *Rules and Regulations*, Part 15, Subpart J. The clock frequency in most computers is above 1 MHz. Many state of the art personal computer clock frequencies run at speeds above 30 MHz.

13.1 CONDUCTED EMISSIONS COUPLING PATHS

A major (and typical) problem in attaining FCC compliance for electronic equipment is the reduction of the conducted emissions that are conducted through the input power lines to the outside world. Switching logic loads can create large amplitudes and wide spectrums of noise on input power lines. Clocking exacerbates the switching loads problem because many gates change state at the same time. Well-designed power input filters are required to attenuate this EMI to acceptable levels. Computers and digital logic circuits often have many high speed I/O lines that may inadvertently conduct EMI to the outside world.

A diagram of a typical computer interface to a peripheral device is presented in Fig. 13.1. Isolated communications between the computer and the peripheral are accomplished with the use of opto-couplers. The major components of the conducted emissions design problem in this example is outlined in the following:

1. Noise sources
    a. Internal clock
    b. Logic changes (signals)
    c. Peripheral switching

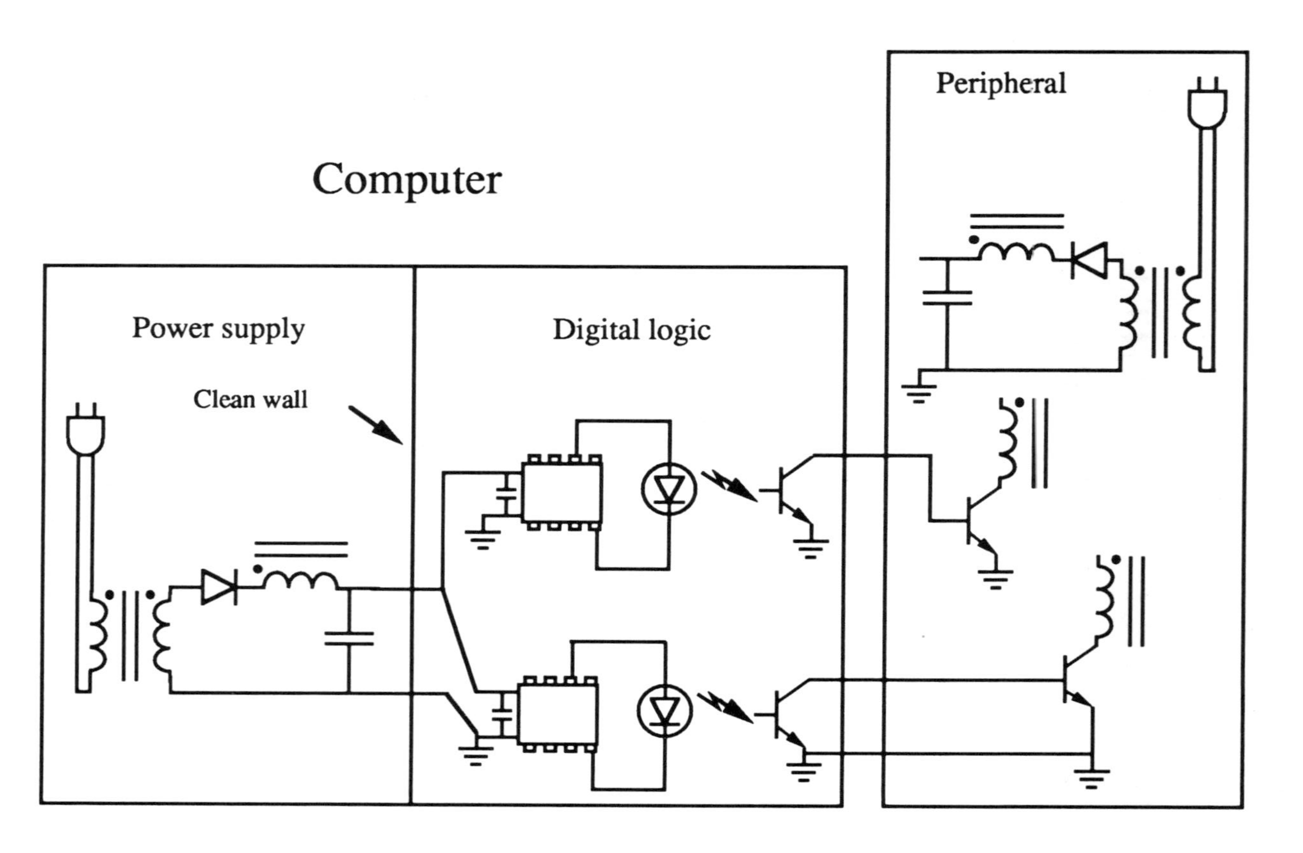

Figure 13.1 Computer interface to a peripheral device.

2. Paths to outside world
   a. Computer power supply
   b. Peripheral power supply

Susceptibility is the inverse problem of conducted emissions. Virtually all of the information about conducted emissions applies to susceptibility. The reduction of circuit loop areas for good layout applies to both problems.

The amount of isolation achieved between the computer and peripheral is far less than we would generally like. Because there are metal electrodes within the opto-coupler, there is a small capacitance of about 5 pF between input and output. At 50 MHz (maximum frequency measured for CE03 conducted emissions) the reactance of 5 pF is 637 Ω. This capacitance could be a very significant design factor in many applications. Magnetic coupling through opto-couplers can also be a significant design parameter. Opto-couplers always have input and output conductive loops. Magnetic fields couple between these loops, which reduces the isolation between circuits.

Magnetic coupling is a major concern for computer and digital logic layout designs and is a major problem in low-impedance circuitry. Of course, low impedance is required for fast operating speeds and is common to most logic families. Electric field coupling is not as much of a problem because of the relatively low operating voltages.

A diagram with front and top views of loop to loop coupling is presented in Fig. 13.2. Current in any conductive loop will magnetically induce a current in any adjacent conductive loop of material. The top view of the two loops shows a representation of the magnetic flux field coupling the two loops together.

A simplified diagram of a single path through the computer and peripheral (Fig. 13.3) has six conductive loops that can couple magnetic energy through the whole system. Let's look closer at the magnetic coupling across the opto-coupler. Peripherals often have switched inductive loads such as motors and solenoids. The switched load circuit loop area, A, in Fig. 13.4 couples to the loop area B. The induced current is circulated in the peripheral input circuit. The magnetic field from loop area C couples to the loop area D of the logic output circuit. The resulting voltages from induced current in loop D may cause logic upsets in the computer. In this case, the conducted emissions come from the peripheral and upset the computer. In order to meet FCC requirements on con-

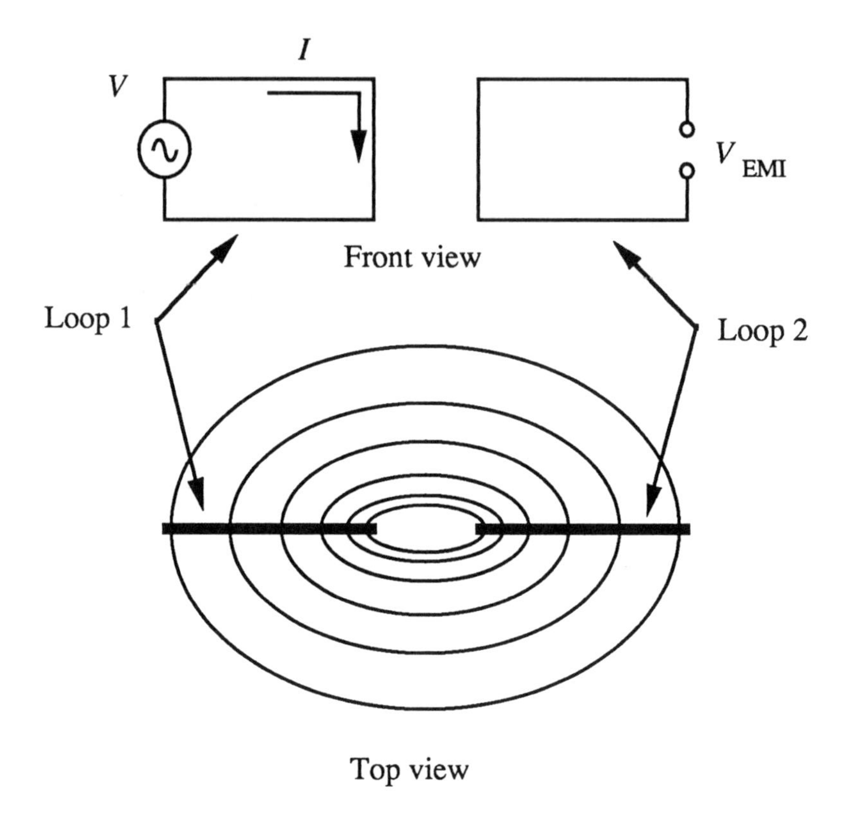

Figure 13.2 Loop to loop coupling.

ducted emissions, the peripheral manufacturer must reduce the circuit layout loop areas. The reduction of circuit loop areas is not always easy because of the size, shape, and mounting methods of the electronic components. Sometimes extraordinary design and/or manufacturing efforts are required to mechanically control the conduction paths so that the desired electrical performance can be achieved.

13.2 SEQUENTIAL LOGIC AND CLOCKS

Sequential logic circuits have outputs such that their state (1 or 0) depends upon the previous states of the circuit inputs. Many sequential circuits are used in computers and digital logic electron-

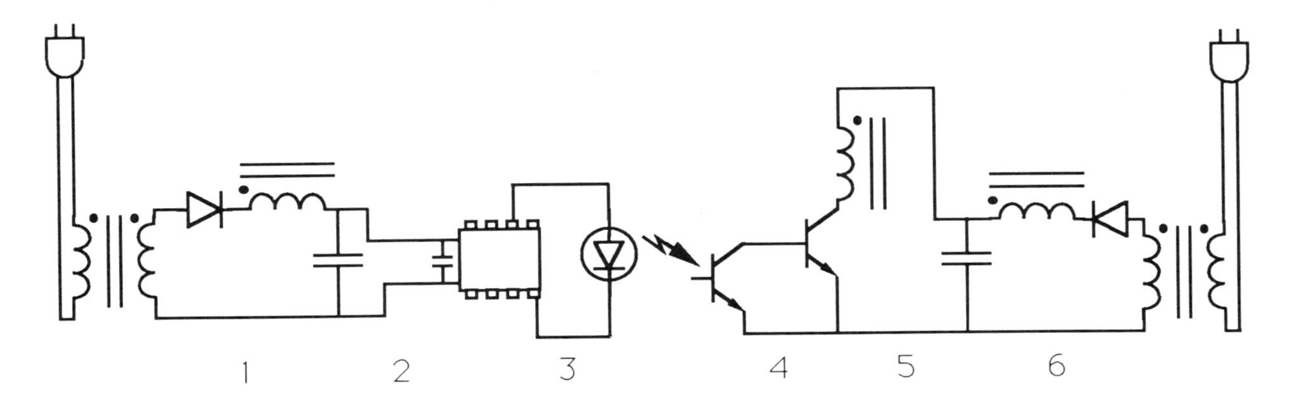

Figure 13.3 Conducted path through computer and peripheral.

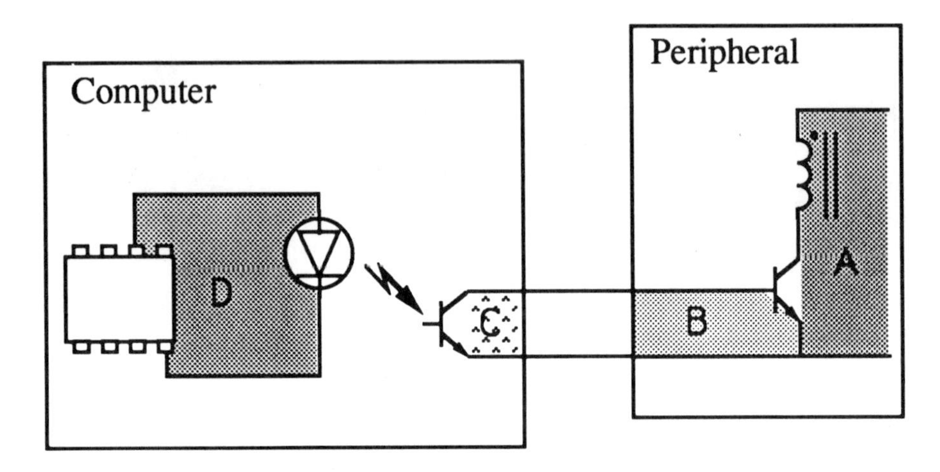

Figure 13.4 Magnetic coupling through opto-coupler.

ics. Sequential logic circuits are commonly said to have "memory." This type of circuitry is used extensively throughout most computers and digital logic circuitry. EMI can cause sequential logic circuitry to change state randomly, or upset, which in some applications can cause catastrophic failures to attached equipment or the loss of data, which can be very devastating and/or expensive.

Clock circuits are generally the largest internal EMI threat to computers and digital logic circuits. Clock circuits are low impedance and have extremely fast rise and fall times. Filtering high speed clock signals or I/O lines is difficult because the filtering slows down response times. The increased rise and fall times cause higher power dissipation. The susceptibility of the circuit also increases when the logic signals are in the linear region too long. Contamination of the V_{cc} bus with clock harmonics will be exacerbated if inadequate bypassing or decoupling is used. Three controllable design parameters for the reduction of conducted emissions caused by clock circuits (and signal lines) are:

1. Reduce clock circuit loop area
2. Use lowest possible clock frequency
3. Use maximum acceptable clock rise and fall times
4. Clock and signal lines should be designed to match characteristic impedances of drivers and loads

Conducted emissions design is important for design related problems as well as for passing the requirements of EMI standards. We will analyze the bypass problem in digital circuit design next. The form of this design problem is very similar to the form of the problem of meeting emissions limits.

13.3 EXAMPLE OF INTERNAL CONDUCTED EMISSIONS

Key Idea 1

> The conducted emissions problem is generally in the form of an electronic unit causing electronic noise on the power source, which results in a noise problem for the other equipment using that power source.

Figure 13.5 shows a diagram of a LSTTL gate U_1 driving another LSTTL gate U_2. A bypass capacitor C_1 bypasses or decouples the driver U_1 from the inductance of the incoming power circuit layout. The general rule of thumb is that for every two digital IC packages, at least one bypass capacitor should be located as close as possible to the packages. Most computer and digital circuitry manufactured today includes at least one bypass capacitor for every digital IC, and some bypass every IC in a design. The reasons for bypassing will be apparent after study of the following example analysis. We will only simulate LSTTL gate turn on (transition from 0 to 1 state). The simulation for the

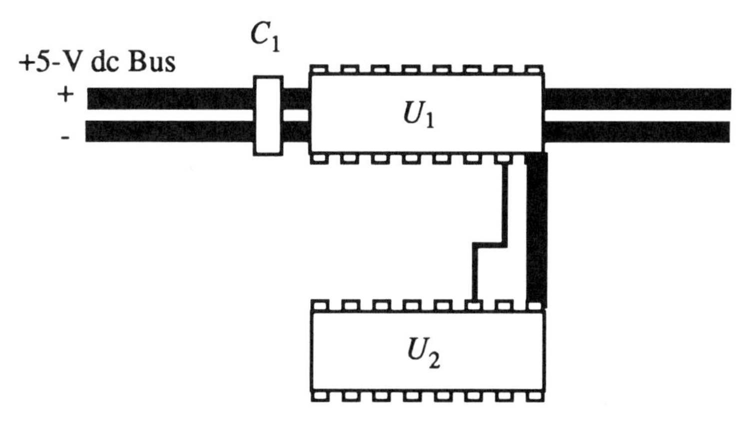

Figure 13.5 LSTTL gate driving another LSTTL gate.

transition from the 1 state to the 0 state could be performed in a similar fashion.

A schematic of the LSTTL example is presented in Fig. 13.6. The battery V_1 models the +5-V dc power supply bus, providing power to the driving LSTTL gate U_1. The driven LSTTL gate is connected to a wholly separate +5-V dc power bus but is referenced to the same ground potential.

A SPICE model of the LSTTL example is presented in Fig. 13.7. The +5 V dc power bus is modeled with V_1, L_1, and L_2. The inductors L_1 and L_2 model the circuit layout inductance of the power and return traces or wires. The bypass capacitor is modeled using C_1, R_1, and L_8. In this design case, the parasitic elements of the capacitor must be modeled because they affect circuit performance. Inductors L_3 and L_4 model the circuit layout trace inductance between the bypass capacitor and the IC package terminals. Inductors L_5 and L_6 model the internal inductance of the LSTTL driver gate caused by the circuit layout inductance from the gate terminals to the gate IC chip. The resistor R_2 models the quiescent dc in the driver gate. The combination of R_3 and C_2 model the current pulse associated with the switch in states of the gate. The load (driven LSTTL gate input) is modeled with L_7, C_3, and R_4. The element C_3 models the input capacitance of U_2 and R_4 models the equivalent input resistance. The inductance L_7 models the circuit layout inductance of the interconnecting circuit traces between the output of U_1 and the input of U_2. The voltage source *A* is used to model the turning on of the load or driving the input of U_2 from the 0 state to the 1 state. The rise time of voltage source *A* is set to 18 nS as specified for LSTTL. This source is shown as sinking current instead of sourcing current for convenience in working with SPICE.

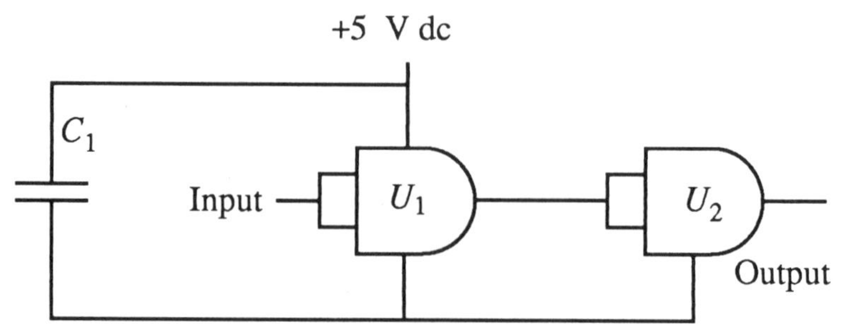

Figure 13.6 Schematic of LSTTL gate driving another LSTTL gate.

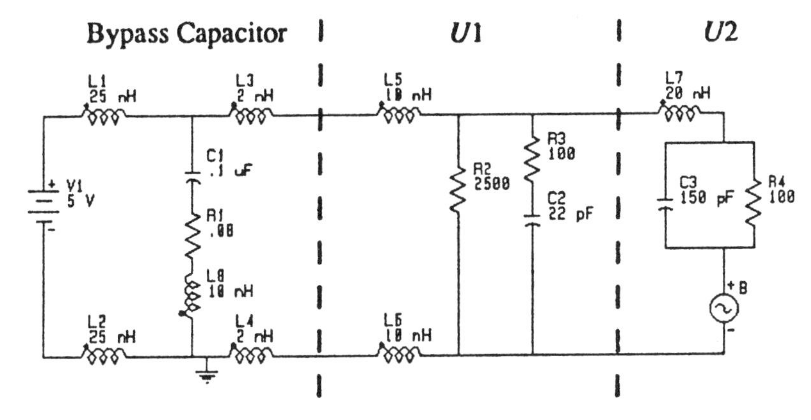

Figure 13.7 Model of LSTTL gate driving another LSTTL gate.

The circuit layout traces are shown in Fig. 13.8. The area labeled Area 1 causes the inductances L_1 and L_2 to be significant in this problem. The area labeled Area 2 causes the inductances L_3 and L_4 to be significant in this problem. The area labeled Area 3 causes the load inductance to be significant. A diagram of the area causing the inductances L_5 and L_6 to be significant internally to U_1 would be similar to Fig. 13.8. The wiring area is created by the traces from the package terminals to the chip mounted inside. The

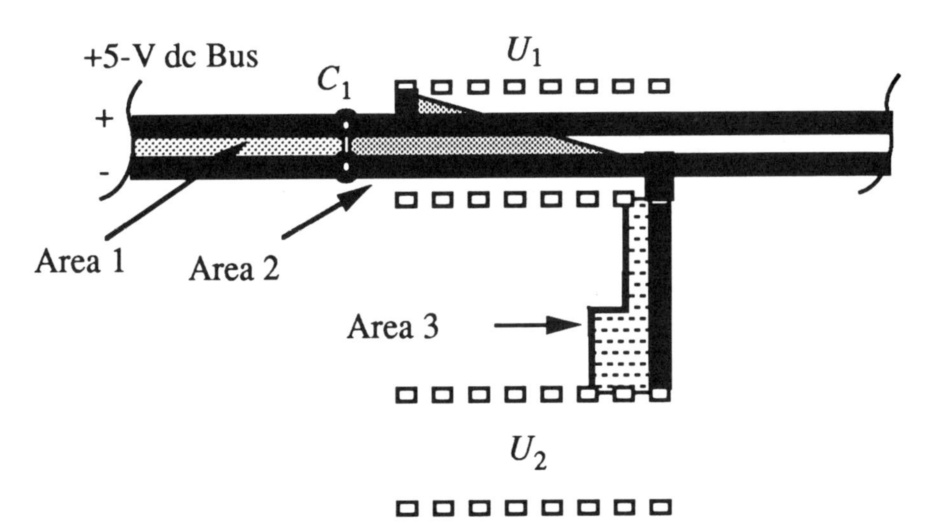

Figure 13.8 Loop areas that contribute to circuit inductances.

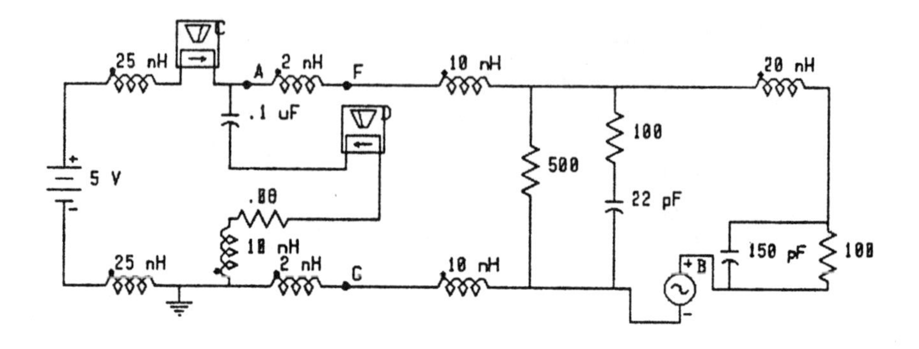

Figure 13.9 Model component values.

manufacturer specifies IC performance measured at the pins of the package. The manufacturer guarantees the performance specified and so must account for the internal IC inductance.

Typical circuit values are assigned to the model as shown in Fig. 13.9. The load capacitance C_3 is set to 150 pF representing a fan out of 10 gates with an input capacitance of 15 pF each. The simulation data are presented in Fig. 13.10. The voltage on the bypass capacitor C_1 and the voltage at the +5 V dc power input pins of U_1 are plotted. The difference between the two plots is the voltage across the layout inductance from the bypass capacitor to the chip inside the IC package. The +5-V dc at the pins of U_1 is pulled down 225 mV.

The noise immunity for the +5-V dc power inputs on LSTTL is generally not specified. The drop in input voltage is directly transferred to the output in the 1 state. The noise immunity for LSTTL inputs in the 1 state is 0.7 V, and is 0.3 V for the 0 state. As a rule of thumb, the +5-V dc power supply voltage at the LSTTL IC pins should be designed to change less voltage at transition times than the smallest noise margin for the input pins. The circuit as modeled and simulated should function cleanly (no upsets from noise).

Key Idea 2

> The purpose of a bypass capacitor is to function as a local low-impedance energy storage tank to bypass the inductive effects of the incoming wiring or circuit traces.

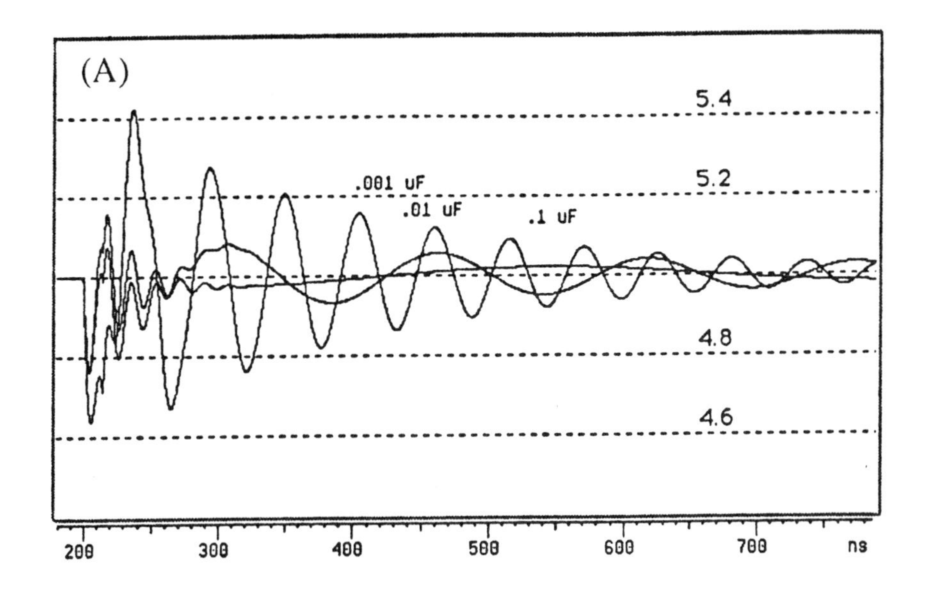

Figure 13.10 Capacitor and logic gate during transition to 1 state.

During a transient change (going from the 0 state to the 1 state on the output) current D is supplied by the bypass capacitor. Current C comes from the source. These currents in L_1 and C_1 are plotted in Fig. 13.11. There are two dominant resonant tanks in this circuit. The low-frequency tank at 2 MHz is the loop with C_1, L_1, L_2, and L_8. Inductances L_3, L_5, L_7, L_6, and L_4 do not act in parallel with the low-frequency tank because the resistance of R_4 is very large compared to the reactance of all of the inductors at 2 MHz. The reactance of C_3 is even larger yet, so that the loop with L_3, L_5, L_7, L_6, and L_4 is a relatively open circuit when compared to the reactance of L_1, L_2, and L_8. The LSTTL rise time is approximately 18 ns. The ringing of the natural frequencies of the circuit continue for a time after the transient output change. The high-frequency (56 MHz) tank consists of the loop of L_3, L_5, L_7, C_3, L_6, and L_4. At 56 MHz C_1 is a relatively short circuit and so completes the loop of the high-frequency tank. During the transient, the peak current in the bypass capacitor C_1 is 95.7 mA. The peak current in L_1 (the current coming from the +5 V dc source) is only 30.2 mA. More than 75 percent of the current for the change of logic states (output change from 0 to 1) is supplied from the stored charge in the bypass capacitor. After the initial effects of the transient have diminished, the bypass capacitor is recharged by the +5-V dc source. Without the bypass capacitor, the power input voltage to the IC packages will be greatly affected by the normal transient operation of the logic changing states.

13.4 WHAT IS THE BEST BYPASS CAPACITOR?

Obviously there is no single answer to this question. The physical size and the amplitude of the voltages and currents will change the bypass capacitor needs. Bypass capacitors should be made of very high-quality dielectrics for satisfactory performance. The capacitor package design must minimize the inductance effects caused by the loop formed by the leads and plates of the capacitor. The size and layout of the package must also facilitate the mounting of the package in a way that will minimize the inductive effects of the circuit layout. For ICs (especially digital ICs) the most common choice is ceramic capacitors in the CKR06 or surface mount device (SMD) style. The most common IC bypass capacitors are 0.1- and 0.01-μF ceramic capacitors. In some cases, 0.001-μF ceramic capacitors are used.

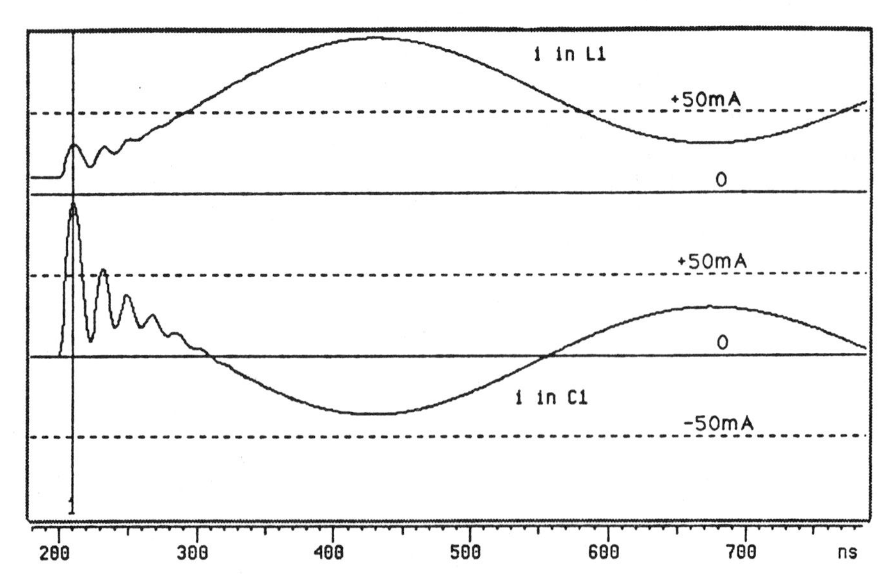

Figure 13.11 Current in L_1 and C_1.

Typical impedance curves for these three capacitors are presented in Fig. 13.12. The inductance of the three different components is relatively the same as can be seen in the right half of Fig. 13.12, where the capacitors are beyond their resonant frequency. The three resonant frequencies are in the ratio of the square root of 10; the capacitance values are the main reason for the different resonant frequencies. The ESRs primarily comprise the plate and lead resistance in series with the equivalent dielectric losses. Basically the larger plate areas of the larger capacitors are like parallel resistances that reduce the total resistance. The 0.1-μF capacitor has the lowest ESR and the 0.001–μF one has the highest ESR of the three capacitors.

Three simulations were run to determine the best bypass capacitor value for this example circuit. The simulation data plots are presented in Fig. 13.13. The three different in circuit resonant frequencies of the different capacitors are evident in the plotted

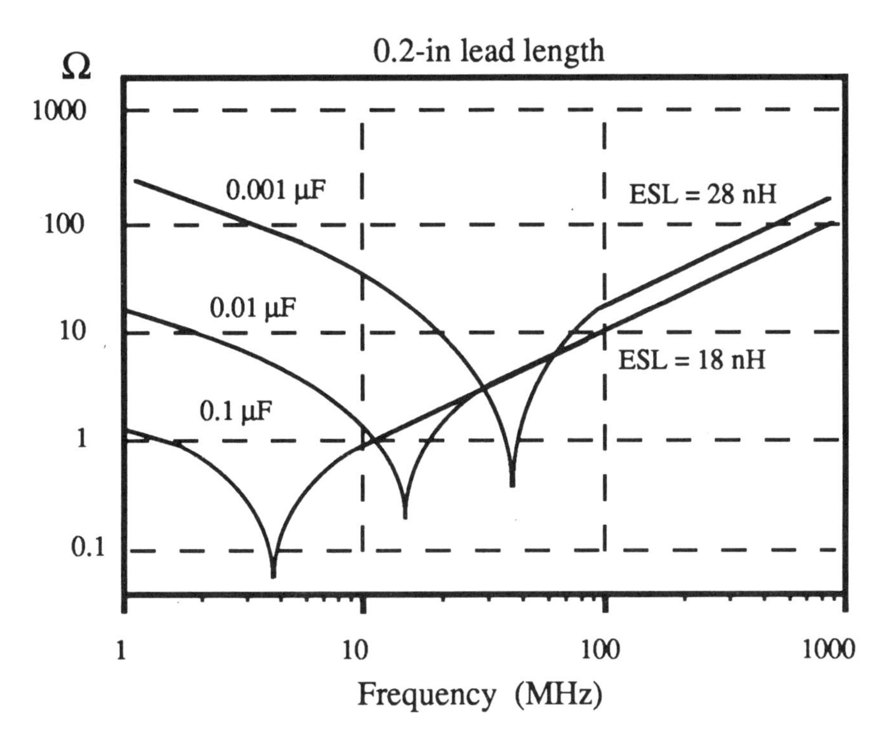

Fig 13.12 Typical impedance curves for ceramic bypass capacitors.

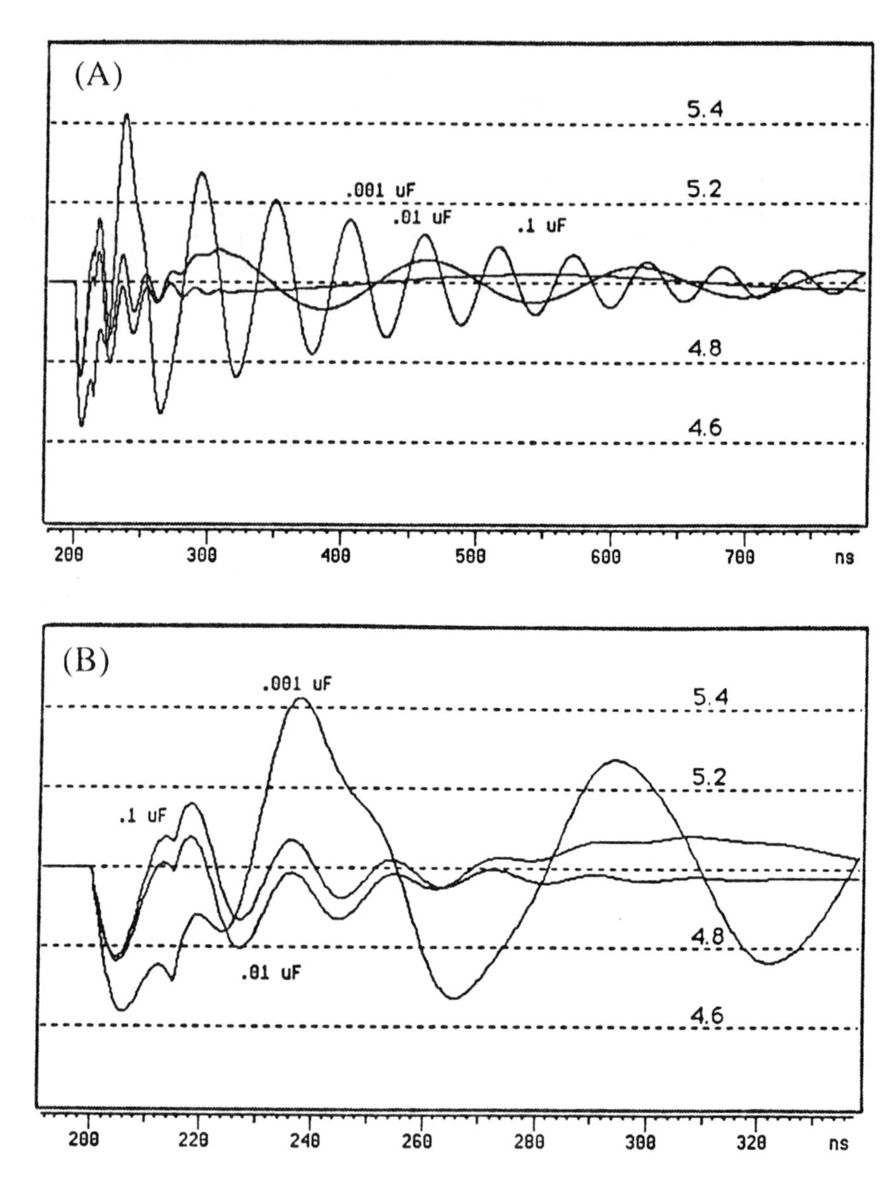

Fig 13.13 Performance comparison for three bypass capacitor values.

data. In this example, the 0.1-µF bypass capacitor has the smallest change from the nominal +5-V dc. The 0.001–µF plot exceeds a 400-mV voltage delta from nominal. The envelope of the 0.1-µF bypass capacitor transient response contains much less energy than the 0.01- or 0.001-μF bypass capacitor cases. The plot in Fig. 13.13B is an expanded plot of the time near the initiation of the transient. Separate plots of the 0.001-, 0.01- and 0.1-μF simulations are shown in Fig. 13.14. Because the ESRs are modeled to be approximately the same for the 0.01- and 0.1-μF bypass capacitors, the initial voltage fall after the transient initiation is approximately the same. In similar circumstances, the bypass capacitor

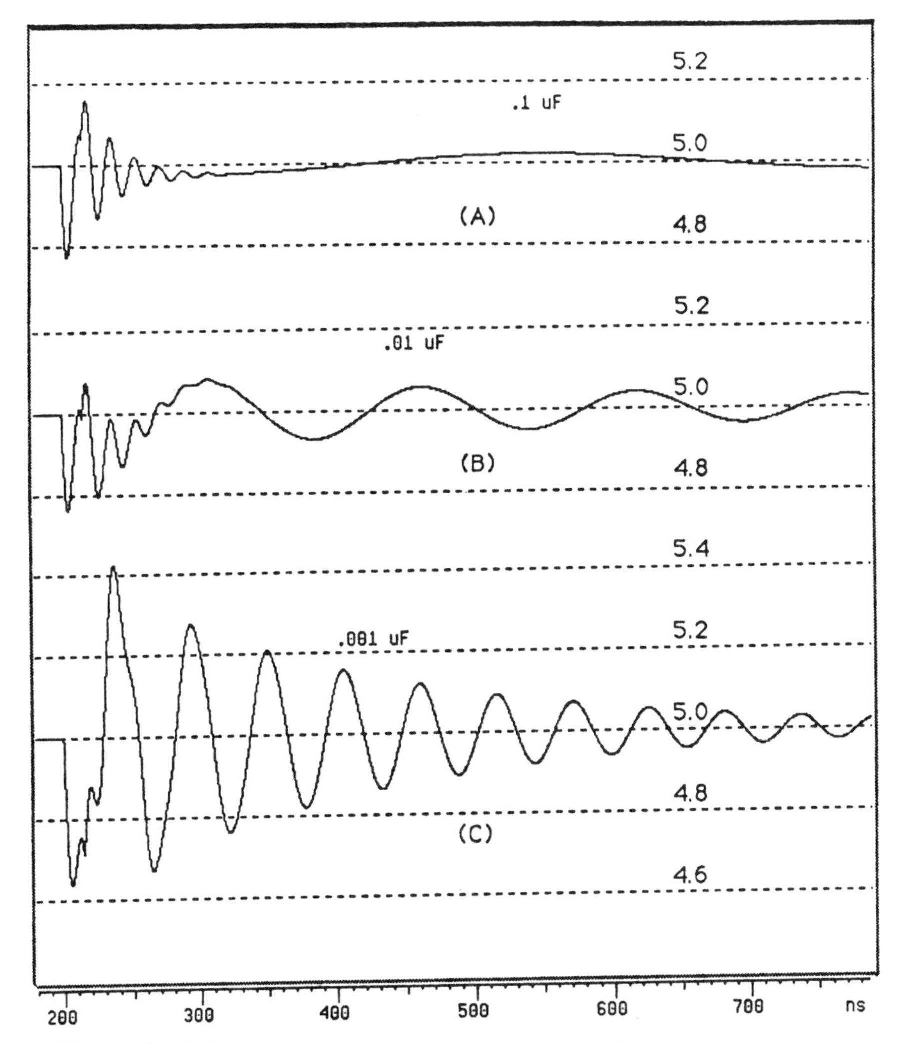

Figure 13.14 Separate plots for bypass capacitor performance comparison.

with the lowest ESR will respond with the least excursion from the nominal voltage.

The possibility exists that a particular layout for a special part or for some other reason may have an unavoidable inductance, which may ring with the 0.1-μF capacitor and cause problems. It may prove to be a better design choice to use the 0.01-μF capacitor if the resulting higher resonant frequency presents less problems than the use of the 0.1-μF capacitor. Another possibility exists that the +5-V dc power supply may have a requirement to be at a nominal value within a certain time limit for fast power up of a computer. In this case, the designer may want to use the smallest value of bypass capacitors possible for fast power up but still have enough bypassing to maintain a solid +5-V dc for fault-free logic operation.

Key Idea 3

> The transient load affects the amount of bypassing necessary for a given technology.

The load on the driver gate greatly affects the response of the +5-V dc power supply to transient changes. A few more example simulations will be performed to get a better feel for the bypassing design problem.

A simulation with the same circuit but the internal IC package inductances (L_5 and L_6) and the inductance of the circuit output (L_7) layout have been changed to 2 nH. A plot of this simulation with 0.1-μF bypass capacitors is shown in Fig. 13.15. The voltage excursion from nominal is now approximately 400 mV. The better the layout for the output circuit (for lower impedance and faster response), the more rigorous are the demands for bypassing. The internal inductance of the package and the output layout inductance isolate (somewhat) the output transient effects from causing transients on the input power to the logic IC package. If the circuit operation was very slow, these inductances could be larger, which would reduce the bypassing needs.

The original simulation, shown in Fig. 13.9, was run with a fan out of 10 driven gates. A simulation with C_3 set to 15 pF, for a fan out of 1, was performed and the resulting data are presented in Fig. 13.16A. The scale of this plot is the same as for the previous plots. It is apparent that the reduction of the load capacitance reduces the amount of needed bypass capacitance. The data plot in Fig.

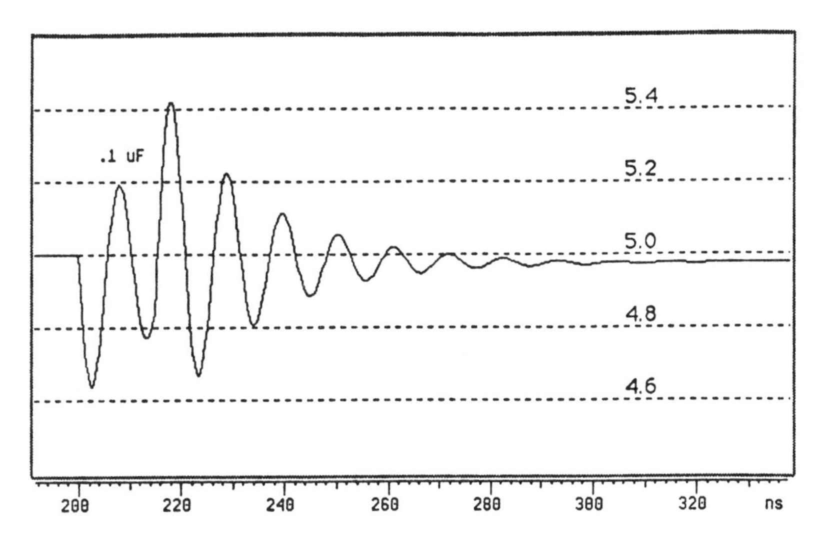

Fig 13.15 Simulation data plot with 0.1-μF bypass capacitor, reduced IC package inductance, and reduced load inductance.

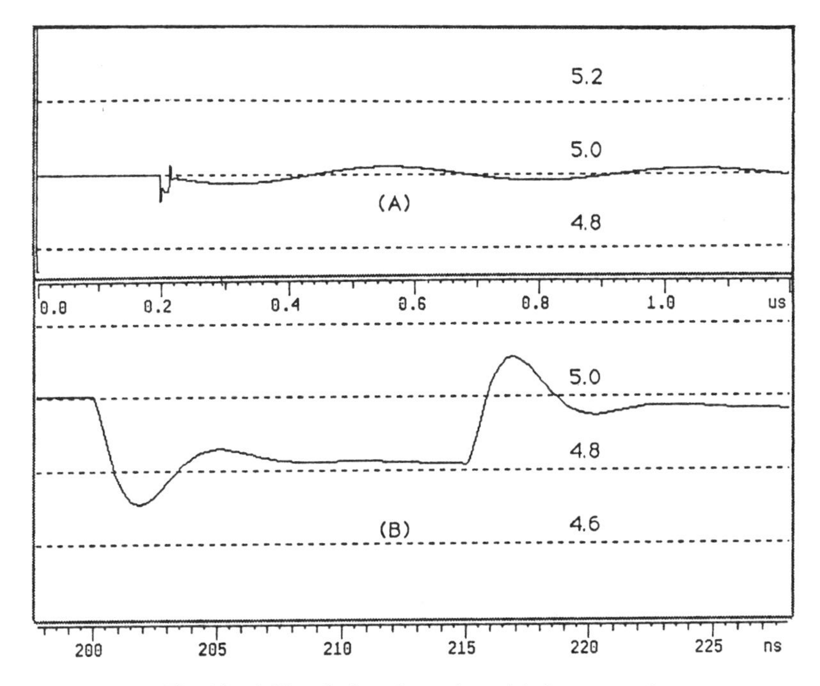

Fig 13.16 Simulation data plot with fan out at 1.

13.16B is an expanded plot of the time near the initiation of the transient. The vertical amplitude scale is expanded. The initial voltage fall from nominal is now only about 75 mV as compared to 225 mV for the case of a fan out of 10.

Once again we return to the original simulation shown in Fig. 13.9, but this time we change the rise time of the LSTTL gate from 15 to 1 nS. The results of this simulation are shown in Fig. 13.17. The voltage fall from the transient is in this case more than 1.1 V. More bypassing is needed in this case to reduce the output transient effects on the input power pins on the IC package. It is interesting to note that when using a particular technology such as LSTTL, as the fan out increases, the output rise times tend to slow down. The offsetting effects of more fan out and slower output rise times cause the bypassing needs to be somewhat constant throughout a design.

The lower the impedance of the technology of ICs, the lower the needed impedance of the bypass capacitor. The characteristic impedance of the bypass capacitor is the square root of the layout inductance and capacitor inductance divided by the capacitance:

$$Z_c = ((L_1+L_c)/C)^{1/2} \quad (13.1)$$

$$= ((L_3+L_4+L_8)/C_1)^{1/2} \quad (13.2)$$

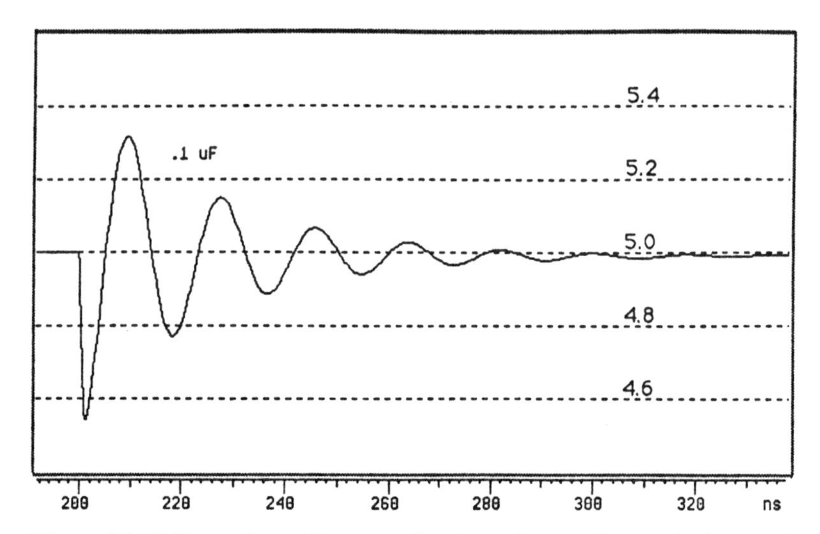

Figure 13.17 Data plot with output rise time changed from 15 nS to 1 nS.

Further, for good transient response by the driver U_1 and for good power supply pin bypassing for U_1 the following condition must be satisfied:

$$Z_c < 0.1\, Z_L$$

where $$Z_L = (L_7/C_3)^{1/2} \qquad (13.4)$$

The best layout for V_{cc} bus decoupling is presented in Fig. 13.18. A +5-V dc power plane and a return plane on adjacent layers provide the lowest impedance possible to the V_{cc} bus. By using surface mount capacitors and vias to connect to the power and return planes, the lowest inductance path is provided between the local energy storage source (the bypass capacitor) and the logic IC package.

The bypass capacitors are not only necessary to make high speed electronic circuits functional, but they also reduce the emissions conducted back into the power source. A list of parameters affecting the amount of bypass capacitance needed follows. The amount of bypass capacitance can best be analyzed

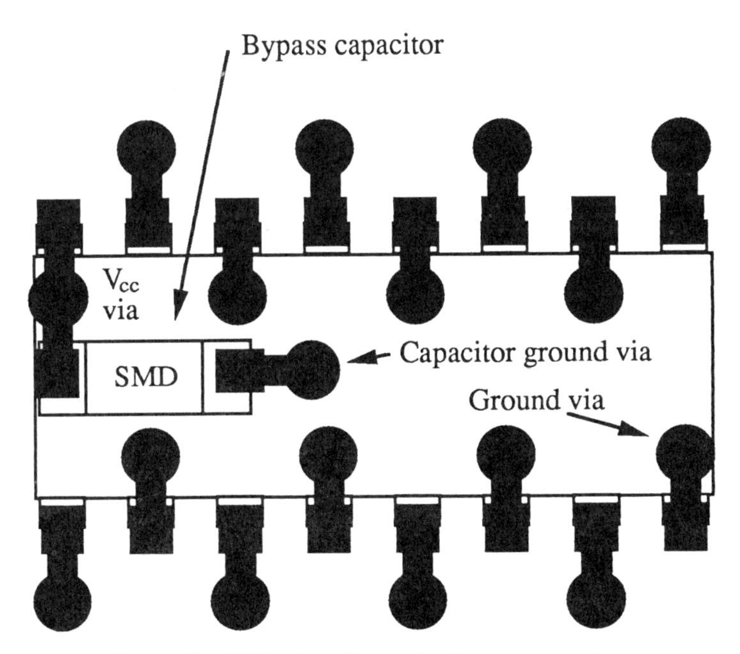

Figure 13.18 The best layout for bypass capacitors.

by its characteristic (driving) impedance, which includes the circuit layout parasitics:

1. Voltage transition time
2. Bypass capacitor layout inductance
3. Package inductance
4. Output layout inductance
5. Fan out

13.5 POWER ENTRY CAPACITOR

Power entry capacitors should be used on every printed wiring assembly (PWA) to bypass the incoming wiring for the same purpose as for the smaller IC package bypass capacitors but at a larger scale. Power entry capacitors must be located on each PWA at the point where the +5-V dc input power enters the PWA. These power entry capacitors provide bypassing for the whole PWA, which lowers the effective impedance of the +5-V dc bus on the PWA. Generally speaking power entry capacitors are 4.7 to 47 μF and are usually tantalum electrolytics. Metallized polycarbonate or multilayer ceramic capacitors can also be used for power entry.

QUESTIONS

13.1 Contrast the conducted emissions problems for computer circuitry and for switching power supplies.

13.2 What are the typical conducted emission paths for a computer?

13.3 What are some conducted emission problems internal to computers that are functional problems but are not measured and compared to standard limits?

13.4 What is the best bypass capacitor? What factors affect the amount of capacitance required to reduce the transient voltages at an IC power input pin to acceptable levels?

13.5 Why are the terms "decoupling" capacitor or "bypass" capacitor used for the capacitor that functions as a local energy source for each IC?

13.6 What kinds of circuits need bypassing or decoupling?

13.7 Does a power entry capacitor do the same function for a whole circuit board assembly as a bypass capacitor does for each IC? Explain.

13.8 How can EMI propagate from input to output if the input and output circuits are isolated with opto-couplers?

14

WHAT THIS ANALYSIS METHOD IS NOT

14.1 DIAGNOSTICS

The purpose for discussing diagnostics at this point is twofold. First, understanding conducted emissions requires the understanding of interrelated phenomena. Conducted and radiated energy are immersed together in the broader field of electromagnetics. Human engineers, for academic reasons, categorize physical phenomena into subsets of information that under the conditions of the subset are able to simplify approaches to problem solving. An example would be the treatment of near-field and far-field radiation. Certain assumptions can be made about a problem in electromagnetics if the designer determines that the distance to the source compared to the wavelength of the signal is far field or near field. We even say "near-field radiation"or"far-field radiation," yet the electromagnetic radiation is the same, but differing assumptions can be made relative to the distance to the source of radiation. A similar type of discussion can be made for conducted versus radiated emissions. After all, we call the current in a capacitor "conduction" when internally the current is passed through near-field radiation from capacitor plate to plate (or between electrodes).

Secondly, the diagnostic skills presented are important in the initial stages of solving a design problem. Field engineers should use these diagnostics to find the nature of a functional problem in electronic equipment to focus in and isolate the problem. For a design engineer the diagnostic skills may be used when a breadboard seems to exhibit EMI problems. These diagnostic skills should also be used in the initial stages of a design to predict and solve the problems up front at an early stage in the electronic design process. The diagnostic skills can help an engineer to identify the EMI design criteria for a particular application.

14.2 FIELDS

The impedance of electromagnetic energy (EMI energy) in air at far field is 377 Ω. If the impedance is determined to be significantly greater than 377 Ω the EMI energy can be treated as capacitive near field (electronic). If the impedance is determined to be significantly less than 377 Ω then the EMI energy can be treated as "inductive near field" (magnetic).

Key Idea 1

Impedance level	Type of coupling
Z > 1K	Capacitive noise
Z < 50	Inductive noise

The transition point of near field to far field is defined as when the distance to the source is greater than the wavelength divided by 2π, or approximately λ/6. Radiative far-field coupling requires a distance between source and receiver greater than the λ/2π.

14.3 RADIATION

Significant amounts of energy can be radiated from electronic circuits when the physical size of the circuit conductors is greater than the wavelength divided by 2. The conductors will work under these circumstances as a whip or dipole antenna.

Key Idea 2

D (maximum length of circuit conductor) = λ/2, which works well for an antenna.

There is no radiation if $d << \lambda/2$.

Electromagnetic (including EMI noise) coupling can be divided into four categories as follows:

1. Conductive	(direct contact)	Metallic path	
2. Capacitive	(electric field)	Near field	$\delta v/\delta t$
3. Inductive	(magnetic field)	Near field	$\delta i/\delta t$
4. Radiative	(electromagnetic)	Far field	$d>\lambda/2\pi$

There are five facts necessary to decide which of the four coupling mechanisms are present:

1. Metallic connection
2. Voltage transition
3. Current transition
4. Distance
5. Frequency (wavelength)

By inspection, if a metallic connection exists between two circuits, the transfer of energy between the circuits will be conducted. This, of course, assumes that the conducted path is of much lower impedance than any possible radiated path.

If the EMI noise includes a dc component (waveform is unsymmetrical), it is probably conductively coupled, although dc components can be caused by RF detection.

The voltage and current transitions of a waveform can be compared so as to determine the approximate impedance of the waveform:

$$Z \approx (\delta V/\delta t)/(\delta I/\delta t) \tag{14.1}$$

If the voltage and current transitions occur over equal time periods, the impedance becomes simply:

$$Z \approx (\delta V)/(\delta I). \tag{14.2}$$

The waveform is:

Capacitive if $Z > 377$
Inductive if $Z < 377$

If the noise waveform changes when a hand or body part is placed close to an electronic circuit, the coupling is capacitive. The body increases the capacitance locally about the circuit so as to change the impedance and natural frequencies of the path. Since the human body is primarily constructed of nonmagnetic materials and is of relatively high impedance, the presence of human flesh does not significantly affect magnetic fields. These capacitive effects are often observed when using an oscilloscope for voltage measurement.

14.4 CHARACTERISTIC IMPEDANCES OF COMMON PAIRS OF CONDUCTORS.

It is often useful to know the characteristic impedances of common configurations of pairs of conductors. Table 4.1 should be memorized and used for a rule of thumb guide.

Table 14.1 Characteristic impedances of common pairs of conductors.

Conductor configuration		Impedance $Z_c = (L/C)^{1/2}$
TV cable wires about 1/2 inch apart	≈	300 Ω
Parallel traces on a PWB	≈	10 Ω
Parallel planes on a PWB	≈	1 Ω

14.5 SHORTCOMINGS OF EMI TEST SIMULATION AS DESCRIBED HEREIN

Figure 14.1 is a model of an inductor wound on a permeable core. The elements in this model are linear and *not* frequency dependent. The purpose of the many elements used to model an inductor component is to create a model that will emulate the frequency dependent performance of the inductor being modeled. These elements are lumped in the model but in the actual physical real world are spatially distributed. Each of these actual distributed elements in themselves are frequency dependent to some extent. The elements' values chosen for the model are averaged or "worse cased" so that the model will represent the actual performance for a specific simulation purpose. Each use of the model should be tailored to the specific problem at hand.

The typical real world frequency dependent effects on the elements used in the inductor model of Fig. 14.1 are diagrammed in

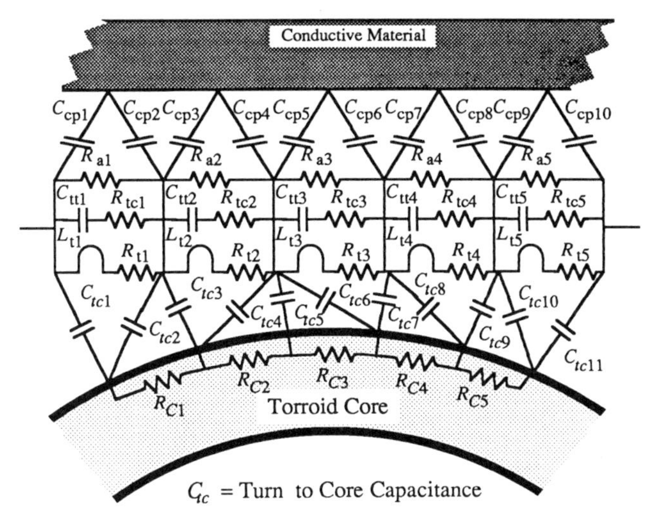

Figure 14.1 Lumped elements modeling a distributed field.

Fig. 14.2. But SPICE does not (directly) have means to vary element values with changing frequency. The mu of a typical inductor core, for instance, actually increases (starting at low frequency), reaches a maximum, and then decreases with increasing frequency. For the frequencies of interest for EMI, the mu of the core gets smaller with increasing frequency. A high level of accuracy for these models is achieved at only one frequency. We try to achieve better accuracy over a wider range of frequencies by adding elements that come into play in certain frequency ranges to improve the accuracy of the model.

The parameters that are not specifically (directly) simulated in SPICE are:

1. Models
   a. Frequency dependence of the elements
   b. Lumped elements modeling distributed effects
   c. Temperature effects
   d. Altitude (atmospheric pressure)

2. Simulation
   a. Electrical fields between components are generally not simulated
   b. Magnetic fields between components not simulated

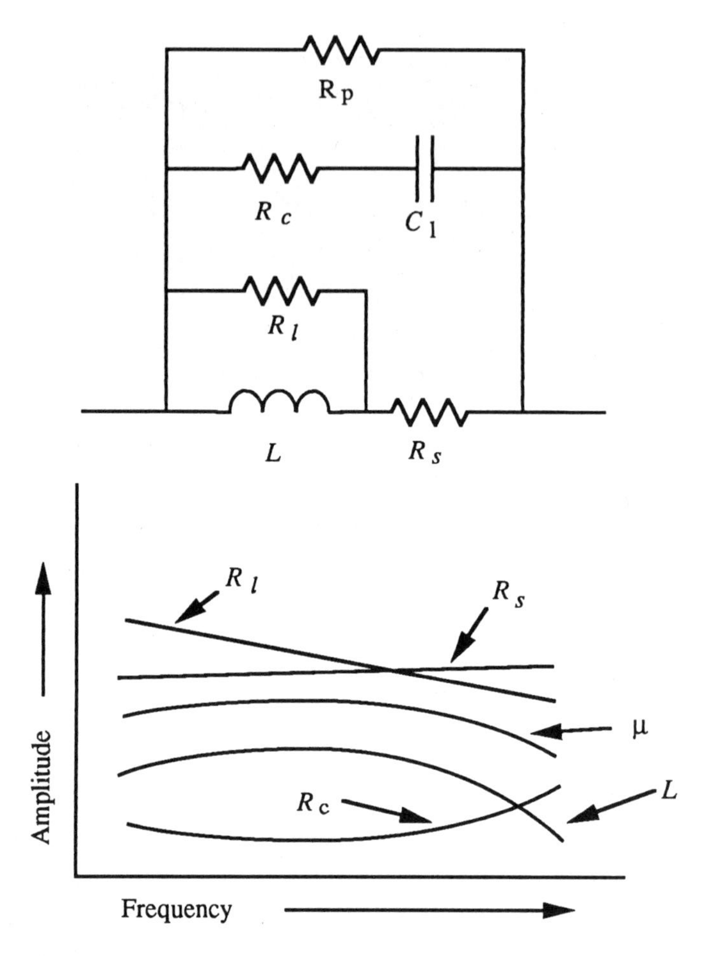

Figure 14.2 Frequency dependent aspects of inductor models.

Items c and d of the models (temperature and altitude effects) are not the subject of this book but can be accomplished easily with most electronic analysis software. The simulation methods presented attempt to extend the ability of simulation to include the effects of items a and b of the models by the use of lumped elements.

Each of the frequency dependent elements is modeled with multiple elements to create a second level of frequency dependence so as to refine the simulation to be more accurate when necessary.

Multiple elements can also be used to simulate distributed effects. As the number of lumped elements increases to infinity, the model approaches an exact simulation of a distributed element real world circuit. This, of course, assumes that the modeling by addition of elements is done properly.

A computer simulation can be used to find the values of parameters that are not easy to measure in the laboratory. The first step is to measure or calculate all of the model parameters possible. If only a relatively small number of unknown parameter values remain, they can be determined by educated guess and refined empirically by performing the simulation and adjusting the empirical parameters until the expected results are achieved. The confidence level will not be as high as it would be if all of the parameters were measured, but the results of empirical data can be very useful if the data are gathered carefully. We would, of course, prefer to measure and calculate all parameters when possible.

There are other software analysis programs that include the modeling of fields. The input parameters include the physical aspects of the components and layout. This type of simulation is definitely going to be used more and more in the future. Some of the developments necessary for these more complicated programs to be useful in the future are:

1. Development of knowledge and skills
2. Development of improved software
3. Faster and more powerful computers

Current software analysis programs that include fields are cumbersome and require a relatively large amount of input for an equivalent simulation. The models take more time to build. Powerful computers (number crunchers) are required to run these programs and the simulation runs typically require longer run times than do the lumped element oriented software electronic analysis programs such as SPICE, SABRE, ECA, and MCAP.

A designer using lumped element oriented software electronic analysis programs may attempt to model all of the magnetic fields with coupled inductors, coupling every element to every other element, and attempt to model the electrical fields with capacitances from each element to every other element. This would be very time consuming and messy. Most of the added elements would not be players and would be superfluous. Also the run times for all of the known lumped element oriented software electronic analysis programs increase tremendously with the addition of each magnetically coupled component. The addition of coupled magnetics causes the analysis convergence to take much more computer CPU time.

When electronic analysis programs that include field modeling become more available, all of the information learned here and in the use of the presented methods will apply to these more advanced electronic analysis programs.

The simulation methods using lumped element electronic analysis programs as described in this book are sufficient to get useful results and the resulting data are sufficiently close to the measured data, so that successful design decisions can be made based upon the simulation data. This design tool is extremely valuable in the early stages of the design process. It does not preclude the need for the other available design tools but should be used in conjunction with them for the best design effort possible. Whenever possible, we would prefer to use all our four main design tools, which are:

1. By hand calculations
2. Computer simulation
3. Laboratory experiments (breadboard measurements)
4. Previous design experience

When all of these items are correlated and give the "same" answer, our confidence level in our design can rise greatly and we can sleep at night, without worrying about unattended design details. Design details, if not addressed, will come back to haunt the designer when the equipment goes into production or into use by the customer out in the field.

QUESTIONS

14.1 Does current pass through a capacitor by metallic conduction or radiation?

14.2 What kinds of fields are modeled in SPICE? What kinds of fields are not practical to model in SPICE simulations?

14.3 What effect does the lack of field modeling in SPICE simulations have on the output data?

14.4 Why is it important to distinguish between conduction and radiation?

14.5 Are there any examples of electronic radiation that occur without any related current conducting source? If so, what are they?

14.6 Why is it important to know the limitations of the analysis methods performed for design purposes?

14.7 Explain what is meant by lumped elements and distributed elements in a model.

15

MAGNETIC SATURATION MODELING

In this chapter a background in magnetic saturation is reviewed to set the basis for the characterization of magnetic saturation. The parameters of minimum and maximum magnetic path length and "porosity" of domains, crucial to proper nonlinear inductor modeling, are identified. Almost all models of magnetics depend on curve fitting. The advantages of parametric models over curve-fitting type models are identified.

15.1 THE POLARIZATION OF MAGNETIC DOMAINS

Consider a voltage source applied to an inductor. Fig. 15.1 shows waveforms of current and voltage and pictures of a model of the polarization of the magnetic domains in different stages of excitement, as a result of the applied voltage waveform. There is no history of previous voltage, but one may see remnance in Fig. 15.1A. There is a remaining bias of polarization of magnetic domains from the last applied voltage pulse in the positive direction. The same state of remnance is in Fig. 15.1E.

At time t_2, all of the magnetic domains are polarized in full alignment with the applied field. In this saturated state the inductor has the inductance of its coil in free space ($B = H$). The current rises very fast when in saturation. Usually the rise in current looks like a short circuit but is still limited in rise time by the coil's free space inductance. If the associated electrical circuit isn't current limited, large currents will flow and some components will fail. However, no damage to the magnetic core is caused by saturation. Although, if the large currents in saturation aren't designed for, the magnet wire of the coil may char due to heat

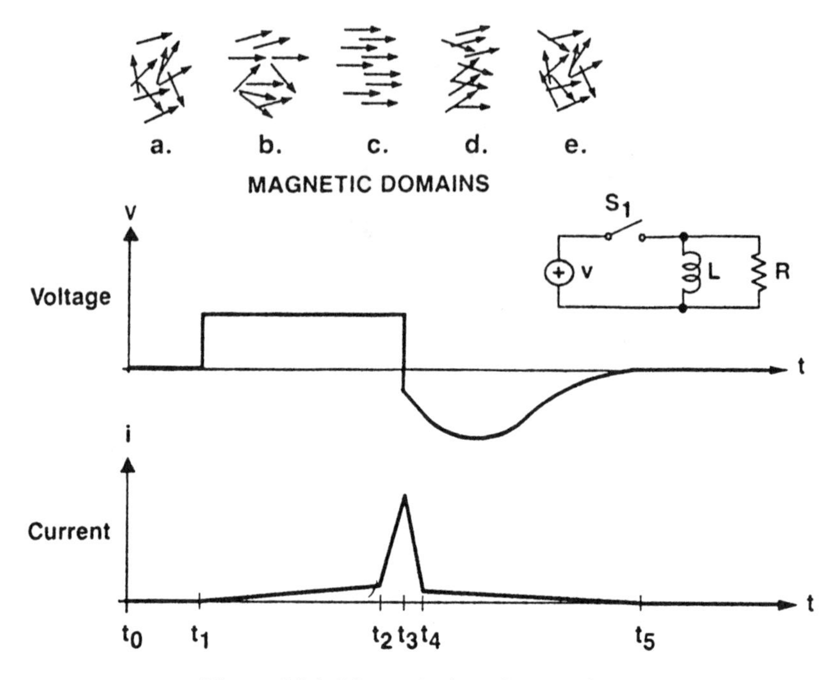

Figure 15.1 Magnetic domain snapshots.

from the heavy current flow. At time t_4, the current is slightly less that at time t_2, because of the energy dissipated in doing the work of polarizing the magnetic domains. This effect is known as hysteresis. It is our intention to avoid any hysteresis in our models because the saturation model must cause only saturation effects or there will be uncontrolled hysteresis side effects from the saturation model.

At time t_5, a polarization of the domains is still left as a remnant of the applied voltage. A negative voltage must be applied to reverse the polarity of the remnance.

15.2 DEVICE, CORE, AND MATERIAL PROPERTIES

The distinction between device, core, and material properties is shown in Fig 15.2. It is interesting to note in Fig. 15.2C that saturation will be abrupt if there is no porosity of magnetic domains in the material. The loops in frames a and b will virtually always saturate more softly because the core geometry will usually

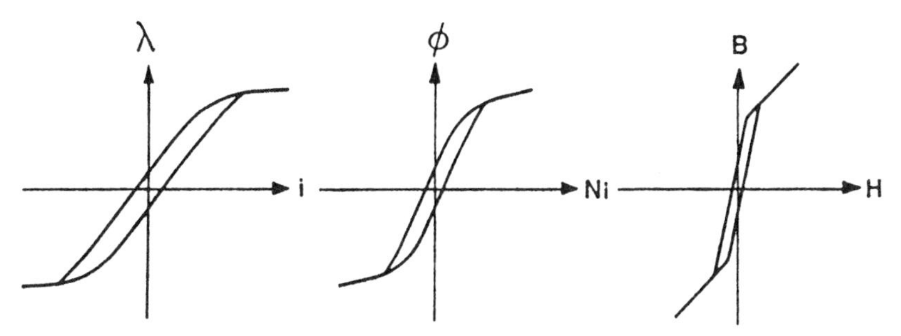

Figure 15.2 Magnetic properties.

not be filamentary. In all three frames of Fig 15.2, hysteresis is evident. Here we will be plotting the device properties to gain information about the material properties. The constants in the proper ratio, shown in Fig. 15.3, are used to pass from the device properties to the material properties' scaling.

To move from device properties (windings) to core properties (geometry) λ is divided by N, and i is multiplied by N. This is a turn to geometry conversion "so to speak."

To move from core properties to material properties the flux is divided by the area of core to get B (flux density). Ni (ampere turns) is divided by the magnetic path length to get H, the magnemotive force necessary to create the field of flux density B.

15.3 CORE GEOMETRY EFFECTS

The flux path will follow a circular path in torroidal cores as shown in Fig. 15.4. For a coil of wire wound on a core and a current forced through it, a magnemotive force of $H = NI/l_m$ is established in the core. For the flux path of l_{min} the magnemotive force is much stronger than for the path of l_{max}. As more magnemotive force is applied, the core saturates from the inside out. Saturation will be softer as the ratio of l_{max}/l_{min} gets larger. To model saturation one needs the input parameters of at least the area of the core, mean magnetic path length, and ratio of l_{max}/l_{min} so that the geometric effects may be modeled. The E core, shown in Fig. 15.4, also has this geometric effect. Only a filamentary core will not exhibit these characteristics.

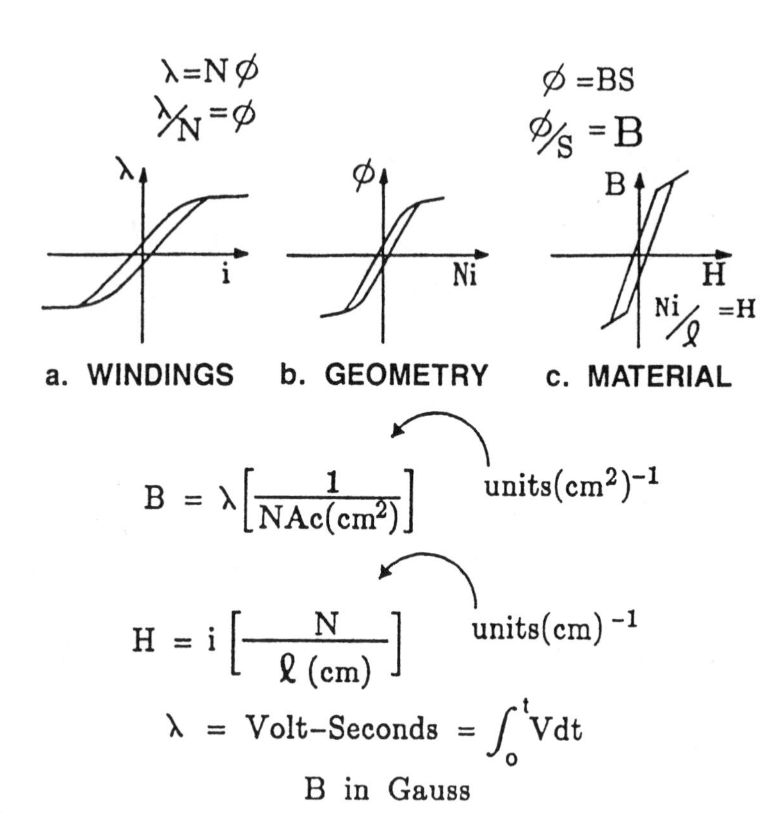

Figure 15.3 Device to material conversion factors.

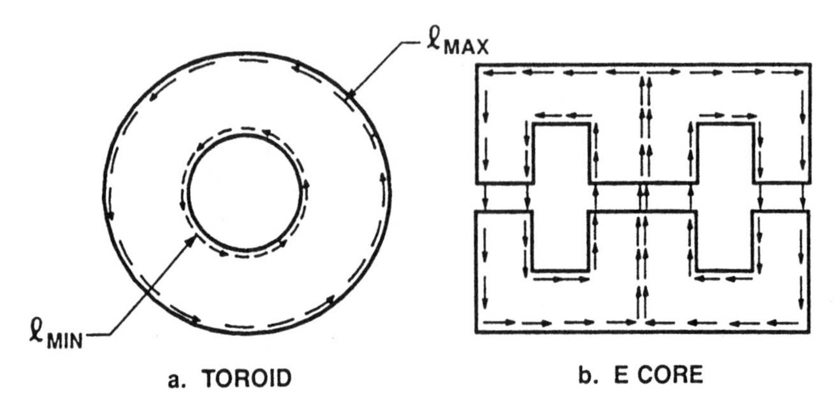

Figure 15.4 Core flux paths.

15.4 EFFECTS OF CORES MADE OF TWO DIFFERENT MATERIALS

Cores made of two different materials will saturate at different levels of flux, as shown in Fig 15.5. The combination of materials smooths the transition into the saturation region.

Cores that are approximately filamentary (l_{max}/l_{mean} = 1) can be wound to observe the core material's properties. The only softness effects on saturation are caused by the amount of distributed air gap, or the "porosity", of magnetic domains of the chemistry of the material being investigated.

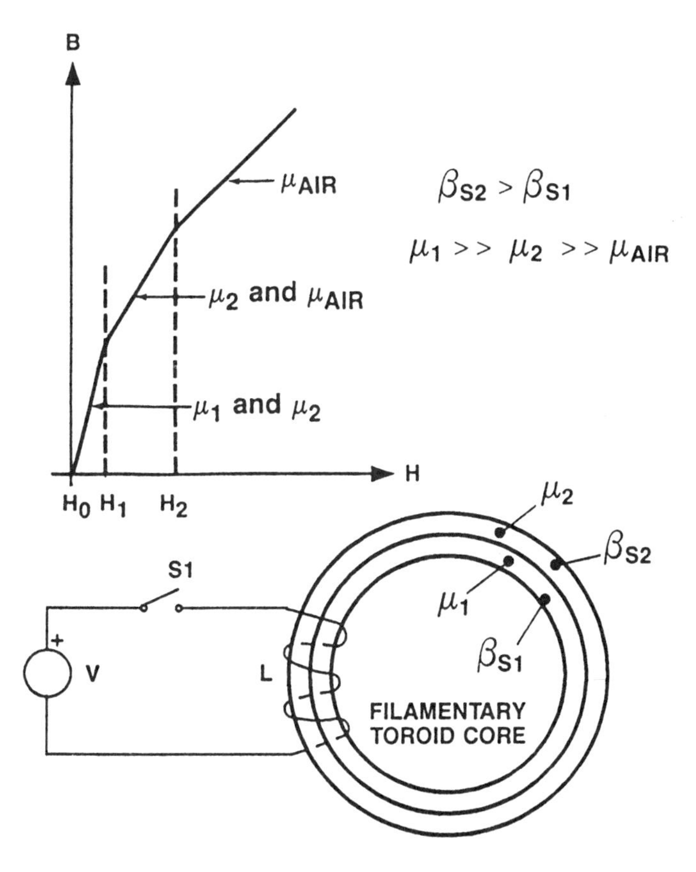

Figure 15.5 Saturation in cores with two materials of different saturation flux densities.

The action of air gaps in materials is similar to the situation of two different materials. These gaps cause a smoothness in the transition into the saturation region.

Mu-metal is relatively continuous with magnetic domains. There is very little molecular space between crystals in the molecular structure of ferrous metal. The *B-H* loop will exhibit abrupt saturation in mu-metal. Powdered iron cores have distributed gaps caused by the binder and fillers used to control the permeability. The saturation threshold will be different for some paths, as shown in Fig. 15.6, causing a soft, or less abrupt, transition into the saturation region.

15.5 SOME CRUCIAL PARAMETERS TO MODEL SATURATION

To model saturation accurately a porosity factor must be a parameter input to the model. Porosity is a chemical property of the magnetic material. Some crucial magnetic parameters for modeling saturation are:

1. Area of the core
2. Maximum magnetic path length
3. Minimum magnetic path length
4. Chemical porosity factor of core material

The saturation model must monitor the flux level in the core and change the rate of current rise (or the inductance) at the threshold of saturation. This implies the use of a software flux sensor and a software switch (Fig. 15.7) to choose one of two inductances on the fly.

15.6 METHODS OF INTEGRATING VOLTAGE

As previously mentioned we plot the integral of the voltage with respect to time versus the current in the core to be able to see the corresponding *B-H* loop of the material being investigated. There are basically only scaling constants involved to move from the device characteristics to the material properties of the core.

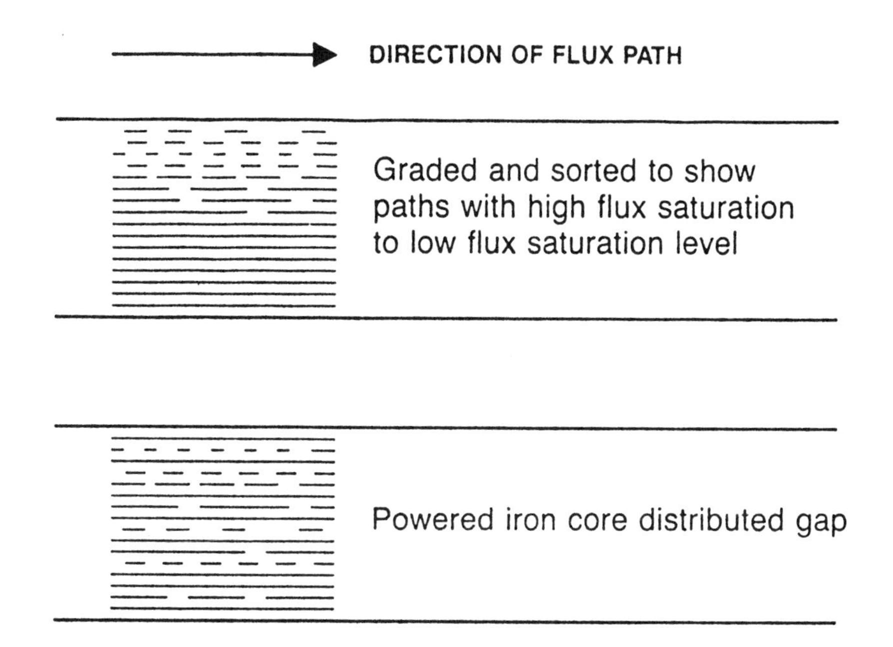

Figure 15.6 Description of softness in saturation because of the porosity of the magnetic domains.

One method of *B-H* loop measurement uses the setup of Fig. 15.8. If the *RC* time constant is large compared to the switching period and $X_c << R$, capacitor *C* will charge in a linear fashion and can be used to integrate the voltage with the associated scaling factors considered. The oscilloscope is used in the *X-Y* mode with the integral of the voltage on the *Y* axis and the current on the *X* axis. For modeling, we really don't want the effects of the *RC* inte-

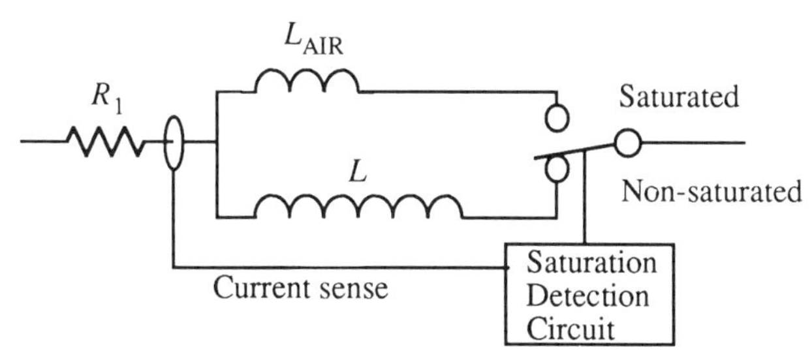

Figure 15.7 Model of change in inductance when passing from linear to saturation mode.

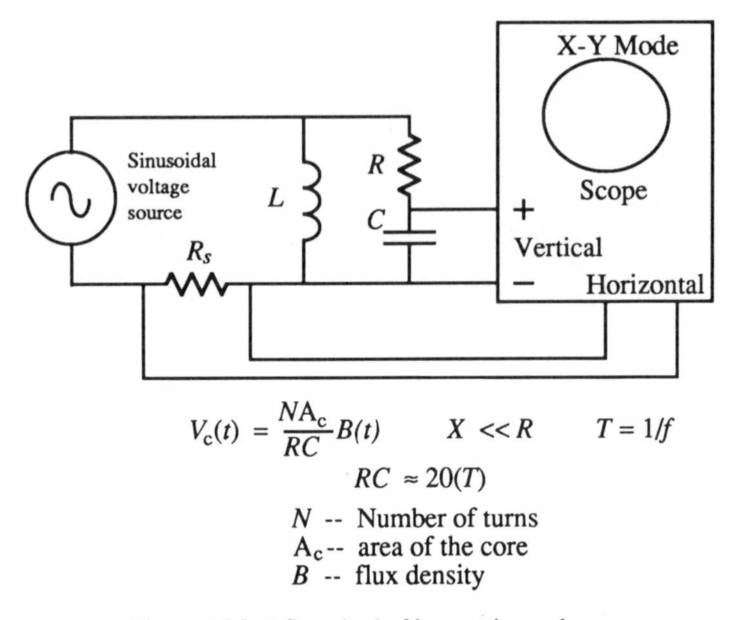

Figure 15.8 *RC* method of integrating voltage.

grator altering the response of the inductor model. This undesired effect can be eliminated by the use of a voltage controlled voltage source, as shown in Fig. 15.9, to isolate the unwanted effects. This method can be used without any processing done external to SPICE. Many new versions of SPICE can plot voltage versus current.

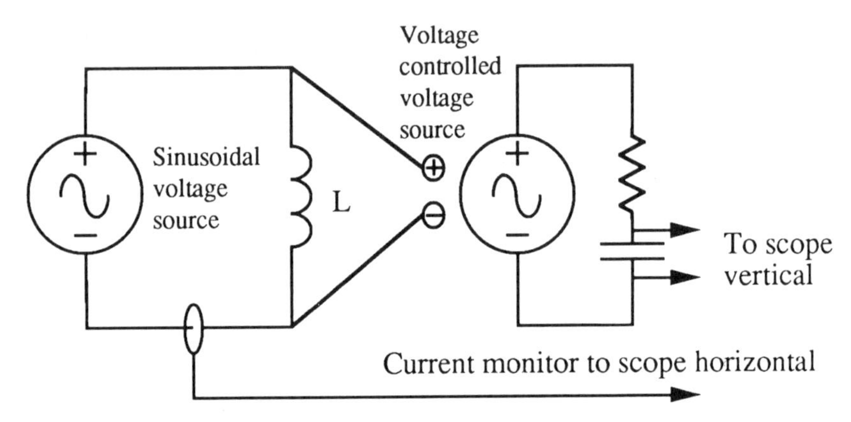

Figure 15.9 Isolation of *RC* integrator by use of controlled source.

Another method to integrate the voltage is to dump the voltage and current data into a file, after the SPICE run, and process the data in FORTRAN, BASIC, or some other appropriate language. This method was used for this material. The integral of the voltage with respect to time, $\int V(t)$ is calculated as follows:

$$\int V(t) = V(t - 1) + [(V(t) + V(t - 1))/2]\ \delta t \qquad (15.1)$$

The δt chosen must be small enough that the residual error in integration is very small when compared to the applied current. With δt too large, the error will cause the model to demonstrate false hysteresis even when no hysteresis is supposed to be in the model. This, of course, is an unwanted effect.

15.7 DR. LAURITZEN'S SATURATION MODEL

Dr. Peter Lauritzen's saturation model in his paper *A Computer Model of Magnetic Saturation and Hysteresis for Use on SPICE2* (IEEE 1984) is suited for implementation in public domain SPICE with use of the poly sources. A simplified version of the model is shown in Fig. 15.10. An actual SPICE implementation of the model is shown in Fig. 15.11.

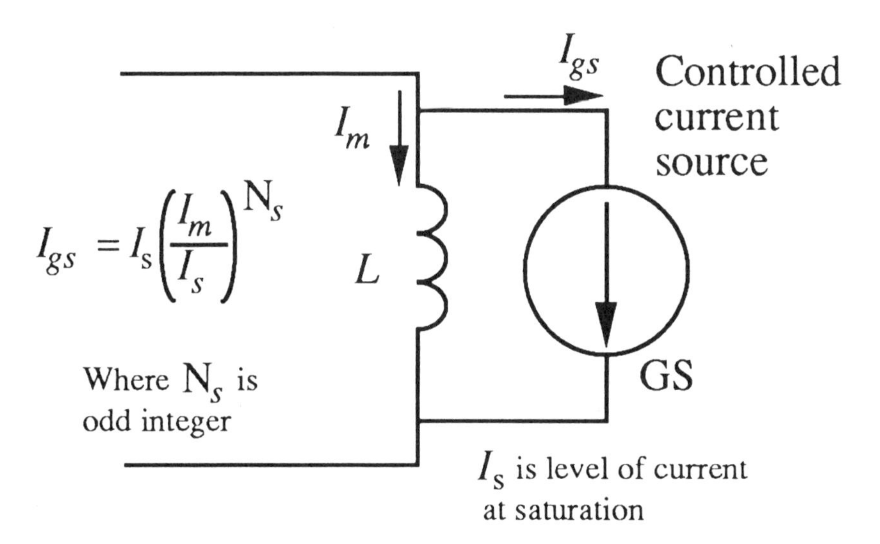

Figure 15.10 Basic saturation model.

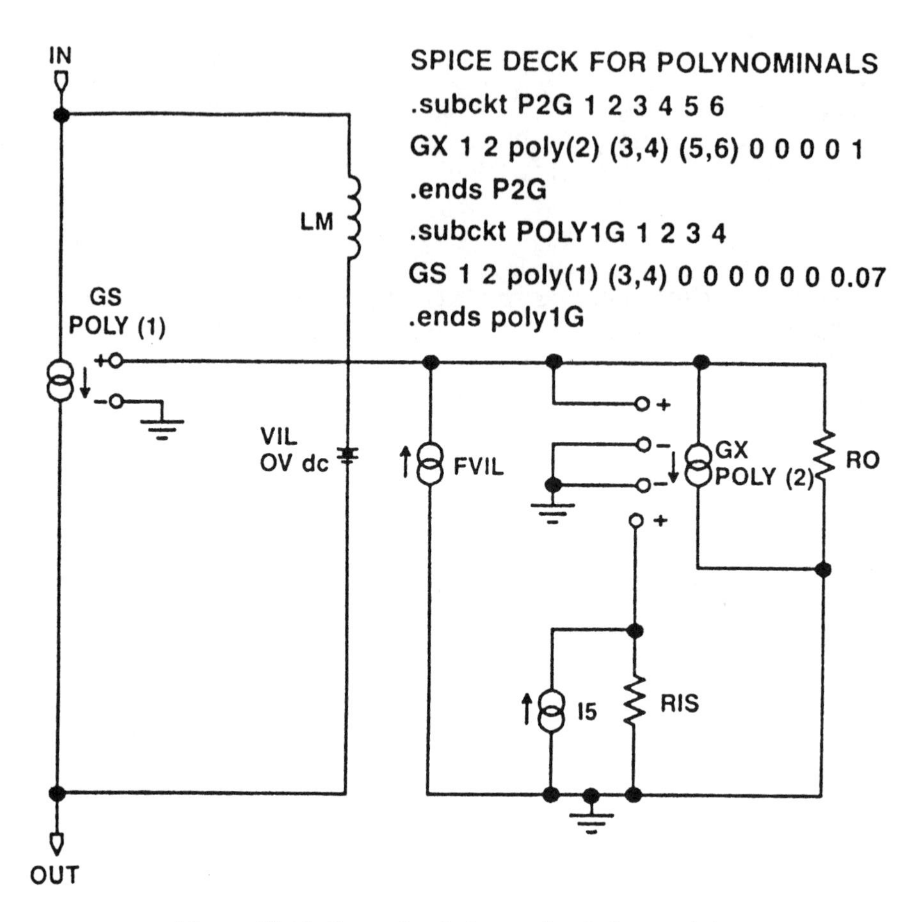

Figure 15.11 Saturation inductor simulation model.

The exponent N_s controls the curvature of transition into the saturation region of the *B-H* loop. The input of FVIL monitors the current in the inductor. GX is part of a feedback circuit that creates a voltage at the positive input of GS that represents I_m/I_s. GS then raises this input to the required exponent N_s and shunts the saturation current across the inductor. N_s must be chosen by curve fitting in the transition area of saturation (See Fig. 15.12). A good feature of Dr. Lauritzen's model is that the current, when $(I_m/I_s) <$ 1, gets small very fast even with the small exponent of 3, which gives a large curvature, but the hysteresis, caused by the minute saturation current for values of $(I_m/I_s) < 1$, is fortunately not perceivable.

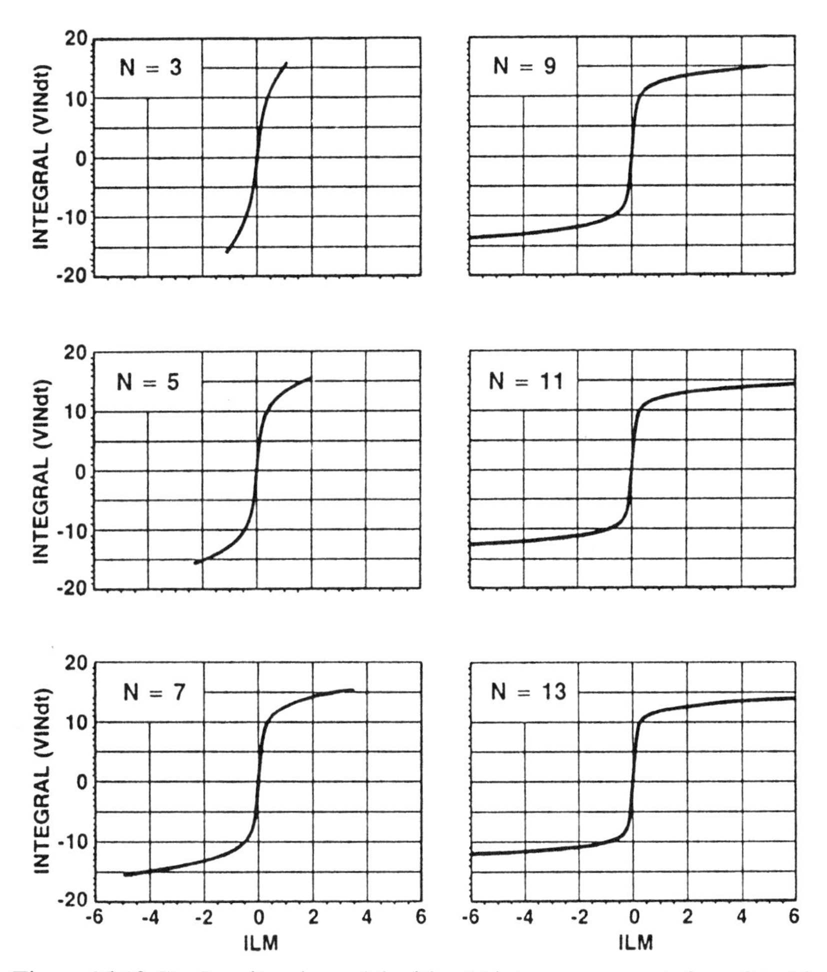

Figure 15.12 Dr. Lauritzen's model with odd integer exponents from 3 to 13.

In this model, the final slope of current rise is different for all values of N_s, yet all materials finally saturate at the slope of permeability of air. This is a shortcoming of the model that we would like to improve. For *B-H* loops the slope in saturation is always 1 (relative permeability). For flux versus ampere-turn loops the slope is equal to the area of the core divided by the magnetic path length (A_c/l_m). A typical slope would be 1/5. For flux linkage versus current loops the slope in saturation is the area of the core divided by the product of the magnetic path length and the square of the number of turns $[A_c/(l_m N^2)]$. This ratio will be approximately zero in nearly all cases. This model usually works satisfactorily for models of inductors or transformers but would not be accurate enough for close modeling of saturable reactors when the amount of current flow in the saturation region is not accurate.

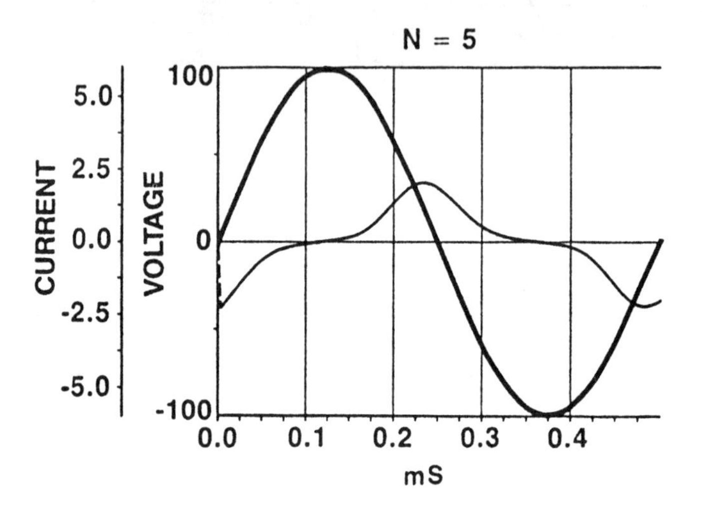

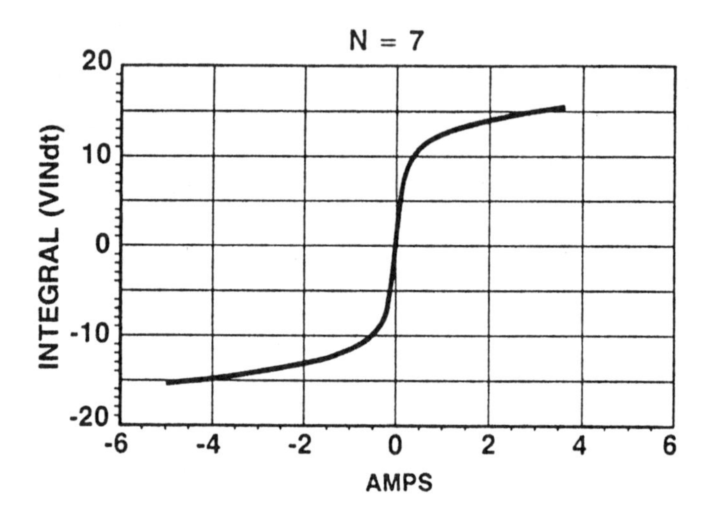

Figure 15.13 Sinusoidal voltage source applied to saturable inductor.

Figure 15.13 shows the outputs from a SPICE run of inductor saturation from a sinusoidal voltage source. The *B-H* loop shape is what we would expect. Also, there is no hysteresis caused by the saturation model. The saturation model was also used in a model of a forward converter, shown in Fig. 15.14. The output voltage and currents are shown in Fig 15.15. At about 3 μS, the core saturates and the current rises to nearly 25 A. During this high rise of current one may see that the FET switch is coming out of electronic saturation as the core saturates magnetically.

15.8 THE CORE GEOMETRY AND MATERIAL POROSITY REGION OF THE *B-H* LOOP

The bold part of the waveform in Fig. 15.16 is not accurately modeled unless the appropriate aspect ratio of the routine is exactly the same as the proper parameter of porosity of magnetic domains

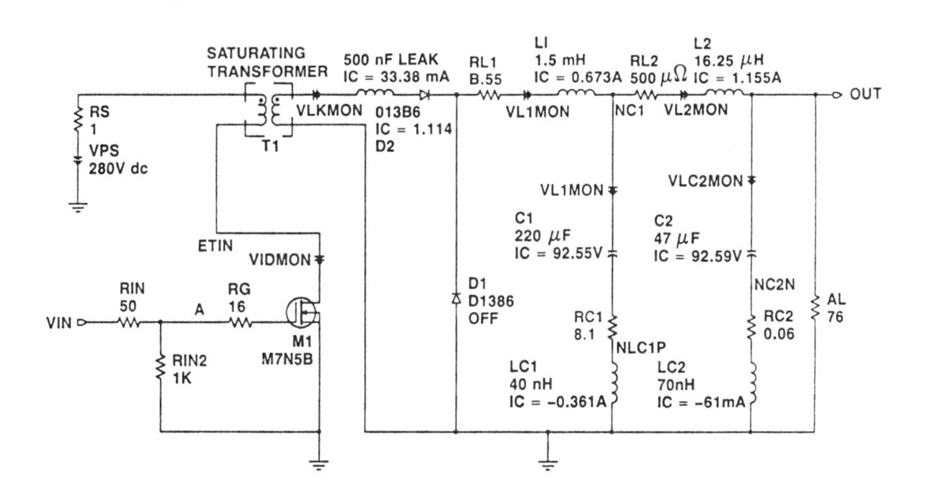

Figure 15.14 Forward convertor model.

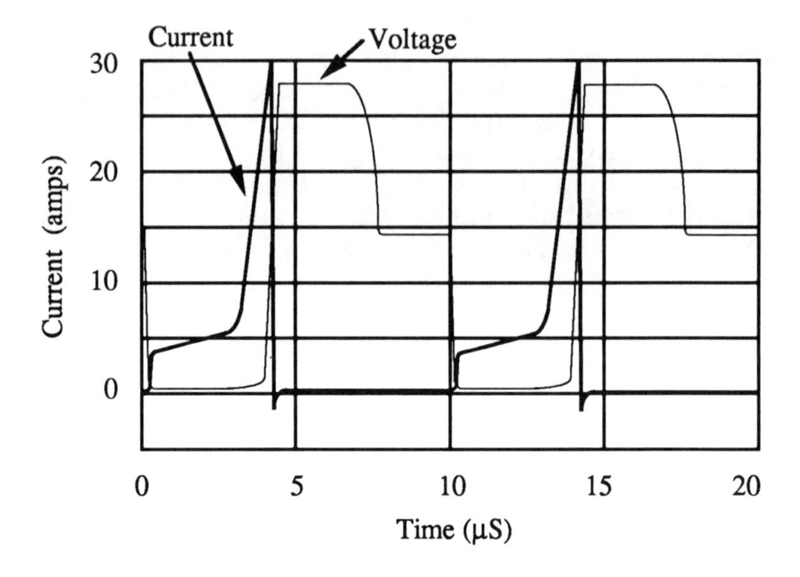

Figure 15.15 Saturated forward convertor output.

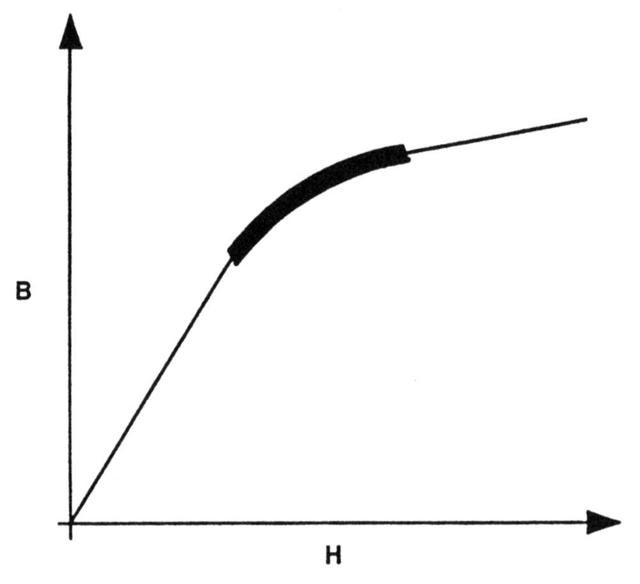

Figure 15.16 Core geometry and porosity of magnetic domains region of *B-H* loop.

and the core geometry. This is highly unlikely. Some high-level SPICE versions still lack one or two of the two needed parameters, which are:

1. Maximum and minimum magnetic path length
2. Chemical porosity of magnetic domains in core material

15.9 CURVE FITTING VERSUS PARAMETRIC MODELS

Curve fitting is a mathematical approximation by means of a algorithm that is adjusted to approximate some known data. The curve fitted model may or may not respond in dynamic situations in any way like the responses of the real object being modeled.

Parametric mathematical models that model the constituents of the object being modeled to a fineness of resolution such that all static and dynamic test responses of the model are the same as the object being modeled are by far more powerful and efficient to apply. The parametric models carry a much higher level of confidence in giving an appropriate response. Parametric models are easier to implement since the curve fitting portion of the work to set up a model is changed to curve checking. This is a much faster and more straightforward approach.

Dr. Lauritzen's model could be used, by curve fitting, for the section of the loop in Fig. 15.17 from I_0 to I_1. For currents greater than I_1 the model is switched, with a software switch, as in Fig. 15.7, to another model with a different inductance. The new inductor L_{air} has the magnitude of inductance of the same coil of wire with no magnetic material in proximity to the area of the coil. The initial condition for the inductor L_{air} will be I_1 in Fig. 15.7. The choice of I_1 is somewhat subjective but should be about at the point where there is no more hysteresis. After all of the domains are polarized, no physical work can be done on them. In the saturation region $\mu_r = 1$ and $B = H$.

SPICE versions that offer FORTRAN capability would be ideal to make a parametric model in the form of Fig. 15.7. IG SPICE allows access to the FORTRAN source code, so a programmer could manipulate the source code to include the needed parameters of maximum and minimum magnetic path length and porosity of domains.

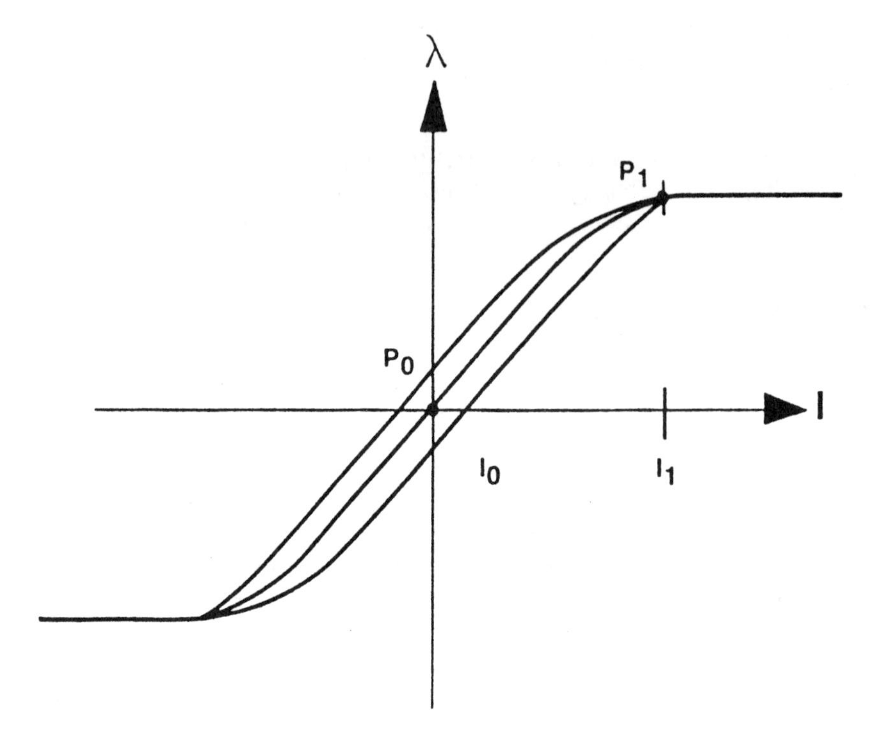

Figure 15.17 *B-H* loop.

15.10 CONCLUSION

The parameters of core geometry and porosity of domains were identified and their effects on *B-H* loops were described. Dr. Lauritzen's model depends upon curve fitting but can be implemented on public domain SPICE versions. An implementation of Dr. Lauritzen's model was demonstrated. His model could be greatly improved for modeling saturable reactors by using a software switch to be able to fit the *B-H* curve in the saturation and linear regions.

IG SPICE, using FORTRAN, offers the possibility of making accurate parametric saturable inductor models by using the magnetic model in IG SPICE but with the needed parameters of maximum and minimum magnetic path length and the porosity of domains added to this model. This model also needs the ability to model gaped cores. A list of definitions follows:

N	Number of turns
λ	Flux linkages
B	Flux density
H	Magnemotive path length
S	Area of the core
l	Magnetic path length
$\int V(t)$	Integral of voltage
ϕ	Flux

QUESTIONS

15.1 Why do you think magnetic saturation modeling may be important in some EMI analysis situations?

15.2 What is the difference between a curve fitted model and a parametric model as defined by chapter 15?

15.3 Does magnetic saturation ever directly cause catastrophic failure?

15.4 At current levels where the core in an inductor saturates, what will be the effective inductance of the inductor?

15.5 What causes the curved portion of the hysteresis loop to be curved?

15.6 What would be the results of an unexpectedly saturated EMI filter inductor on the EMC performance? Are there other functional situations in which magnetic saturation would affect EMC performance?

15.7 When a balun saturates, what happens to its coupling coefficient? How can a balun saturate?

APPENDIX

BASIC FFT

```
10 REM ****PROMPTED RECTANGULAR WAVE INPUT ***
20 REM
40 PRINT "RECORD LENGTH MUST BE A POWER OF 2."
50 INPUT "INPUT RECORD LENGTH.";N
60 PRINT "WHAT IS THE TIME PERIOD OF THE WINDOW WIDTH?"
70 INPUT "PERIOD IS";WW
80 M=LOG(N)/LOG(2)
90 DIM INPUTRE(N-1),INPUTIM(N-1),BUFFERRE(N-1),BUFFERIM(N-1)
100 DIM MAG(N-1) , MAGLOG(N-1)
110 DIM TFRE#(N/2-1),TFIM#(N/2-1),TMP(N-1)
120 REM  *******RECTANGULAR WAVE PARAMETER PROMPTS *********
130 INPUT "WHAT IS AMPLITUDE A ?";A
132 INPUT "WHAT IS AMPLITUDE B ?";B
140 INPUT "WHAT IS PULSE WIDTH?";PW
150 INPUT "WHAT IS PERIOD OF WAVEFORM?";T
160 INPUT "WHAT IS RISE TIME ?";TR
170 INPUT "WHAT IS FALL TIME ?";TF
180 REM ******* CREATE RECTANGULAR WAVE  ****
190 RRR = 0
200 RR = RRR
210 TS = WW/N
220 TTT = T/TS
230 TT = TTT
240 NC = WW/T
250 XXX = TR/TS
260 XX = XXX
270 SLOPETR = A/XXX
280 YYY = PW/TS
290 YY = YYY
300 ZZZ = ((PW+TR/2+TF/2)/TS)-1
310 ZZ = ZZZ
312 SLOPEHAT = (B-A)/(YYY-XXX)
320 SLOPETF = B/(TF/TS)
340 FOR L = 0 TO NC - 1
```

```
350 FOR J = RRR TO XXX
360 INPUTRE(J) = (J-TTT+TT)*SLOPETR
370 NEXT J
380 FOR J = XXX TO YYY
390 INPUTRE(J) = ((J-TTT+TT)*SLOPEHAT)+A
400 NEXT J
410 K = 1
420 FOR  J = YYY TO ZZZ
430 INPUTRE(J) = B-(K*SLOPETF)
440 K = K + 1
450 NEXT J
460 FOR J = ZZZ+1 TO TTT-1
470 INPUTRE(J) = 0
480 NEXT J
490 RRR = RRR + TT
500 XXX = XXX + TT
510 YYY = YYY + TT
520 ZZZ = ZZZ + TT
530 TTT = TTT + TT
540 NEXT L
550 REM  *******************************************
560 REM     **** INPUT DATA FOR XFORM HERE--REAL DATA INTO ARRAY**
600 REM ***** SCALES SCREEN TO DISPLAY INPUT DATA *****
610 MAX = 0
620 MIN = 0
630 FOR J = 1 TO N-1
640 IF INPUTRE(J) > INPUTRE(MAX)  GOTO 680
650 IF INPUTRE(J) < INPUTRE(MIN)  GOTO 700
660 NEXT J
670 GOTO 720
680 MAX = J
690 GOTO 660
700 MIN = J
710 GOTO 660
720 PRINT "MINIMUM IS "INPUTRE(MIN) "     MAXIMUM IS "INPUTRE(MAX)
730 REM  ************* PLOT INPUT DATA *****************
740 INPUT TYPE$
750 CLS
760 KEY OFF
770 SCREEN 2
780 WINDOW (0,INPUTRE(MIN)) - (N+8,INPUTRE(MAX))
790 FOR J = 0 TO N-2
```

```
800 LINE (J,INPUTRE(J)) - (J+1,INPUTRE(J+1))
810 NEXT J
820 REM ****** PROMPT FOR FORWARD OR INVERSE FFT ***********
830 INPUT TYPE$
840 CLS
850 PRINT "IS THIS AN INVERSE OR FORWARD"
860 PRINT "TRANSFORM?  ENTER I OR F."
870 INPUT TYPE$
880 IF TYPE$="I" THEN SIGN=1
890 IF TYPE$="i" THEN SIGN=1
900 IF TYPE$="F" THEN SIGN=(-1)
910 IF TYPE$="f" THEN SIGN=(-1)
920 IF SIGN=-1 THEN 1090
930 REM
940 REM ****REORDER DATA FOR INVERSE FFT****
950 REM
960 FOR I=0 TO N/2
970 BUFFERRE(I)=INPUTRE(I+(N/2-1))
980 BUFFERIM(I)=INPUTIM(I+(N/2-1))
990 NEXT I
1000 FOR I=0 TO N/2-2
1010 BUFFERRE(I+(N/2+1))=INPUTRE(I)
1020 BUFFERIM(I+(N/2+1))=INPUTIM(I)
1030 NEXT I
1040 FOR I=0 TO N-1
1050 INPUTRE(I)=BUFFERRE(I)
1060 INPUTIM(I)=BUFFERIM(I)
1070 NEXT I
1080 REM
1090 REM ****GENERATE TWIDDLE FACTORS****
1100 REM
1110 PI#=3.141592653589795#:PI2#=2*PI#
1120 FOR P=0 TO N/2-1
1130 TFRE#(P)=COS(PI2#*(-P)/N)
1140 TFIM#(P)=(SIGN)*SIN(PI2#*(-P)/N)
1150 NEXT P
1160 REM
1170 REM ****** COMPUTE FAST FOURIER TRANSFORM ******
1180 REM
1190 FOR I= 1 TO M
1200 L=0: H=0
1210 G=(N/2^I)
```

```
1220 FOR K=0 TO (N-1) STEP G
1230 TFI=0
1240 TFIFLAG=(-1)^(L+1)
1250 FOR J=0 TO (G-1)
1260 TFI=J*2^(I-1)
1270 R=K+J:S=J+H:T=J+G+H
1280 IF TFIFLAG>0 THEN 1320
1290 BUFFERRE(R)=INPUTRE(S)+INPUTRE(T)
1300 BUFFERIM(R)=INPUTIM(S)+INPUTIM(T)
1310 GOTO 1360
1320 TEMPRE=INPUTRE(S)-INPUTRE(T)
1330 TEMPIM=INPUTIM(S)-INPUTIM(T)
1340 BUFFERRE(R)=TEMPRE*TFRE#(TFI)-TEMPIM*TFIM#(TFI)
1350 BUFFERIM(R)=TEMPRE*TFIM#(TFI)+TEMPIM*TFRE#(TFI)
1360 NEXT J
1370 L=L+1:H=INT(L/2)*G*2
1380 NEXT K
1390 FOR II=O TO N-1
1400 INPUTRE(II)=BUFFERRE(II)
1410 INPUTIM(II)=BUFFERIM(II)
1420 NEXT II
1430 NEXT I
1440 FOR I=O TO N-1
1450 INPUTRE(I)=INPUTRE(I)/N
1460 INPUTIM(I)=INPUTIM(I)/N
1470 NEXT I
1480 REM
1490 REM ****BIT REVERSAL ROUTINE TO UNSCRAMBLE FFT RESULTS****
1500 REM
1510 FOR I=O TO N-1
1520 INDEX%=I
1530 IOUT%=O
1540 FOR J=1 TO M
1550 TEMP%=1 AND INDEX%
1560 IOUT%=IOUT%*2
1570 IOUT%=IOUT%+TEMP%
1580 INDEX%=INDEX%\2
1590 NEXT J
1600 BUFFERRE(I)=INPUTRE(IOUT%)
1610 BUFFERIM(I)=INPUTIM(IOUT%)
1620 NEXT I
1630 IF SIGN=1 THEN 3360
```

```
1640 REM
1650 REM   ***** ORDER FFT OUTPUT FOR NEG. FREQ. AT O TO ************
1660 REM   ****** N/2-2, DC AT N/2-1, POS. FREQ. AT N/2 TO N-2,*******
1670 REM   ******* AND NYQUIST AT N-1.****
1680 REM
1690 FOR I=O TO N/2
1700 INPUTRE(I+(N/2-1))=BUFFERRE(I)
1710 INPUTIM(I+(N/2-1))=BUFFERIM(I)
1720 NEXT I
1730 FOR I=O TO N/2-2
1740 INPUTRE(I)=BUFFERRE(I+(N/2+1))
1750 INPUTIM(I)=BUFFERIM(I+(N/2+1))
1760 NEXT I
1770 REM
1780 REM    ****FFT OUTPUT IS HERE IN ARRAYS INPUTRE FOR REAL
1790 REM    PART AND INPUTIM FOR IMAGINARY PART****
1800 REM
1810 KEY OFF
1820 FOR L=0 TO N-1
1830 MAG(L) = (INPUTRE(L)^2+INPUTIM(L)^2)^(1/2)
1840 NEXT L
1850 REM ***** FIND MAX AND MIN TO SCALE LIN VS LIN PLOT ***
1860 MAX = 0
1870 MIN = 0
1880 FOR J = 1 TO N-1
1890 IF MAG(J) > MAG(MAX)  GOTO 1930
1900 IF MAG(J) < MAG(MIN)  GOTO 1950
1910 NEXT J
1920 GOTO 1970
1930 MAX = J
1940 GOTO 1910
1950 MIN = J
1960 GOTO 1910
1970 PRINT "MINIMUM IS "INPUTRE(MIN) "    MAXIMUM IS "INPUTRE(MAX)
1980 REM ***** PRINT MINUMUM AND MAXIMUM **********
1990 REM ********************************** ***************
2000 SCREEN 2
2010 REM **********   EXPAND MIDDLE OF PLOT ? **********
2020 INPUT TYPE$
2030 CLS
2040 PP = 0
2050 PRINT "EXPAND CENTER OF PLOT  Y/N  ?"
```

```
2060 INPUT TYPE$
2070 IF TYPE$ = "N" THEN 2120
2080 IF TYPE$ = "n" THEN 2120
2090 CLS
2100 INPUT "SCALE 0 TO .499 ? ";PP
2110 WINDOW (N*PP,MAG(MIN))-((N-1)*(1-PP),MAG(MAX))
2120 FOR L=PP*N TO (N-1)*(1-PP)
2130 LINE (L,0)-(L,MAG(L))
2140 NEXT L
2150 INPUT TYPE$
2160 FOR L=N/2-1 TO N-2
2170 IF MAG(L) = 0 THEN MAG(L) = .000001
2180 REM *************** AVOID LOG OF ZERO ERROR *************
2190 REM MAGLOG(L) = 8.68589 * LOG(MAG(L))
2200 REM **************** SCALE FOR NAT LOG TO LOG10 ***********
2210 MAGLOG(L) = 8.68589 * LOG(MAG(L)/.000001)
2220 REM *************** SCALE TO DB/UA OR DB/UV **************
2230 NEXT L
2240 REM **** FIND MAX AND MIN TO SCALE LOG VS LOG PLOT ****
2250 MAX = 0
2260 MIN = 0
2270 FOR J = N/2-1 TO N-2
2280 IF MAGLOG(J) > MAGLOG(MAX)  GOTO 2320
2290 IF MAGLOG(J) < MAGLOG(MIN)  GOTO 2340
2300 NEXT J
2310 GOTO 2360
2320 MAX = J
2330 GOTO 2300
2340 MIN = J
2350 GOTO 2300
2360 PRINT "MINIMUM IS "MAGLOG(MIN)" db     MAXIMUM IS
"MAGLOG(MAX)" db"
2370 REM ************************************
2380 INPUT TYPE$
2390 CLS
2400 REM WINDOW (0,MAGLOG(MIN))-(8.68589*LOG(N-
1),MAGLOG(MAX)*1.2)
2410 WINDOW (0,0)-(8.68589*LOG(N-1),MAGLOG(MAX)*1.2)
2420 FOR L=N/2-1 TO N-2
2430 HOG = 8.68585 * LOG(L-N/2+2)
2440 REM ************* FREQUENCY MARKERS **********************
2450 REM  ************* 10,20 AND 30 ARE THE *****************
```

```
2460 REM  *******************SELECTED HARMONICS *********
2470 IF L = (N/2-1)+10 GOTO 2510  REM **** 10,20 AND 30 ARE THE *******
2480 IF L = (N/2-1)+20 GOTO 2510
2490 IF L = (N/2-1)+30 GOTO 2510
2500 GOTO 2520
2510 LINE (HOG,7*MAGLOG(MAX)/8)-(HOG,MAGLOG(MAX))
2520 LINE (HOG,0)-(HOG,MAGLOG(L))
2530 NEXT L
2540 INPUT TYPE$
2550 REM *** CREATE Y AXIS GRADICLES ********
2560 REM LINE (0,0)-(N-1,0)
2570 REM *** THIS ONE DESTROYS FREQUENCY TICS BUT CAN BE USED *
2580 LINE (0,20)-(N-1,20)
2590 LINE (0,40)-(N-1,40)
2600 LINE (0,60)-(N-1,60)
2610 LINE (0,80)-(N-1,80)
2620 LINE (0,100)-(N-1,100)
2630 LINE (0,120)-(N-1,120)
2640 REM ****** LABEL THE "Y" AXIS **********
2650 INPUT TYPE$
2660 LOCATE 7
2670 PRINT"   120 db"
2680 LOCATE 10
2690 PRINT"   100 db"
2700 LOCATE 13
2710 PRINT"   80 db"
2720 LOCATE 16
2730 PRINT"   60 db"
2740 LOCATE 19
2750 PRINT"   40 db"
2760 LOCATE 22
2770 PRINT"   20 db"
2780 REM  FREQUENCY LABELING CAN BE DONE BEFORE A SCREEN PRINT
2790 REM  AND IT WILL NOT EFFECT THE PROGRAM RUN
2800 REM **********************************************************
2810 REM ** REMOVE THE FIRST 30 & 60 HARMONICS **
2830 INPUT TYPE$
2840 CLS
2850 PRINT "REMOVE LOW ORDER HARMONICS FOR BETTER RESOLUTION"
2860 PRINT "OF HIGH ORDER HARMONICS   Y/N  ?"
2870 INPUT TYPE$
2880 IF TYPE$ = "N" THEN 3130
```

```
2890 IF TYPE$ = "n" THEN 3130
2900 CLS
2910 INPUT "HOW MANY LOW ORDER HARMONICS REMOVED ?";HARM
2920 FOR L=N/2-1 TO N-2-HARM
2930 HOG = 8.68585 * LOG(L-N/2+2)
2940 LINE (HOG,0)-(HOG,MAGLOG(L+HARM))
2950 NEXT L
2960 REM  *************** REMOVE FIRST 30 HARMONICS **********
2970 INPUT TYPE$
2980 CLS
2990 PRINT "REMOVE LOW ORDER HARMONICS FOR BETTER RESOLUTION"
3000 PRINT "OF HIGH ORDER HARMONICS   Y/N  ?"
3010 INPUT TYPE$
3020 IF TYPE$ = "N" THEN 3130
3030 IF TYPE$ = "n" THEN 3130
3040 CLS
3050 INPUT "HOW MANY LOW ORDER HARMONICS REMOVED ?";HARM
3060 FOR L=N/2-1 TO N-2-HARM
3070 HOG = 8.68585 * LOG(L-N/2+2)
3080 LINE (HOG,0)-(HOG,MAGLOG(L+HARM))
3090 NEXT L
3100 REM  ************ REMOVE FIRST 60 HARMONICS ************
3110 REM ** LABEL FREQUENCY DOMAIN OUTPUT DATA TABLE **
3120 INPUT TYPE$
3130 CLS
3140 PRINT "PRINT OUTPUT DATA TABLE  Y/N ?"
3150 INPUT TYPE$
3160 IF TYPE$ = "N" THEN 3320
3170 IF TYPE$ = "n" THEN 3320
3180 CLS
3190 PRINT "IS OUTPUT VOLTAGE OR CURRENT    V OR I  ?"
3200 INPUT TYPE$
3210 IF TYPE$ = "I" THEN 3260
3220 IF TYPE$ = "i" THEN 3260
3230 PRINT "   VOLTS          DB/MICRO VOLT        FREQUENCY"
3240 PRINT "                 "
3250 GOTO 3280
3260 PRINT "   AMPS          DB/MICRO AMP         FREQUENCY"
3270 PRINT "                 "
3280 REM L = N/2-1 TO N-2
3290 FOR L = N/2-1 TO N/2-1+16
3300 PRINT MAG(L) "       " MAGLOG(L) "          "(L-(N/2-1))/WW
```

```
3310 NEXT L
3320 STOP
3330 REM
3340 REM *** SCALE INVERSE TRANSFORM OUTPUT BY FACTOR OF N **
3350 REM
3360 FOR I=O TO N-1
3370 INPUTRE(I)=BUFFERRE(I)*N
3380 INPUTIM(I)=BUFFERIM(I)*N
3390 NEXT I
3400 REM
3410 REM   **** INVERSE FOURIER TRANSFORM OUTPUT IS HERE IN
ARRAYS INPUTRE
3420 REM *** FOR REAL PART AND INPUTIM FOR IMAGINARY PART******
3440 STOP
```

INDEX

Absorption losses 65
Admittance 74
Aliasing 25, 44
Analysis 23, 163

B-H loop 313-314, 316
Balanced circuits 14, 16, 83, 95, 97, 99, 101
Balun 83-87, 91-96, 102, 214
Balun inductance 86-87, 101
Balun model 102
Baluns 132, 163, 204
BASIC 29, 32, 265, 309
Beads 132, 204
Bonding 150
Bonding resistance 132
Broadband 25-26, 232
Broadband emissions 181
Broadband noise 199
Bypass capacitor 275-280, 284-288
Bypassing 49

Cable shielding 159
Capacitance 2, 4, 8-9, 11-12, 15, 18, 51, 66-74, 101-102, 105-106, 110, 116, 124, 129-130, 147, 163, 209-210, 214, 221, 223, 227, 229-230, 237, 239, 281, 292-293
Capacitive coupling 131
Capacitive reactance 56
Capacitors 259
Capacitive coupling path 126-129, 155
Capacitive reactance 167
Capacitor 18, 49, 51, 54, 56, 59, 61, 101, 106, 109, 111, 114-117, 120-126, 130, 144, 163, 199, 210, 227, 276, 279-280, 285, 291, 307, 331
Capacitor impedance 56
Capacitor inductance 52, 286
Capacitor model 53
CE01 175, 233, 265-266
CE03 26, 109, 110, 174, 186, 205, 207-208, 211, 213, 227, 232-233, 242, 265, 267, 271
Characteristic ground impedance 145
Characteristic impedance 8, 96, 115-116, 144, 147, 149, 228, 286, 288, 294
Chassis grounding 133
CISPR 181, 184
Clean area 155
Clock 4-5, 24, 26, 101, 269, 272, 274
Common ground impedance 15
Common impedance 143-144
Common impedance voltage 15
Common mode 5-6, 13, 84-87, 93-94, 98, 100-101, 143-146, 155-156, 160, 213, 222-226, 235-237, 241-242
Common mode balun inductance 93
Common mode capacitance 15, 101, 223, 236-237, 240

Common mode conducted emissions 204
Common mode current 14-15
Common mode elements 204
Common mode EMI filter 197
Common mode filter 95, 151, 205, 239
Common mode flux 85
Common mode impedance 16
Common mode inductor 83, 87, 204
Common mode loops 16
Common mode noise 83, 87, 101
Common mode voltage 14
Component 163
Computer 4-5, 101, 134, 145, 182, 184, 269, 270-275
Conducted emissions 20, 23, 26, 164, 169, 181, 183, 197, 199, 202, 205, 213, 217, 224, 227, 230, 233, 244, 265, 269, 271, 275
Conducted susceptibility 200
Control plan 187
Convertors 39
Core 110, 303
Core losses 67-68
Core materials 71
Coupled inductor 102, 297
Coupling coefficient 87-91, 102
Crosstalk 3-4, 12-13
Current return 133, 135, 137, 140, 147
Curve fitting 310, 315

Daisy chaining 143
Damping effect 122
DFT 24
Diagnostics 291
Dielectric 56, 60, 230
Dielectric absorption 61
Dielectric constant 51, 61, 237
Dielectric losses 56, 67-68, 76, 109, 281
Dielectric strength 51, 61
Dielectric withstanding voltage 61
Differential circuit 3, 6, 98, 239
Differential current 13, 84, 87, 91
Differential elements 240
Differential filters 6, 204
Differential inductance 92
Differential mode 5, 12, 145, 156, 202, 213
Differential mode flux 84
Differential noise 3-4, 12, 14, 16, 200, 205
Differential spectrum generator 224
Diodes 163
Dipole molecules 51
Direct contact 18
Discontinuity 25, 44
Discrete Fourier transform 24
DO-160B 190, 194
DOC 181
Driving impedance 96-97, 116

Earth ground 152, 154
Earthing stake 153
ECA 1, 247, 250-251, 259, 297
Eddy current 68, 161
Electromagnetic radiation 7
Electric dipoles 56
Electric field 18, 271, 293
Electrolytic capacitors 60
Electrostatic field 18
Electromagnetic compatibility 5
Electromagnetic emissions 5
Electromagnetic energy 18, 20
Electromagnetic interference 1
Electromagnetic Interference Control Plan 187
Electromagnetic shield 155
Electromagnetic wave 18

Electrostatic shield 153, 155, 158-159
Elements 75, 163, 294-296
EMC 1, 5, 20, 75, 155, 181, 185, 187
EMI 1, 4-7, 23, 59, 74, 84, 105-106, 116-117, 125, 131, 133-134, 150, 155, 164, 169, 173, 185, 188, 191-193, 197, 199, 204-205, 208, 210, 219-224, 229-232, 233, 235, 239-240, 242, 244, 265, 269, 274, 275, 292, 295
EMI analysis 163-164, 169, 210, 214
EMI filter 110, 131-132, 151, 197, 199, 200, 221
EMI modeling 164
EMI noise generators 175
EMI simulation 205
EMI spectrum 214
EMI spectrum generator 175, 178, 230
Emissions 5, 188
Equivalent resistance 56, 62, 68
Equivalent series inductance 121
Equivalent series resistance 50-51, 56, 167
ESR 56, 61, 117-118, 167, 220, 227, 281

Far field 18, 20, 155, 291-293
Faraday shield 110, 132, 153, 158-159, 197, 205
Fast Fourier transform 24, 265
FCC 2, 109-110, 181-184, 187, 192, 194, 233, 269, 271
Feed-through capacitors 132, 192, 204, 237, 240
Feed-through filter 131
FFT 24-26, 34, 35, 164, 246, 265
Filter 4, 6, 49, 74, 105, 108, 113, 116-119, 124-126, 152, 164, 199, 205, 209, 210-214, 217, 220-221, 223, 227, 232, 242, 249-256, 259
Filter capacitors 220
Filter connector 131, 199
Filter design 260
Filter inductors 209
Flight-critical 134
Flux 68, 87, 91-92, 303-305
Flux density 53, 303
FORTRAN 309, 315-316
Fourier integral 25
Fourier series 25
Fourier trans-form 25
Frequency domain 23, 25, 35, 37, 41, 167, 210, 231, 260, 267
FTZ 184
FTZ/VDE 181
Fundamental switching frequency 214

Ground 7, 133-137, 145, 147
Ground circuit 143
Ground fault interrupter 138
Ground loop noise 13
Ground plane 138-142, 149

Harmonic distortion 76
Hold-up 49, 105, 200
Hysteresis 68, 302, 310, 313

IEC 181
IG SPICE 315-316
Impedance 8, 56, 65, 77, 105
Impedance curve 56, 59, 73, 79-80, 281
Impedance matching 146
Inductance 2, 4, 8-9, 11, 18, 52-54, 70, 84, 90-91, 96, 105, 109, 116, 120, 124-126,

128-130, 147, 150, 160, 163, 202, 204, 229, 275, 277-278, 280, 285, 307, 315
Inductance element 210, 213
Inductive 18, 74, 227, 271, 278, 292-293
Inductive reactance 56, 65, 77
Inductor 65, 67, 70-78, 80, 86-87, 105-106, 109-110, 112-115, 120-126, 130, 144, 163, 199, 202, 221, 224, 227, 246-249, 259, 294-295, 310, 313, 315
Inductor capacitance 72
Inductor impedance 76
Inductor model 102
Inductor parasitics 109, 124
Inductor reactance 65
Inherent parasitics 109
Input to output capacitance 110
Input to output isolation 152
Insertion loss 117-119, 131, 250-251, 257, 259
Insertion loss curve 105-106, 117, 121-122
Insertion loss simulation 251
International Electrotechnical Commission 181
International Special Committee on Radio Interference 181
ISM 184

Key players 166

Layout inductance 286
LC characteristic (or driving) impedance 227
LC filter 95, 106, 108, 111-122, 166, 167, 173, 200, 218, 236, 239, 249
LC tank 131, 214, 217, 227, 228
Leakage inductance 88, 90, 92, 102
Leakage resistance 51, 53
Line Impedance Stabilization Network 123, 183
LISN 123-124, 183, 189, 191, 192
Lossy pi filter 131, 199
LR ladder 118
LSTTL 275

Magnemotive force 53, 303
Magnetic core saturation 87
Magnetic coupling 68, 160, 271, 274
Magnetic domains 68, 76, 302, 305-306, 313-314
Magnetic field 12, 18, 128, 160, 271, 293-294, 297
Magnetic field strength 13
Magnetic flux 271
Magnetic flux density 181
Magnetic loop coupling 3-4, 12
Magnetic saturation 301
Magnetic shields 12, 154
Magnetically coupled noise 12
MCAP 1, 297
Mil-Hdbk 235 188
Mil-Std 461 26, 109, 110, 174, 175, 185-187, 192, 194, 205, 233, 244, 265
Mil-Std 461A 181
Mil-Std 461C 185
Mil-Std 462 181, 185, 187, 208, 242
Mil-Std 462 207, 211-212
Mil-Std 463 181, 185
Mil-Std-461B 208
Military 175, 181, 187, 191, 199, 235
Military bonding 150
Military standard 26, 174

Mutual inductance 87

Narrowband 26
Narrowband conducted emission limits 182, 186
Narrowband noise 233
National Electrical Code 138
Natural response 130
Near field 18, 292-293
Near-field radiation 18, 154, 199
Noise 2-5, 13, 18, 150, 152, 159, 164, 197, 275, 278, 292, 293
Noise and ripple 2
Noise margin 152
Notch filter 122

Parametric models 315
Parasitic 75, 211, 226-227
Parasitic capacitance 67, 106, 204
Parasitic elements 54, 106-107, 121-122, 219, 276
Parasitic inductance 53, 106, 110
Parasitic winding capacitance 110
Parasitics 50, 66, 70, 75, 106, 108, 110, 213, 243, 288
Path 6-8
PCB 148, 160
Penetration 157
Permeability 8, 50, 66, 71, 149, 306, 311
Permeable core 70, 294
Permeable magnetic shield 161
Permeable material 154-155, 160
Permittivity 8, 147, 149, 210, 237
Phase angle control 41
Player 119
Player element 120
Players 94-95, 120, 224, 227
Polarization 302
Power conversion 84, 205-206, 236, 239
Power entry capacitors 288
Power supply 101, 110, 116-117, 135, 140, 152, 164, 167, 197-198, 200, 202-203, 208, 231-232, 240, 260, 271, 276, 278
Power transistor switches 125
Printed circuit board 101, 130, 147, 199, 204
Printed wiring assembly 288
Printed wiring board 147, 209, 210
Pulse width modulation 140
PWA 288
PWB 4, 11, 106, 125, 209-210, 229-232
PWB layout 54
PWM 101, 140-141, 239

Q 76, 115, 130, 249
Qualification Plan 187
Quasiwiring diagram 137

Radiation susceptibility 150
Radiated emissions 125, 181, 291
Radiation 145
Radiative 18, 293
Radiative shield 160
Radio frequency interference 2
Radio Technical Commission for Aeronautics 188, 194
Reactance 49, 54, 65, 94
Rectifier noise 3-4
Remnance 301
Resistance 2, 4, 56, 74-76, 106, 163, 224-225, 229, 281
Resistive losses 67, 124
Resistive parasitic effect 109
Resistors 163, 231
Resonance 56, 66, 76
Resonant frequency 56, 73, 76-77, 95, 101, 106, 116, 130
Resonant LC tank 130

Resonant tank 120, 130
Return plane 11, 149
Returns 136
RF narrowband conducted emission limits 186
RFI 2
Ripple 2-3
Ripple current 62
Routing 147
RTCA 188

S/N ratio 152
SABRE 1, 297
Safety 150
Safety grounding 133, 138
Saturable inductor 312
Saturation 75, 84, 301-313
Self-inductance 87, 149
Self-resonance 66-67, 70-71, 73, 109, 110, 125, 130
Sequential circuits 272
Shared ground impedance 3, 16, 144
Shared return 143
Shielding 153
Shielding effectiveness 155, 159
Signal to noise ratio 152
Simulation 75, 163
Single-point ground 140, 143
Skin effect 75, 109
Skin effect losses 76
SMPS 208, 210, 212-213, 230
Spectrum analysis 260
Spectrum generator 164, 231
SPG 142
SPICE 1, 102, 164, 171-172, 210, 276, 295, 297, 308, 313, 315-316
SPICE ac analysis 169, 173
SPICE EMI analysis 164
SPICE EMI generator 214
SPICE EMI spectrum generator 171
Stakes 153
Stray capacitance 105
Subharmonic 45
Susceptibility 5, 125, 149-150, 181, 187-188, 271, 274
Switching noise 3

Test procedures 187
Test report 187
Time domain 32-33
Timing 49
Trace routing 127
Transformers 125, 163
Transient response 117
Transistors 163
Transmission line 145
Tree 143, 150

Unbalance 14-15, 144
Unbalanced circuit 83

VCCI 181, 188, 192
VCCI RF narrowband limits 189
VDE 109-110, 184-186, 192, 233
VDE0871 184
Victim 7

Wave impedance 20
Windowing 25
Wiring diagram 134-135